DEDICATION

To Jaq

For the steady hand on the tiller when my own course drifted.

For the love that asked for nothing but gave me everything.

For the patience that absorbed the noise of engines, the weight of silence, and the rough edges of a life lived elsewhere too often.

You have been the calm in my turbulence, the warmth when the world felt cold, the laughter that reminded me who I was beyond the uniform.

Every page that follows carries your quiet courage between the lines.

If this story has wings, it is because you gave them to me.

AUTHOR'S NOTE

I'm writing this before I forget the details that come with a life as an aviator. The shrill whine of the Lynx's Gem engines and the beat of rotor blades. the sight and sound of London Heathrow at night.

In between the sensory beatings were quieter moments: the tired silence and the 'brown eye' after long NVG sorties, the chaos of family life squeezed in around rosters and business meetings. Through it all runs a single thread, a life shaped by flight.

So why Unlucky Jim? It started as a nickname, part joke, after a run-in involving a tail rotor and a fence. It stuck like shit to a stick. For years it followed me through flying tours, cock-ups, boardrooms and barstools. But this isn't about bad luck.

When I sit in quiet contemplation now, I know I haven't been the father or the husband I wanted to be. I haven't been the father I wanted to have. I didn't have a father to copy, so I built my own ideas from fragments; some worked, some didn't. My focus has always been to provide for my family and to be a man my children might one day be proud of.

Before he died, my father spoke with bitterness about his own dad, a man I never met. My aunt later told me why. My grandfather was an abuser. She said he was the type who'd watch his nine kids eat soup while he tucked into steak. I swore I'd never be that man, metaphorically or otherwise.

The pictures I've used are an attempt to illustrate key moments. Many are poor quality. I've tried to repair them, but, as with life, there's only so much you can do to fix old and lasting damage.

This isn't just a military story. It's a human one; stubborn, hopeful, occasionally funny. If I've done it right, you'll laugh, wince, maybe swear a bit too. This is my truth, told as honestly as a pilot with an ego can manage.

TABLE OF CONTENTS

PROLOGUE 11

Bessbrook Mill Helipad, Northern Ireland – 21 June 1998 11

PART 1: EARLY LIFE AND TRANSATLANTIC TRANSITION 13

Chapter 1. Warships and Warplanes 14

Chapter 2. Heroes and Heroines 17

Chapter 3. The Osmonds on Tour 19

Chapter 4. Lessons I Never Asked For 21

Chapter 5. From Deli to Drill Square 26

PART 2: JLR, ROYAL ARMOURED CORPS 31

Chapter 6. Thrown In the Deep End: Lad's Army 32

Chapter 7. Cunning Running and Other Surprises 36

Chapter 8. Pranks and Pilgrimages 39

Chapter 9. Shoes and Best Boots 41

PART 3: NOSEDIVE TO LEVEL FLIGHT – LIFE ON THE GROUND 45

Chapter 10. It Shouldn't Happen to a Vet 46

Chapter 11. Gunnery, Goonery and a Wedding 50

Chapter 12. A New Start 52

Chapter 13. Horses and Dogs 53

Chapter 14. Driving, Dogs and Duties 58

Chapter 15. Murder and Medics 61

Chapter 16. Pasta, Peaches and Wine 64

Chapter 17. Driving Miss Daisy 67

Chapter 18. Tank Suits and Toilets 73

Chapter 19. History in the Making 76

Chapter 20. Checkpoint Charlie 81

Chapter 21. Bare Naked Ladies 82

Chapter 22. Running with a Brown Dog 85

Chapter 23. History is Made 87

Chapter 24. Signalling the Future 90

Chapter 25. Sultans and the Sin Bin 93

Chapter 26. Selection and Separation 98

Chapter 27. Chipping Away 100

Chapter 28. Amalgamation and Division 103

Chapter 29. Doors and Departures 104

PART 4: LEVEL FLIGHT TO STEADY CLIMB – LIFE IN THE AIR 108

Chapter 30. Ground School Phase 109

Chapter 31. Basic Fixed Wing Phase 112

Chapter 32. Survival and Aeromedical Training 115

Chapter 33. Basic Rotary Wing Phase 118

Chapter 34. Advanced Rotary Wing (ARW) Phase 127

Chapter 35. Brickbats and Bouquets 134

Chapter 36. A Brave New World 137

Chapter 37. The Para, the Gamekeeper and a Psycho 139

Chapter 38. New Barracks, Old Ways 144

Chapter 39. Promotions and Missiles 147

Chapter 40. Hunting and Heroes 151

Chapter 41. The Hunters become the Hunted 156

Chapter 42. Callsigns 160

Chapter 43. Command and Loss 161

Chapter 44. Flying from a Desk 163

Chapter 45. Transferring Traditions to Habits 166

Chapter 46. Where Eagles Dare 169

Chapter 47. Rescuing Dead Cows 173

Chapter 48. More Pasta, Peaches and Wine 175

Chapter 49. Grief and Loss 180

Chapter 50. Back in the Air 183

Chapter 51. Immersion Suits and Clogs 184

Chapter 52. Leaping with Frogs 189

Chapter 53. Riding on a High 191

Chapter 54. Combat Frogs 194

Chapter 55. Homes on Fire 199

Chapter 56. Why 201

Chapter 57. The Dawn of Tombstone and its Dogs 208

Chapter 58. SA-6 and the Legacy of Scott O'Grady 211

Chapter 59. The Mujahadeen v Monsieur SF 213

Chapter 60. NATO Firepower in all its Glory 215

Chapter 61. Going Dutch 217

Chapter 62. Split 220

Chapter 63. Skyhooks and Right Hooks 224

Chapter 64. Pranks, Skits and Bored Pilots 226

Chapter 65. Low Routes and High Stakes 228

Chapter 66. The General and his Close Shave 231

Chapter 67. Compassion and Minefields 233

Chapter 68. Debrief, Bibles and the Clap 236

Chapter 69. In a Mess 239

Chapter 70. Elections, BATs and PIFWCs 241

Chapter 71. Wine, Squid and Alanis 247

Chapter 72. Tail Rotors, Fences and Ranges 249

Chapter 73. Outlaws, Comanches and Apache's 255

Chapter 74. Running Down 259

Chapter 75. Homeward Bound 262

Chapter 76. Reflections 265

Chapter 77. New Year, New York, New Life 267

Chapter 78. In a Mess. Again 270

Chapter 79. Auf Wiedersehen Pet 273

Chapter 80. The Jock Nag 276

Chapter 81. Flights, Friends and Fuckwits 279

Chapter 82. The Hitman's Wife 281

Chapter 83. Walk Before You Can Fly 286

Chapter 84. On Ilkla Moor Baht 'at 289

Chapter 85. Know Your Place 292

Chapter 86. Options for Change 294

Chapter 87. Clerks, Magazines and Hard Landings 302

Chapter 88. The Festive Season 305

Chapter 89. Schools, Teachers and Vasco de Gama 308

Chapter 90. Belfast and Romance don't Mix 314

Chapter 91. Tankers! 317

Chapter 92. To Drumcree and Beyond 323

Chapter 93. The Long Road to Peace 328

Chapter 94. The Apprentice, without Sugar 339

Chapter 95. Leadership v Management 342

Chapter 96. Tragedy Strikes Again 348

Chapter 97. Hero to Zero: Betrayal 351

Chapter 98. The Twilight Zone 362

Chapter 99. The Aftermath 366

PART 5: A CIVILIAN IN UNIFORM 368

Chapter 100. Settling in 369

Chapter 101. On the Line 374

Chapter 102. Baboons, Babies and Twin Towers 379
Chapter 103. Back to Bedlam 383
Chapter 104. Big Plans 387
Chapter 105. The Writing on the Wall 393
PART 6: BEGAN MARRIED, ENDED DIVORCED: BMED 399
Chapter 106. On the Busses 400
Chapter 107. War Zones and Potential War Zones 403
Chapter 108. The Blonde Bombshell 407
Chapter 109. Blackhawks and Cowboys 409
Chapter 110. Love is in the Air 415
Chapter 111. Fanning the Flames 419
Chapter 112. Belfast Child 422
Chapter 113. The Russians Attack 424
Chapter 114. The Other Side of Divorce 426
Chapter 115. Tehran and Other Great Ideas 428
Chapter 116. Finners 433
Chapter 117. Headaches and Brainwaves 437
Chapter 118. A Star is Born 441
Chapter 119. African Promise 442
Chapter 120. Genesis 444
PART 7: GROUNDED 447
Chapter 121. Pulling Together 448
Chapter 122. Wobbles and Wins 451
Chapter 123. Good Morning Ghana 455
Chapter 124. Gathering Steam 458
Chapter 125. A'hhv Bin ta Leeds 461
Chapter 126. From a Jax to a Queen 464
Chapter 127. New Directions 469

Chapter 128. Testing Times 474

Chapter 129. The Birds and the Bea's 478

Chapter 130. Snow Days and Groundhog Days 484

Chapter 131. Wedding Cakes and Doughnuts 487

Chapter 132. The Three Musketeers 494

Chapter 133. Sing Sing 498

Chapter 134. Home Sweet Home 502

Chapter 135. Boom and Bust 510

Chapter 136. A Tale of Two Cities 513

Chapter 137. Performance under Fire 521

Chapter 138. Lights, Camera, Action 524

Chapter 139. Independence Counts 529

Chapter 140. It Ain't Half Hot Mum! 533

Chapter 141. Bless the Rains Down in Africa 541

Chapter 142. Sandyman to Sandyland 543

Chapter 143. Sandyland to the Land Down Under 548

Chapter 144. Administering the Last Rites 550

Chapter 145. Walk Towards the Light 559

Chapter 146. Far, Far Away 564

ROLL OF HONOUR 567

Family 567

Army Air Corps 567

The Light Dragoons 567

Northern Ireland 567

IN REMEMBRANCE 568

PROLOGUE

Bessbrook Mill Helipad, Northern Ireland – 21 June 1998

It was just another Lynx 5 duty in South Armagh, until Buzzard's voice came over the PA system.

"CRASH-OUT; 5 and 7, CRASH-OUT."

No hesitation. Training took over, boots moving before the brain caught up. The pilots were out of the door, the aircraft 'burning and turning' before commanders were even strapped in. In the back, the doorgunners had the GPMG ready while marshalling the ARF into place, dog, its' handler and all.

Buzzard gave me the bare bones: "*1.5 K's north of Cross on the Blaney Road, patrol in contact*". The rest came as we climbed hard and fast up 'the Tube', Charlie at the controls, max power.

A joint RUC and British Army patrol had stopped a tanker smuggling heating oil. Twice. The second time, the driver deliberately ran over the patrol commander, Corporal Gary Fenton, killing him.

The tanker was now just over the border, close enough to see, too far to touch without permission.

I called for clearance to cross and engage. Denied.

We pressed on anyway, right to the line, the invisible border. The doorgunner was poised, finger curling around the trigger, waiting on my word. I looked at the road ahead, the tanker in full view, and called it off.

I told Charlie to take us back.

I didn't know it then, but I'd spend the next two decades replaying that moment, the muscle memory screaming go, the rules saying no, and me stuck right between the two. Every day since I've lived with being stuck in the middle of that.

Gary's death didn't just drop into my logbook; it anchored itself there. A man I never even met.

PART 1: Early Life and Transatlantic Transition

USA, Pittsfield, Mass: May 1967 - 1975

Chapter 1. Warships and Warplanes

As much as I hate admitting it, I was born in Pittsfield, Massachusetts. My mother, Jane, had met my father, Sam, in early 1966 in Bergen, Norway where my mother was working as an au pair for an English/Norwegian family.

My dad had been drafted into the US Navy because of the Vietnam war and his ship, the aircraft carrier USS Randolph had served as the recovery ship for two Project Mercury space missions, including John Glenn's first orbital flight.

During one Atlantic tour in 1966, my father was summoned to the bridge. He had no idea why, but knew something was off when an officer randomly handed him a bowl of ice cream.

The captain then proceeded to bluntly tell him that his younger brother, also drafted into the Navy, had been killed in Vietnam.

Seaman James M Termini was only 18 years old when he arrived in Da Nang, Vietnam. He had been assigned to the Lighterage Section, Naval Support Activity (NSA) Da Nang, under MACV, the Naval command that handled barge transfers between supply ships moored offshore and shore facilities.

He had been in Vietnam only 2 weeks.

It was an utterly ordinary evening. Bunks, strip lights, quiet chatter. No tension, no drama, just sailors doing what they'd done a hundred times before.

Then one of them didn't check his weapon. Not properly. In fact, not at all.

A round went off at bunk height. An 18-year-old boy was shot where he sat, killed instantly by his own mate. No enemy, no warning, no excuse. Just incompetence, a moment's laziness, and a life gone in a flash of noise and cordite.

That was the sum of his war. No firefights or glory. Just blunt randomness in a place where the margin for error was zero.

After Bergen, the Randolph steamed for Portsmouth. Mom was due home anyway and they had agreed to meet at Portsmouth Naval base. My father was due for discharge from the Navy, so he took his soon-to-be bride back to New England to start a new life.

In mid-1967 the happy couple had a boy and named him James M Termini, after his uncle. Two years later my brother, Edward Robert, arrived, presumably to keep me in check.

We shared a two-family weatherboard house outside Pittsfield, cousins and family everywhere, before work pulled Dad south to Bloomfield, Connecticut.

My dad was an aviation nut to his core. At every opportunity he'd drag us to airshows, narrating the Thunderbirds routines, but saving his real reverence for the Navy's Blue Angels. They were his favourite, and therefore mine.

Back then they flew F4 Phantoms, beautiful brutes that smoked like they ran on burning tyres.

At home he'd glue together Airfix kits of the same jets and carriers. The kitchen smelled permanently of enamel paint, glue, and cigarette smoke.

Half the models bore more of his fingerprints than decals.

I loved building those models alongside him, convinced I was helping, when in truth I probably just glued my fingers to the table. It became one of my favourite things to do. Years later, Ed and I had *Super Flight Deck*, a spring-loaded aircraft carrier game where you catapulted a jet off a deck and then tried, mostly in vain, to land it.

But Dad wasn't there to share it. That toy should have been his crowning glory, the thing he'd spent years rehearsing with Airfix and getting high on glue fumes for. Ed and I played with it for hours, laughing and bickering as brothers do.

At the many airshows we went to, we would walk around the static aircraft displays of C5 Galaxies, C130 Hercules and everything else that was on offer. To this day I still adore the Phantom, its beautiful lines and classic shape.

At the time, Boeing was conducting large scale testing on their new 747; my dad would take us to the threshold of the nearby Bradley International airport, where we would sit right under the final approach and watch in awe as this behemoth's landing lights scorched the roof of our Buick Estate Wagon, with its fake wood panels and acres of chrome, it was more a rolling living room than a car.

Somewhere between airshows and model making, the shouting began. I was seven or eight. Raised voices, slammed doors, nothing explained or understood. The raised voices occasionally became more than that. On one occasion, I was out playing in an area of scrub land with a friend, getting hot, dirty and messy throwing large rocks into a small brook.

We had gone back to his house for a drink of grape Kool-Aid, the kind that stained your tongue purple for about a week, when I was told that my mom was on the phone.

"*Which plane do you want to fly to England, Jimmy, the 707 or the new 747?*" Confused, I replied the 747. I didn't grasp that she was booking one-way tickets, that Dad wasn't coming, and that a chunk of my life was to disintegrate.

I never got to say goodbye to my dad.

We were given a lift to the airport by one of my mother's friends and that was that. I never saw him again for many years and never knew why.

I coped in the way seven-year-olds do: I focused on the airplane ride and my BOAC junior jet club logbook, trying not to cry.

I didn't know it then, but that flight severed more than the tie to my father. It ended a chapter I wouldn't understand until years later.

UK, Ilkley, Yorkshire: 1975 - 1984

Chapter 2. Heroes and Heroines

Arriving at my grandparent's stone terraced house in Ilkley, Yorkshire was an extreme shock to the system. My grandparents seemed positively ancient to me.

My grandfather, Harry, had suffered a severe stroke a few years previously through medical negligence. He could now barely speak, being only able to force out a handful of words that were utterly useless in conversation.

The only words he was able to confidently use were born of frustration and were generally swear words, which of course was simply hilarious to a kid.

His movement was similarly limited, but he did manage to haul himself to the off-licence once a week to buy his Bells whiskey. He would often just stop random cars, without saying anything because he couldn't, and commandeer a lift into the town.

Harry had been somewhat of a war hero during the second world war; he was rapidly promoted through the ranks until receiving a battlefield commission in the Libyan desert.

As I got to know him, I realised he'd once been a very funny man who loved winding my mother up with pranks when she was a kid. Now he was trapped inside a barely functioning shell.

I'd sit on his knee while he tried to show me the soldier he used to be, sketching memories for me in pictures and maps and crude little drawings from his time in the war.

The words were gone, but the stories were still in there somewhere.

I loved him very much, but still managed to take advantage of his disabilities in practical jokes on him. Hiding his whiskey or putting firecrackers in the bowl of his smoking pipe – and erupting into fits of laughter when his pipe exploded.

Shouting *"BLOODY SEE!"* and shaking his one good fist at us with fragments of smoking tobacco all over his beige cardigan, was about all he could do to in return.

I've talked about Harry, but the true hero of the family was Betty, my grandmother. A tall, stylish and elegant woman best described as an English 'gentlewoman'. A military hero in her own right, Betty was commissioned from Warrant Officer in the Auxiliary Territorial Service (ATS).

As a Junior Commander (the female equivalent of a Captain in the Army), Betty was in command of an ATS air defence battery defending the skies above the Midlands of England.

On one particularly heavy raid she was plotting for the guns when word came through that there wasn't an officer in the gun pits. Someone had to give the firing orders. So, she sprinted out of the relative safety of the plotting room, strode straight into the gun line and took charge.

The gunners followed her orders, engaged the target and brought an enemy aircraft down. Family legend insists she became the first female officer with a kill.

The war left my grandmother with a serious tic. She'd been travelling on a train during a blackout, lost all sense of movement and stepped out thinking she'd reached a station. She hadn't, and the fall caused injuries that stayed with her for years.

To two young boys the tic was an open invitation for hilarity. Every time she jerked her arm into the air with a sudden grunt, Ed and I would collapse in fits. She never told us off.

Chapter 3. The Osmonds on Tour

Starting school in Yorkshire, with an American accent and a haircut that made me look like I had a giant mushroom on my head, coupled with being called Jimmy at a time when the Osmonds were the big thing meant a few things. I learned quickly to lose the accent, lose the name, and stick up for myself.

I hated school, far preferring to be up on Ilkley moor with friends, poking around the quarries and rock falls above Heber's Ghyll looking for caves. We found a few tucked under huge granite boulders and kept going back, 'Stig of the Dump' lodged firmly in our heads.

Our house was near the end of a terrace and directly next door to the best Fish & Chip shop in Ilkley. It did mean though that we permanently smelt like old chip fat and haddock.

This was Yorkshire in the time of the miners' strikes. No central heating, no hot water, only what little coal we had to keep the fire grudgingly alive. When that ran out, the cold wasn't just uncomfortable, it was constant, something we lived with and endured.

Baths were a Sunday night ritual. "*Treat*" would be the wrong word; it was more like lowering yourself into two inches of the North Sea in mid-January, sometimes with a squirt of Matey to mask the suffering.

Whoever got in first got it almost warm, whoever was last in was basically rinsing in other people's history, and with a brother – sometimes worse.

Afterwards, towels steaming by the fire, condensation streaming down the windows, and *'All Creatures Great and Small'* on the Waltham portable TV showing us the luxury we apparently lived in, or '*Oh No It's Selwyn Froggitt'* mumbling away in the background, that was the rhythm of our Yorkshire Sundays.

A kind neighbour had long been storing a reclaimed oak parquet floor, he donated it to us in lieu of coal to burn and try and get some warmth into the house, and heat into the back boiler for hot water.

While that lasted, it was luxury. And spectacular when the hardened tar melted and took a hold.

Life for my mother was tough. She worked long hours as a laboratory assistant, testing the quality of woollen garments sent in by the manufacturers. The odd one made it home once testing was complete, including a green woollen Army jumper each for Ed and me. They became standard issue.

Chapter 4. Lessons I Never Asked For

In summer 1978, Ed and I were loaded onto a train in Leeds with a suitcase and a supermarket bag with haslet & cheese sandwiches, cold sausages and gherkins for lunch. We were on our way to Exeter to stay at my mother's cousin's farm, Eastern Hill, for the summer holidays.

'Uncle' John stepped in as a surrogate father without ever making a fuss. He was kind and as hard working as they come. He had us up at 04:00 in the dairy milking cows and dragged us into every part of farm life, including driving tractors long before any sane adult should have let us near one.

I loved every minute of our time on the farm, it felt like I should have always been there, knee deep (sometimes literally) in cow shit, nursing piglets and moving the herd.

Ed and I still laugh about that train journey, and the kind of idiocy only brothers could find entertaining. One of us would hurl a cold sausage out of one carriage window, and the other would lean dangerously out of another, desperately trying to catch it.

Forget cricket, this was the true national sport of Yorkshire, lads armed with leftover pork products.

Looking back, those summers at Eastern Hill did more than just fill a holiday. They gave me a sense of graft, routine and quiet responsibility that I carried long after we left. You learned quickly that if you didn't turn up, the job still needed doing, and that stuck. It was the first time I felt useful, properly part of something, and it set a baseline for everything that came after, whether I realised it at the time or not.

I think I was 12 when my mum took me to a party at a friend's house somewhere outside Ilkley. There were a couple of girls there my age, her friend's daughters, so I was told I'd be fine.

At some point during the evening a man introduced himself as 'David' and struck up a conversation with me, talking to me like an adult. Later, when everyone was pissed up, he asked if I'd ever been to Blackpool, extoling the wonders of the Golden Mile.

He persuaded my mum to let him take me there that night.

On the drive, he told me to get some sleep. I closed my eyes.

That's when it started.

Thinking I was out cold, he began fumbling with me. At some point he must have been driving badly, because we were pulled over on the motorway by the police. I stayed completely still, breathing shallow, pretending to be asleep while an officer shone a torch through the window. He saw nothing wrong.

They waved him on.

As soon as the blue lights disappeared ahead of us, he got bolder. No hesitation now. Just hands where they had no right to be, as if the interruption had given him permission. Every instinct told me to move, to fight, to do something, but I stayed frozen in that awful half-state, pretending to sleep and hoping he'd stop. He didn't.

The car kept rolling through the dark. I lay there counting the seconds.

He dropped me home late the next day.

It was never spoken about. Not by me, not by my mum. We both carried our own version of it and left it there.

Years later, we found out that "David" died whilst in prison for a range of similar offences.

School carried on, life carried on, and like most lads my age I learned to fold things away rather than deal with them.

After turning 13, some of my friends had joined the local detachment of Army cadets, they persuaded me to have a go. I went along on the next drill night in a small wooden hut between the cemetery and the sewerage works.

The Ilkley detachment was affiliated to the 4th Battalion of the Parachute Regiment, a reserve battalion made up of many ex-regular paratroopers who we were in total awe of.

The cadets was a game changer for me. I loved the idea of the uniform, but the reality was something else. The kit looked like it had been stitched together from old sandbags and passed down since the Boer War.

But there was always something happening: camps where we swaggered about like we were on operations and shooting weekends where we felt like snipers. These "boys' own" adventures filled our heads with pride and purpose.

In a world without PlayStations or TikTok, this was it; the original, unfiltered adventure mode.

The cadets gave me more than discipline; it gave me mates, memories, and a first taste of that strange mix of hardship and laughter that would end up defining my life in uniform.

It was now late 1979, I had survived the education system and moved up to the secondary school. Ilkley Grammar School was an old Victorian building on the very edge of the moors.

My best friend John and I would meet at the top of my road and cross the railway yard via a long, narrow footbridge. As we turned off the road towards the stone stairway that led up to it, we stopped. The scene that met us that morning was horrific in its scale and meaning.

There were police officers milling around, streaks of plastic police tape, and piles of sand everywhere soaking up vast pools of blood. A 21-year-old woman had been attacked by a man wielding a hammer on the footbridge, but a passerby had intervened, preventing a fatal outcome.

Later, the name of the attacker surfaced: Peter Sutcliffe, the "Yorkshire Ripper." The same man who haunted headlines and kept women indoors at night.

Not long after, I accidentally found a knife under my mother's pillow. At the time I didn't understand why it was there, so I thought nothing of it.

I understand now.

In spring 1982 our cadet detachment headed out to Hohne in West Germany for a two-week camp. It was one of those trips that sounded far more impressive than it probably was; a bunch of teenagers being let loose on a British Army training area.

Once in the barracks we met up with other detachments and spent time with our hosts, probably annoying the life out of them, but having enormous fun.

On the way to Luton airport, we passed through London Kings' Cross railway station, dressed in our cadet uniforms and wearing the cap badge of the Para's, passing blackboards with:

"ALL MEMBERS OF THE PARACHUTE REGIMENT ARE TO RETURN TO BARRACKS IMMEDIATELY"

It was of course the start of the Falkland's war. Watching old ladies staring at these boys in uniform running through London's main station wearing that cap badge made us all feel very important!

The cadets had a 'star' system; whereby accumulated skills resulted in the presentation of a star to wear, denoting your status. The highest level was '4-star. I asked if Ilkley had had a 4-star cadet; they hadn't. John and I decided we would go through the 4- star system together and become the first.

We volunteered for everything going, completed all the aspects of community service we needed to and even organised a group to trek along the Dales Way, a 130-kilometre-long footpath in Northern England, traversing both the Yorkshire Dales and Lake District National Parks. We enjoyed it so much, we did it twice.

Eventually, we scraped together the required skills. Progress came in small wins and sheer bloody-mindedness. We got knocked back, corrected, and occasionally bollocked, but we stuck at it. When we finally earned our scarlet red badges, it wasn't just relief.

A small red badge, earned through stubbornness, became our first lesson in grit.

In the background, my mother had been working and saving hard; soon enough she was able to afford a larger house. Still a terraced house, just around the corner, but with two bedrooms plus the attic room. This time she took the attic room, and Ed and I each had our own bedrooms for the first time.

At the end of our terrace was a large family, you could say feral, but the kindest, most natural people you could hope to meet. The Dinsdale's. Straight out of old seaside postcards, the mum, Winnie, was larger than life and a proud Yorkshire woman with a heart of gold.

They had 3 sons. The middle son, Graham, was 18 or 19 and a gamekeeper at a local farm.

I got to know the family well; Gerry would take me down to the River Wharfe and taught me how to fly fish with an old split cane trout rod. He even taught me how to tie flies, the 'ginger quill' becoming my go to fly. I became good friends with Graham, known simply as 'Dinny'. Dinny introduced me to his work as a gamekeeper at a farm up the road near Bolton Abbey.

From then on, I would head up there and help him with the daily grind of rearing pheasants. That meant everything from hatching and raising the birds to releasing them into the pens and later working as a beater on organised shoots.

School wasn't going well for me, largely because I would much rather go up to the farm with Dinny. When it came time for GCE and/or CSE exams, I hadn't prepared at all and had less than zero interest in them. My intention was to go and work full-time at the farm.

I didn't even bother to attend a couple of the GCE exams; the first Mom knew of it was when the exam board sent her a fine for my no-shows, suffice to say, she was not amused. My highest qualification was a grade 2 CSE in French, which never did a single thing for me. As we'll see.

This was all a great plan; it would have been an even better plan if the farm's owner, the man who would have been paying me, had known anything about it.

Sadly, he hadn't been consulted.

Then when he did find out, he didn't share my enthusiasm for the arrangement. Apparently, his vision didn't include a fifteen-year-old dreamer (read idiot) with no qualifications and a head full of plans.

And so, just like that, my countryside life ended before it had even started. At fifteen, with zero useful exams to my name and worse prospects than a meat raffle at a vegan festival, I had no choice but to go and find what everyone else called "a proper job."

Chapter 5. From Deli to Drill Square

After just under a year working in a local deli, wearing a white jacket and wellies, the realisation that I was going nowhere fast finally slapped me around the chops, like one of the fresh salmon on display. The Army was the only door left open.

I caught a local train into Bradford, found the Army Careers Office, and laid it all out to the recruiter.

He told me I'd left it too late to apply for the Junior Leaders Regiment (JLR). Which was basically a full-time, paid cadetship designed to give young soldiers a head start.

That meant waiting at least another year, and even then, I'd be joining as a regular soldier.

Then he mentioned that the civilian Youth Training Scheme (YTS) had just been extended to the JLR, and the age limit had been removed. The YTS was a Thatcher government scheme designed to get kids off the streets and into work. It paid £25 a week and ran for twelve months.

I took the bait and signed up, with my mother's full agreement, for the JLR Royal Armoured Corps.

Ilkley's moors and the miners' strike had already taught me how to endure cold, hunger, and the sound of my own pulse in the dark. On 10 January 1984, I swapped the daft wellies, a white coat, and the moors for a Dorset barrack square, where the cadets had laid the groundwork for my military life.

1 Harry aged 25, with his sister

2 Betty aged 23

3 Harry aged 35

4 WO2 HQ Lievesley, Alexandria, 1941

5 SN James M Termini, aged 18, my uncle

6 SN Samuel R Termini, aged 18, my father

7 Mom & Dads Wedding Day, 1966

8 In Vermont with my Grandmother

9 Me with Nan & Granddad

10 Extended family, me sitting next to Amy

HELICOPTER CLUB

This is to Certify That James Termini
Having remained motionless in space: flown backward, forward, sideways and vertically in a Bell 47G2 Helicopter, is hereby designated a genuine
HELICOPTER WHIRLYBIRD

9-26-71
Date

Chief Rotorhead

Helicopter Charter, Inc.
Municipal Airport
Shelby, N. C.

11 Prophetic Certification

12 Me and Ed at Pease AFB Airshow

Part 2: JLR, Royal Armoured Corps

UK, Bovington Camp, Dorset: Jan 84 – Dec 84

Chapter 6. Thrown In the Deep End: Lad's Army

I arrived at Wool, Dorset train station in the early afternoon, swimming inside an old twill suit that had been lent to me by my grandfather; it was at least 2 sizes too large and enjoyed that damp, musty old wool smell.

Joining me were a dozen other nervous lads.

We were 'met' by a Sergeant wearing a Regimental Provost armband who immediately hated each of us with a passion normally reserved for paedophiles.

The trip to Stanley Barracks in Bovington Camp in the back of a 4-ton truck only took about 15 minutes. Bovington Camp is the historic and spiritual home of The Royal Armoured Corps (RAC), and the world-famous Tank Museum. Nearby is Clouds Hill, the home and death place of Lawrence of Arabia.

Dismounting from the truck under a hail of screaming and 'encouragement' to be quicker, we met our new parents in the form of the Permanent Staff who had been assigned to 'A' Squadron (Alamein YTS) troop.

Our Troop Leader was an officer from the Royal Military Police (RMP). Lieutenant Ivan Smith was perfect for the role, he always portrayed a humourless, rigid officer who saw things in black and white and was intimidating from day 1.

The Troop Sergeant was Sergeant Jacob Price, also RMP – the real power in the place. Unrelenting, trigger-happy with punishment, and openly fond of a few favourites.

Having been given all of three minutes to find a bedspace in the two-storey, wartime brick block, I ended up on the second floor in a 16-man room. Each floor had an ablutions room and a so-called drying room. The drying room was a massively overheated space, the Army's answer to tumble dryers. It smelt like years of old socks and baked sweat, which, in fairness, was exactly what it was.

In the same room were other boys, soon to be men, who became the best of friends. Miguel 'Mig' Gomez was a unit of a bloke, of Spanish heritage and at the age of 17 was already an award-winning weightlifter and 'gym bunny'. Mig and I, despite our physical differences rapidly became the best of friends.

Bedded next to Mig was Tom. Tom was from Portsmouth and was another sportsman, playing tennis at a high level and went on to represent the Army in various sports.

There was a group of lads from the Northeast of England who had all joined the Army with the intention of going to the same Regiment, the $15^{th}/19^{th}$ The King's Royal Hussars. The 15/19H only recruited from its area, the Northeast of England.

My local Regiment was the $4^{th}/7^{th}$ Royal Dragoon Guards, which I wasn't overly keen on joining as I had experienced them through the Army cadets. I was unsure of which Regiment I would try and apply to.

On the YTS scheme, we had twelve months to prove ourselves and earn a place in the real Army.

The first term was effectively the most basic of basic training, but as boys we still had to attend educational sessions. The Army cadets had prepared me for the military side of training and so basic training was no shock, but I really did not want to be back in a classroom with teachers.

I also had to work hard in physical training, I was never at the front of the run or selected for teams, despite volunteering for everything. As a result, I became a target for both the troop staff as well as the Physical Training Instructors (PTIs).

One PTI, a small, stocky Scotsman, known as 'Jockey', was a particularly nasty bully, one of his favourite pastimes in the gym involved the trainees laying flat on the floor in a long row, with arms and legs in the air whilst he would jog up and down the line, his rubber soled 'daps' twisting and grinding the flesh from ribs under our PT gear.

We spent hours, late into every night preparing our room for never ending inspections from the Troop staff. Using gallons of orange floor wax that radiated a unique oily smell and an ability, despite it being wax, to never shine as it was expected.

We scrubbed and waxed until our hands stank of orange polish and defeat.

It never met the standard required.

Bedding was transformed every morning into the infamous 'bed block'. The perfect bed block was two blankets and two sheets folded and layered inside a third blanket that wrapped around the sandwich.

Not a crease, fold or any other imperfection was permitted. A counterpane was stretched out over the Army issued, anti-piss foam pad (mattress). The counterpane was a garishly coloured, striped polyester number that any Woodstock hippy would have been proud of wearing.

Serious failings included, but not limited to, gaps between the layers of the bed block, any difference in the size of the folds exceeding 2mm, any slack in the counterpane or anything else that was an excuse to launch it out of the second-floor window.

It became common practice to gently lift a good bed block (one that had not been ripped apart) off the bed after an inspection and store it, ready for the next inspection. Then roughing it on the floor for the night. The staff were wise to this and soon put an end to the practice by destroying even the good bed blocks.

In the classroom, 'English' became 'communication skills', maths became 'military calculations' and we had to learn German, as most of us would probably end up posted there at some stage. I wasn't a star pupil, but I surprised myself such that the teachers in Ilkley Grammar School would have been impressed.

One teacher, a Mr G Tipping wrote in an end of school report that *"Jim will be lucky to be considered as a road sweeper"*

The Army teachers were ancient Majors who had seen a lifetime of service and would often regale us with their stories and experiences. One old boy, whose name I don't recall, was apparently the Ugandan dictator, Idi Amin's Sergeant Major in a different era.

For pastoral care we enjoyed the Regimental Padres talks and support; Major Basil Pratt, a former Olympic canoeist was a larger-than-life character who never once rammed religion down our throats. He would ride at speed around Stanley barracks on a Vincent Black Shadow motorcycle in a leather jacket bidding everyone a good day.

For the first 2 weeks we were only allowed to wear Army green coveralls, not yet deemed fit to wear a proper uniform. Having progressed from coveralls to uniform, the next stage was to 'pass off the square', a test of drill, both foot drill and weapons, carrying the immortal Sterling sub-machine gun.

This test, conducted by the Regimental Drill Sergeant, determined whether we were now further fit to wear a beret and the RAC cap badge. Our drill instructor was Warrant Officer Harry Shearer of The Scots Guards, probably, no, definitely, the most impressive human I had ever met.

Mr Shearer was the stereotypical Guardsman, permanently immaculate in every respect with a booming Scots accent. His unique ability to inspire, frighten, berate and compliment a young junior leader was as impressive as it was rare, even Jacob Price was a different person in his presence.

Held to the same, if not higher standard than the regular junior soldiers, we passed off the square at the first attempt and received our hard-earned berets. We also earned the right to wear a 'stable belt'. Bizarrely, soldiers were expected to purchase their stable belts, but as we were only being paid £25 per week, with £10 deducted for food and accommodation, we were given them.

Chapter 7. Cunning Running and Other Surprises

At the end of the first term, having passed off the square, we were given a week's leave. The troop excitedly bomb burst into a hundred directions, and I returned to Ilkley a very confident, fit young man.

It did come as a shock though to find my bedroom had been let out to a mature student from Ghana; this information would have been useful before I ended up on the sofa for a week.

Without much money there still wasn't a lot to do in Ilkley.

Thankfully, I managed to see my grandfather. I marched into his front room in uniform and snapped out the smartest salute I could muster. A tear ran down his cheek, his face split into a grin, and with a booming "*BLOODY SEE!*" he pulled me into the biggest hug of my life. No one before or since has been that proud of me.

He died a few months later, but his influence never left me. I only wish I'd known the pre-stroke version of this remarkable man. He was a figure of enormous character, the kind who led from the front, earned respect without needing to raise his voice, and never asked for more than he was willing to give.

Resilience and grit were in his DNA. I think we would have got on very well, trading stories over a beer, both of us too stubborn to admit when things were getting rough.

Back in barracks, having returned a day early, I found our room empty and couldn't wait for the other lads to get back, especially Mig.

The second term built on everything we had learnt in the first. Training stepped up. Signals, night navigation, and longer exercises in the field. Map reading lessons started to include night navigation, in and out of vehicles, and our field craft was advanced to multiple days and nights in the field.

PT also advanced at a pace with swimming becoming a dreaded activity. 10 lengths butterfly, GO! was something you never wanted to hear. I continued to volunteer for every team, whether I liked the sport or not.

We were not excluded from the Army ritual of 'milling', a sport only enjoyed by sadists like wee jockey, the PTI. Milling is a primitive form of boxing where the fighters are deliberately mismatched and provided with giant, unwieldy gloves that resembled, and smelt like, sacks of straw and liniment.

The objective is to knock your opponent over as many times as possible in 2 minutes. Those that try and box according to rules invariably lose and end up bleeding heavily.

I donned the pillow-like gloves and was just thankful that Jockey hadn't put me against Mig. I fired up the windmills and just swung as hard and as fast as I could, making contact several times with my opponent's nose. I won and was elated.

Afterwards I volunteered for the boxing team, firm in the knowledge that I was now a serious contender.

After the practice bouts of actual boxing to determine suitability and weight category, I was told to *'fuck off and never try again for the team'*.

I was determined to get on a team though, so I volunteered next for the orienteering team, not because I was a good runner, I wasn't, but because I could navigate and map read well. I liked the sound of '*cunning running*'. I took part in the trials and, unbelievably, came first by some margin.

For the first time, I was the one at the finish line, cheering others in. It felt brilliant. I had made a team. For once, I wasn't the last man standing when it came to selection.

During the second term we were taken to an adventure training camp in Devon. Renney Lentney was an old naval gun battery from WW2, now repurposed for Junior Leaders. The old Nissan huts housed us in the most primitive of conditions. Placed on the edge of cliffs we were regularly run up and down them to the freezing waters of the English Channel for some hyperthermic swimming.

Whilst we did do some fantastic adventurous training activities, they were interspersed with torturous 'challenges'. I found the leap off an old Devon stone humpbacked bridge into a dubiously shallow stream no problem, but many struggled.

The most difficult challenge was known as 'escalation'. Escalation was simply a race over 6 miles of Dartmoor ending at the top of a 'tor' (a rocky peak), dragging a Land Rover axle with a 40-gallon drum of water lashed to it.

Two ropes extended from the front, which were never long enough to let everyone on them, meaning to be a team player, you had to get stuck in and push it from behind. Broken bones were common.

Around this time, we were ordered to pick hobbies. We had to choose at least two. I picked bird watching, which earned much ridicule, and stables, which made far more sense thanks to my time on the farm with Dinny. Being around animals was familiar.

The stables sat within the camp, tucked away behind the world-famous Tank Museum opposite Stanley Barracks. The museum had long carried a reputation for ghosts and haunted tanks. The story of "Herman the German", a Tiger tank driver said to still roam the place, was ingrained in all of us.

Those stories suddenly took on a bit more weight just before we were assigned night guard duties in the museum. Confidence was high right up until lights out, at which point even the bravest started questioning their views on life after death.

Wandering around the museum with nothing more than a pickaxe handle was no fun, a Ghostbusters proton pack complete with neutrona wand would have been more effective.

I reported to the stables, along with the others, to Corporal Major Pete Bishop, a Warrant Officer from the Household Cavalry seconded to Bovington to run the place. There were five retired cavalry blacks for us to learn on, older horses that were placid enough to tolerate a group of enthusiastic but largely useless beginners.

Chapter 8. Pranks and Pilgrimages

'Soldier' Magazine came to Bovington to write an article about us YTS soldiers. They followed me over to the stables and I featured in the centre pages of the magazine resulting in much hilarity, due to Pete Bishop commenting to the reporter that *"his love of the beast is something else"*!

I had been working late over at the stables one Saturday and arrived back in the room with the place in total darkness and 15 lads asleep. Smelling horribly of horse shit and burnt hoofs I went for a shower.

Tired, I went to get into bed when a steel bedside locker door slowly creaked open by itself.

Unnerved, I became instantly very awake.

The curtain behind me twitched open. A steel drawer from another locker screeched slowly open; panicking with memories of 'Herman the German', wandering the halls of the museum in eternal unrest, fresh in my mind, I called out to Mig, *"We're being spooked! He's here!"*

No reaction from Mig or Tom, or anyone else, the creepy noises and movements kept on coming, a curtain twitched open.

Then a snigger from someone that couldn't hold it together anymore, the lights flashed on, and the room erupted in fits of laughter. Only then could I see the spiders' web of Army green string!

Not all our formative experiences were pranks, though; some were profoundly humbling.

Religion has never been a part of my life and it's fair to say it never will be. So, it came as a surprise to be called into the Squadron offices and told: *"As a Catholic, you've volunteered to go on a charity trip to Lourdes in the South of France."*

A charity called 'The Across Trust' was seeking help transporting disabled and terminally ill people on a pilgrimage to Lourdes. They requested Catholic volunteers and according to my Army records, I qualified.

The Trust had a giant bendy-bus, or "Jumbulance," fitted with stretchers and medical equipment for the journey. My job was simple: be the muscle, helping move people in wheelchairs around the very steep parts of Lourdes.

I was paired with a little girl named Layla, around 8 years old, who had severe, life-limiting conditions. She never complained. Not once. She enjoyed the sunshine, the trips around the Stations of the Cross, and being surrounded by other children who didn't make her feel different.

She was dying, and she knew it, but she never let that define her. Her fierce drive, not to live forever, but to live well. She refused to be pitied, never asked *"why me,"* and never stopped making plans, even when time was clearly running out.

She showed me that strength isn't loud or showy. It's getting up, choosing your attitude, and cracking on anyway.

Her determination stayed with me long after she passed, just a few weeks after we returned to the UK.

I've never forgotten her. I doubt I ever will.

For our second period of leave, Mig and I arranged for rail warrants that started in Bournemouth, where he was from, and then up to Ilkley before coming back to Bovington.

Chapter 9. Shoes and Best Boots

In our third term we were told whether we could transfer to a regular Army contract. Most of us that wanted to be, were successful, some chose to leave. Others were found unsuitable.

I elected to apply to the Royal Army Veterinary Corps (RAVC) for training to be a farrier.

I was sent to Melton Mowbray, the HQ of the RAVC, to go through a selection process as it was a sought-after trade with few vacancies. I used my time at the stables to practice making horseshoes in an open forge with a lot of help from Pete.

I passed the selection process, forging a horseshoe from straight bar metal against the clock.

Back in Bovington, we spent more time in ever advancing training. We sat our education exams which I passed with good grades. It gave me satisfaction knowing that the grades and reports received from school, and Mr Tipping had been wrong.

We were on the square with Mr Shearer at least once a day, preparing for our final pass-off. We looked good, we were very fit, and we felt it. Most of all, we were proud.

Taken off the street by a government scheme, we had become young soldiers, as good as, if not better than, the regular entrants.

The Junior Army Orienteering Championships took place in Epping Forest during my final term. I finished second in my category, and our team won the championship overall. I also managed to grow six inches in height. More remarkably, I received the only sports medal of my entire life.

About a week before the passing out parade we all received our Regimental cap badges and accoutrements. I received my RAVC badges. Mig elected to go to the Royal Engineers, where he excelled attached to the Royal Marines. We would meet again in very different circumstances.

My mother came down to Bovington for the passing off parade and was a proud Mom, I knew the lad in the oversized suit had finally found his place.

Or so I thought.

Bovington's Youth Experiment Proves to be a Success . . .

13 The Infamous Soldier Magazine Article Pt 1

'We have ended up with some very good soldiers'

Story: Graham Smith Pictures: Les Wiggs

14 The Infamous Soldier Magazine Article Pt 2

15 Simon McVey, Steve Humphries, Tom and Mig

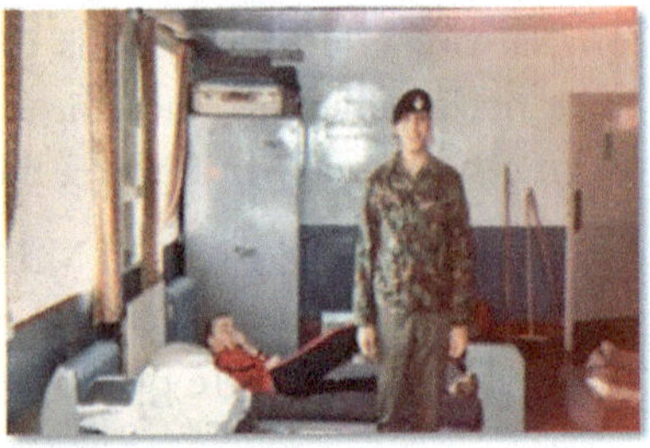

16 Mig, Tom on bed

17 Practice events for the Junior Army Orienteering Competition

18 Me in Vest

19 Finishing Second in Epping Forest

20 Proud Mom at the JLR Passing Out

21 Alamein Troop, A Squadron, Junior Leaders Regiment, Royal Armoured Corps

Part 3: Nosedive to Level Flight – Life on the Ground

UK, Melton Mowbray, RAVC: Jan 85 – Apr 85

Chapter 10. It Shouldn't Happen to a Vet

We passed off in December 1984. Our contracts converted to the regular Army. We received 12 months of back pay to reflect what we would have earned had we originally been on regular pay. I went home on Christmas leave with over £3000 in my bank account, my report date for the RAVC in Melton Mowbray was, ironically, the 10th January 1985.

We'd joined as boys, left as young men.

My first stop was the motorcycle shop in Ilkley. The learner laws had recently changed, new riders were now limited to 125cc machines instead of 250cc, so the shop had a good number of used bikes of that size. One was begging to be bought and ridden home.

I bought it, along with a helmet, and rode it home, much to my mother's horror. I talked her into getting on the back while I rode it up to my grandmother's house, I hadn't seen her since my grandfather died during my second term in Juniors.

On the way home a car pulled out directly into us. We were hit on the left-hand side; both of us thrown violently from the bike. My Mom slid across the road on her face, the full-face helmet she had worn absorbed the impact and she was badly shaken, but largely unhurt.

The impact of the car caught my ankle which was shattered between the car bumper and the bikes engine block. After being taken to hospital, my ankle was reset and cast in plaster.

The driver of the car inexplicably tried to blame me for the accident; my mother took immediate offence and countersued him. Luckily the police apportioned full blame for the accident to him, but I was prosecuted for driving with no licence, no insurance and an illegal pillion passenger. Mom was prosecuted for *'aiding and abetting'* the same offences. I was not popular.

I arrived at Melton, as ordered, on 10th January but had to report sick immediately so that the Army could take over my recovery.

Melton Mowbray Camp was in an unusual place, the main gate was placed between two houses, and it felt oddly civilian after a year of training having just completed 12 months of basic training. I had an

8-man room to myself and could move furniture around to create a sort of studio flat.

After a period of physiotherapy my ankle had healed enough to have the screw removed. I was taken to an RAF Hospital near Swindon for the operation, but as soon as the wounds had healed, I was anxious to get down to the forge to start RAVC trade training.

Before I was allowed near the forge, I had to undergo the mandatory RAVC soldier training. This involved veterinary assistant training, guard dog handling, and a riding course.

Whilst posted to the veterinary hospital to learn some of the skills required, I had my first experiences of working with the girls that were already qualified as veterinary nurses.

My general tasks involved checking piles of horse shit in the paddocks for signs of worms, cleaning everything in the hospital 12 times per day and shovelling shit from the dog kennels. Oh! And making brews.

There was no actual training going on, but I stuck at it.

The Army manages its defence animals as it would any other piece of kit. Dogs and horses have numbers. When they are no longer useful, they are 'cast'.

Euthanised.

On paper it's a process. In reality, it's the quiet end of a relationship with a colleague who never felt like equipment.

It was heartbreaking, especially the dogs. I hated it.

Whilst I didn't enjoy working at the vet hospital, I was beginning to enjoy the social life that came with working with the girls. I was going down to the town at the weekends, the local club was called Tubes, and it was a dive.

One Friday, one of the girls, Sophie, told me her younger sister, Emily was visiting for the weekend, I would like her, and we should go to Tubes together. I was just turning 18, and all the girls were a lot older than me, it was an opportunity not to be missed.

Although she was just over 3 years older than me, Emily and I got on well, we liked each other from the start and stayed in close touch after she had gone home. The fact that she had a 2-year-old little boy, Matthew didn't phase me at all.

When my time at the vet hospital came to an end, I was loaded onto a guard dog handlers' course with soldiers from other units.

There were a lot of Irishmen from the Ulster Defence Regiment (UDR) and other units that used dogs of all kinds in Northern Ireland. I would listen to their stories of the truth of the conflict and what they had to deal with daily.

At the same time, my 1 man, 8-person room became a 2-man room. Joe 'Banksie' Banks, a skinny lad from Darlington was on his 3-month long class 2 farrier course. Banksie was in the stables troop of the 15/19H who had recently taken over as the centre Regiment back in Bovington.

We hit it off straight away, a bit older than me, he was worldly wise, funny and great to be around in my eyes.

Together, we were either in the pubs of Melton or sweating our tits off in Tubes, punching the air to The Communards' *"Don't Leave Me This Way"* or swaying awkwardly to Billy Ocean's *"When the Going Gets Tough."*

We were even roped in by the management of Tubes to do a strip'o'gram act. Wearing very little other than leather chaps and homemade white cardboard collars & cuffs, we performed the most ridiculous act ever seen in Leicestershire; but we enjoyed it, even if no one else did.

If Emily hadn't come down for a weekend, Banksie and I were busy getting into trouble, which wasn't going unnoticed.

I passed the guard dog handler's course. I'd done everything they asked. So, when they told me I wasn't staying in the RAVC, it didn't make much sense at all.

I was angry.

All that effort, the horses, the forge, the hours, gone in a sentence. I walked out of that office convinced I'd failed. Not the course. Myself. And yet, somewhere between disappointment and stubbornness, a little voice kicked in: screw this. I'm not done yet.

Stunned and shocked, Banksie took me out for a consolatory pint. He told me not to worry about it as the 15/19H were looking for farriers and it would be a simple matter to transfer to them and fuck the RAVC.

My choice was simple. Transfer to the 15/19H or leave the Army. Gutted, I applied to transfer to a place where I had just left, knew no one apart from Banksie and was left disappointed with myself, being told I wasn't good enough after everything I'd worked for crushed me.

But instead of feeling defeated, I realised how much I had been wasting my own potential.

UK, Catterick: The 15th/19th The King's Royal Hussars

Chapter 11. Gunnery, Goonery and a Wedding

I knew about the 15/19H from my days in JLR, their name as an outstanding reconnaissance Regiment, their fierce internal pride and their reputation as fighters. On transfer I reported to Catterick Garrison to undergo trade training as a tank crewman. The training centre was manned by the Queens Own Hussars (QOH).

The QOH had the opposite reputation to the 15/19H, they were viewed at the time as scruffy and ill disciplined, but they also had a stable's troop. This troop was manned by soldiers who hadn't fitted elsewhere and were largely incapable of being tank crewmen.

The QOH had a gentle, giant grey drum horse called 'Dettingen', named after one of their battle honours. Dettingen would be paraded at every passing off parade held in Catterick Garrison. He would be turned out immaculately with two large silver kettle drums to lead the bands.

This meant that while I was holding for a gunnery course, I was seconded to the QOH stables troop, still wearing the RAVC cap badge. I was instantly tarred with the same brush as the resident troop of misfits, and it proved a hard perception to shake off.

I really enjoyed my Chieftain tank gunnery course and passed with the highest grade on the course. The gunnery instructors were appalled and vocal that I was going to a stables troop given my results; they tried to talk me out of it, but I had made a commitment to Banksie that I would keep given he had helped me.

Catterick was a stone's throw from Emily's house in Stockton-on-Tees, and I saw even more of her. Living with her parents, I had no excuse not to meet her mum and dad. Arthur, her father, was a very quietly spoken, tall, distinguished gentleman with an impressive range of degrees in engineering. Bobby, her mum, was the opposite.

Bobby and I took an instant dislike to each other, although I tried to hide it. I found her rude and overbearing, her opinions were usually extreme, and I saw a nasty streak in her. She was however a very talented artist.

Being so close to Emily, in a large garrison where I didn't know anyone, was ostracised anyway for being in the stables troop, meant we saw a lot of each other and became closer. Despite her mother being very difficult to get along with, let alone like, she was overheard referring to me as *'just a fucking squaddie'*, it was funny and still makes me laugh.

With a posting back to Bovington, at the other end of the country imminent, we decided that the only option available to us was to get married so that we became entitled to Army housing. On the 13th July 1985 we married at Bradford Registry office to muted celebrations from both mothers.

We cared for each other, sure, but in truth it was a shotgun wedding with the Army holding the shotgun in the form of a house.

Ironically, we had tickets for Live Aid. The one where half the world tuned into watch the world's biggest gig. But we didn't go. Instead, we got married. While Freddie Mercury electrified Wembley, we were signing a registrar's book in Bradford.

By now, my mother had sued the driver of the car that had smashed into us on the motorbike, and I was awarded £4,500 in compensation. Using this money, we bought a second-hand Ford Sierra from a dealer in Stockton and Emily drove us down to Bovington, as I still didn't have a licence.

UK, Bovington Camp: 15th/19th Hussars

Chapter 12. A New Start

We were allocated an Army house in Bovington Camp. Our first neighbours were Kenny and Joanne. Kenny was a smaller bloke from Middlesbrough with a character twice his size. Kenny was lovingly known as Rommel.

Kenny and Joanne never judged anyone. We became friends quickly and they introduced us to their friends in the Regiment.

On my first day in the stables troop 15/19H I met the whole team. The officer in charge was Captain Lew Godfrey, a lanky officer who had been commissioned from Regimental Sergeant Major. Lew took to me and I liked him from the outset. He was always kind and helpful, I later found out that it was Lew that sponsored me into the Regiment, not Banksie.

Lew's wife, Marge, was a softly spoken woman with the gentlest of Northeastern lilts. Like Lew, she was kind, thoughtful, and quietly protective of this young couple trying to find their feet. Lew drove a 200-series Mercedes that looked every inch the German taxi it probably once was. It was his pride and joy; spotless, polished, and maintained with reverence. But he was notorious for driving it like Hoke from Driving Miss Daisy, slow, steady, and utterly impervious to urgency.

Chapter 13. Horses and Dogs

The Troop Sergeant was Duffy Fox. Duffy was also the master farrier for the Regiment and was built as you would imagine. Shoulders designed for flogging errant soldiers and an attitude that meant he would enjoy it. In his troop were three or four troopers. Then there was me, the lowest-ranked hangover from another troop of misfits.

Working at the stables was physically demanding work, day in, day out. Being the duty groom would mean a dark start and finish. Throughout the polo season, we would spend days away from home being grooms for the officers. Whilst they would be larging it, literally with Royalty, we would be living in the back of a horse truck, roughing it in a sleeping bag.

Some were dismissive, arrogant, and blind to the effort we put in. But not all. Others were superb, generous with their time and quick to acknowledge hard work.

I met a few great officers, the kind you could imagine having a drink with, in another universe. One stood out: Captain Patrick Darling; a gentleman through and through, and an officer I greatly admired.

Years later, in 2017, I visited Patrick at his estate in Herefordshire. He still had the blanket box I'd made for him out of a gearbox packing crate, tucked away but clearly treasured. I stayed overnight, and he was exactly the man I remembered: warm, sharp, and full of stories.

During this period the Regiment based in Bovington would host the annual "*Bovington Battle Day*", a massive display of armour through the years in action at the display area, just east of TE Lawrences' former home in Clouds Hill. That year the selected theme was '*Cavalry Through the Ages*'.

The idea was that Stables Troop would start the day with a dashing mounted display, using the Cavalry Blacks, of the old cavalry sport of '*tent pegging*'. Tent pegging used a sabre, a lance or a pistol to remove the pegs holding the enemies tents up in a bid to cause chaos and confusion.

Four of us were chosen; an Officer, Duffy as the Sergeant, Banksie as a Lance Corporal and me as a Trooper. Four red balloons (representing tent pegs) were to be placed equally between two large bushes, each of us would then canter or gallop to demonstrate bursting the balloons, first with a sabre, then a rapid change to a lance, burst the second run of balloons on the return charge and depart to much applause and cheering for the dashing young cavalrymen.

On the charge out the first three all hit their targets as intended; I missed. On the return charge I was determined to join the suave and sophisticated trio ahead and kill the balloon with my lance. The lance is a bamboo pole, about 12 feet long with a sharp spike on the dangerous end and a large steel counterbalance on the other end. The approved technique was to raise the lance so that it travelled under the right armpit, take aim, drive the sharp end through the target and then allow the shaft to smoothly glide through your grasp behind you.

I'm afraid to say that in my determination to hit my target, the approved technique was the first thing that died that day. Although I did hit the balloon, I retained a firm grip on the lance shaft and as a result was spectacularly pole vaulted over Wadden's head.

As I reached my zenith, just about directly above his ears I dropped the lance and grabbed the bridles brow band in a desperate attempt to remain mounted. The combination of the lance dropping near his face and me perched on his ears caused him, quite rightly, to slam the anchors on and stop in his tracks.

I, however, continued and landed about 6 feet in front of him, de-bridling him as I flew forward. I stood up, bridle in one hand, my busby in the other and pride in the mud, watching Wadden disappear into the distance. I felt the full weight of 10,000 spectators erupt into cheers and clapping. I'm just thankful cameras and cell phones weren't the thing they are today!

Emily would often walk down to the stables, about a 30-minute walk, pushing Matty each way to bring me some lunch, otherwise I wouldn't eat. The cookhouse had decided to cordon a section off and surround it with plastic plants due to complaints about the smell from us. I wasn't going to be humiliated and the marmite and cheese sandwich always tasted good.

One day, in a moment of genius (or heatstroke), someone proposed a round-robin of mounted wrestling. The rules were unencumbered by Health & Safety rules: bareback, bridles only, and the objective was simple; knock your opponent off their horse before they knocked you off yours. No points for style, this was 'mounted milling'.

I drew George for the first round. He was compact, wiry, and looked like he'd grown up riding puppies. I fancied my chances right up until the moment he smiled. I missed the grin that said he'd been waiting his whole life for an officially sanctioned excuse to commit GBH on horseback.

I rode up to him, got him in a passable headlock, and braced to twist him off sideways. That was the plan, anyway. George, unbothered, gave his horse a casual nudge and trotted off, taking me with him like a bit of laundry snagged on a door handle.

I hit the ground hard. Elbow first. Something popped, which wasn't part of my strategy. My shoulder had vacated its socket entirely, and my right arm locked up in a posture I can only describe as 'yoga for idiots'. My right arm was curled up and over my head, fingers touching my left ear like I was trying to eavesdrop on myself.

Duffy bundled me into his ancient, steel-sprung Land Rover and rattled us all the way to Poole Hospital. I felt every pothole, stone, and ripple in the road like a punch to the shoulder. A trainee doctor had a go at resetting it and only made things worse.

Then came the final humiliation.

Straight from the stables, I absolutely reeked of horse. The nurses on the ward thought I'd shit myself and promptly gave me a bed bath to deal with the complaints from other patients. No one ever mentioned it again, but then they didn't need to.

If it wasn't for Kenny and Joanne, we'd have gone stir crazy. They brought friendship, laughter and a lifeline that made Army life bearable. With burnt corks, daft party games and proper belly laughs, they helped turn routine into something that felt like living.

Emily fell pregnant soon after arrival in Bovington and we had our first daughter in September 1986; I was delighted and couldn't wait for the new arrival.

Alexandra Louise arrived like she'd always been there. She didn't cry much, she just stared at the world with these wide, calm eyes. I was smitten, but not in a movie way. It was something deeper. Like gravity had shifted and the centre of everything had quietly moved. I wasn't just a soldier anymore.

I was a father, a husband, and someone who had to keep fighting not just for himself, but for them.

Soon enough, the Regiment was off back to West Germany, this time to Detmold on Chieftain tanks. Before we left, they asked for volunteers for the guard dog troop, part of the Regimental Police. Training, promotion, and real responsibility were on offer. It sounded like a decent deal.

I jumped at the chance to get away from the stables. Already a qualified dog handler, I was a natural fit and was duly selected.

I would assume the rank of LULC on arrival in Germany. It was only then that I found out that LULC stood for 'Local, Unpaid, Lance Corporal'. A total con trick that meant I had to purchase an expensive Junior NCO's mess kit and pay mess bills, but without the perks of any extra pay.

We bought a cheap roof rack for the Sierra, from Halfords, along with an RAC road map of Europe and made our own way to Detmold. All we knew was that our bank would be the Sparkasse and we had been allocated a married quarter in an estate known as '*Hakedahl'*, or simply *'The Hak'*. On the 10th January 1987 we packed everything we had, including the kids and set off.

22 Captain Patrick Darling

23 Sergeant Duffy Fox

24 L/Cpl Joe Banks Mounted, Capt. Darling Standing

25 Me and Archie

W. Germany, Detmold: 15th/19th Hussars, 1987 - 1992

Chapter 14. Driving, Dogs and Duties

The drive to Detmold was an adventure. We made it to Dover without too many stops and got in line for the Sally Lines ferry crossing to Dunkirk, the cheapest crossing available. The weather wasn't great and once out of the car, we went upstairs to one of the grotty lounges on the aging ferry.

The second we sailed past the calming harbour walls the ferry transformed into the stereotypical vomit comet.

Whilst the two very young kids remained blissfully unaware, like smug little tourists on a disaster cruise, both Emily and I were as sick as *Sea-Sick Steve* after a rough night on the town. Every movement set off a fresh wave of nausea. Changing nappies in that condition? That wasn't childcare, it was biological warfare. Eyes watering, stomach flipping, gag reflex doing backflips, and the baby just slept.

On the way back to the car as we neared Dunkirk, we chatted with a friendly German couple who asked where we were headed. The Germans realised we didn't have a clue about directions or the European motorway network, despite the RAC road atlas, and suggested we follow them for a good portion of the drive.

We stuck on their tail up to the point we arrived at their house, not realising we were supposed to leave them at a large junction many miles before. They fed us some lunch and provided us with written instructions and hand drawn maps on how to get to Detmold.

As it was starting to get dark, we arrived in Detmold, and looked for signs for Hakedahl, no one had told us that it was called '*Heberhausen*' in German. After hours of driving, we decided to look for the Regimental barracks and ask for directions from there.

We set off looking for Lothian Barracks. No one had told us that it was called '*Klute Kaserne*' in German.

A theme of unpreparedness had developed that was biting us squarely in the arse.

We stumbled across the main gate after what seemed like hours, both of us more than a tad irritable with two crying kids in the car. We were both sick to death of '*Now That's What I Call Music 8'*, the only cassette we had.

We finally got accurate directions and found our new home in Londoner Strasse. Our direct neighbours in this block of apartments were from the Army Air Corps (AAC). Ken and Sarah Marsh.

Ken was a Falklands War veteran; he was one of the first Army corporals to pass pilot selection and earn his pilots' wings.

Ken and I became good friends and, unbeknown to either of us at the time, he would become a major factor in my story.

Dog Section of the Regimental Police was led by Corporal Kit Gibson, a big bloke from inner Newcastle upon Tyne with an accent so thick you could have built the foundations for the Tyne bridge from it, I genuinely couldn't understand a word he said for the first week, which was nerve wracking as he was a heavy weight boxer with a reputation for not taking any shit, at any time, from any person.

The section came together with three other LULCs.

As it turned out, I was not just the only person in the dog section that wasn't from the recruiting area, I was one of a very select few across the Regiment, but this wasn't my first rodeo being a chameleon. In very short order, and subconsciously, I picked up the Geordie accent very quickly and ironically, my name reverted to Jimmy.

Our first duty on forming the dog section was to go to Sennelager, the RAVC centre, to meet our guard dogs. I was issued an older, experienced German Shepard Dog called Beach. He was the most intimidating dog up for issue, but I bonded with him straight away. Beach was probably the best dog in the section, but then I would say that.

A standard duty for us lasted 24 hours. We would conduct patrols of the whole barracks irregularly at night and train with the dogs during the day. It was easy to get volunteers to dress in the heavily padded '*baiting suit*' and then to act as an intruder.

Our dogs were classed as '*right arm true*', meaning their primary target was, unsurprisingly, the right arm. They were trained to latch on and then drag the unlucky person to the ground.

On an evening, we would relocate from the kennels to the main guardroom, where the barrack guard would be, tethering the duty dog to a post on a long leash; just long enough to scare the ever-loving shit out of anyone that got too close, but hopefully not long enough to launch a successful attack.

Drunks returning to the barracks after a Friday night in Felixs' Bar, or the Schwarze Witwe (Black Widow) night club occasionally got too close and suffered instant regret along with a sore hangover.

Each year, the RAVC would run an International Canine Biathlon, a competition for dog handlers from all NATO nations. It involved a 10km run as a dog team with your personal weapon. After 5km the teams arrived at a shooting range where a range of targets were presented.

A further 2km run to the finish line culminated in a release of the dog against a simulated intruder. Points were awarded for the time taken on the first 5km run, accuracy on the shooting range and finally the technical aspects of handling skills for the interdiction.

As a section, we did well, I scored highly on the shooting, but the release didn't go quite so well. As I entered the final arena for the judged release, I hurled, largely due to the party the night before! While I was bent over double, fertilising the arena, Beach did what he did best and took care of the intruder with zero control from me.

As with the Bovington Battle Day, the sound of applause was deafening.

Life in the dog section was better than the stables but was still not a career path in an armoured Regiment, it wasn't a place to shine or to stand out. The endless duties and time in the barracks away from home meant Emily and the kids were still barely seeing me.

In early 1987 Emily was pregnant again and in September, exactly a year after Alex was born, Jessica Ann arrived in a whirlwind, partially in the bath in the house, with a full arrival in the back of the ambulance on the way to the local German hospital.

Jess was born with a full head of long, dark hair and an attitude to match. Alex and Jess, only 12 months apart, were as close as twins.

Jess had a sparkle in her eyes, when you could see them under her long fringe, that said *"Here, hold my beer, watch this."*

Chapter 15. Murder and Medics

In late 1988, a member of our Royal Electrical and Mechanical Engineer (REME) detachment, Sgt Geoff Massey, drove into camp through the main gate. The IRA threat level was high and as such every vehicle coming in was subjected to a thorough search, including an under-body search with a mirror.

The guard reported an object under the car as the boot was opened. In the boot was a load of muddy and blood-stained clothes.

The guard commander called the RMP to attend and it soon became clear that the under-vehicle object was a hoax bomb.

What later emerged was as unbelievable as it was tragic. Massey was a loner that had an interest in battlefield archaeology, several WWI unexploded grenades were found in his room in the mess, along with other battlefield finds.

Prior to the hoax bomb, Massey had apparently seen a car for sale in the force's newspaper, the Sixth Sense. The person selling the car was a female Army doctor in the nearby city of Bielefeld, Massey arranged to view the car and asked for a test drive.

During the test drive he abducted the doctor and drove her to a prepared location deep in a German forest on a British Army training area near Sennelager. He'd found the site during a previous exercise in the area.

Having tied up the doctor, he lit a fire to cook her some food; unfortunately, the local German forest ranger saw the smoke and with his son, went to investigate.

Massey had a self-made pistol with him and shot the ranger and his son dead on the scene.

According to reports at the time, the doctor was a psychologist and talked Massey into letting her go, and then for him to hand himself in. The hoax bomb was an attempt to get rid of the evidence in the hope that his car would be destroyed by the bomb squad.

What is certain is that two innocent men lost their lives and Massey was convicted of murder in a German court and was sentenced to life imprisonment.

Life in Detmold carried on, but the atmosphere had shifted.

By then, the 15/19H had established a fourth Squadron of 14 Chieftain tanks, plus associated armoured fighting vehicles. 'D' Sqn was commanded by Major Aidan Moonie, a highly respected officer whom I had met whilst in the stables.

Speaking to him in Detmold, I was virtually begging for him to take me into 'D' sqn, in any position available.

Maj Moonie came up with the goods and found me a position in the 'D' Sqn headquarters troop, known as Fighting HQ, or FHQ for short.

I jumped at it and was now the driver of an armoured ambulance., armoured personnel carriers that had been converted to become a field ambulance.

The ambulance was crewed by a driver and a medic, the latter doubling as the vehicle commander. I was paired with Ray Hutt, the Squadron medic. Day to day, Ray worked out of the medical centre in a separate barracks on the edge of town, near the Hak, which meant I had free rein when it came to maintaining and servicing the ambulance.

As a crewman by trade, I had completed the radio operator training and my Chieftain gunnery course in Catterick; I just had to complete a driving course to become a qualified class 3 crewman. I was allocated a driving instructor from within the sqn; Cpl Les Stoker got me through the driving course on the ambulance.

Crewing the ambulance was a great job; I enjoyed it very much. On exercise Ray and I would be together with the two sqn command tanks and the Squadron Sergeant Major (SSM), but we would be tasked to operate between the Regimental Aid Post (RAP), field hospitals and the sqn.

The SSM when I joined 'D' sqn was Graham Poole, a great man but he was replaced shortly afterwards by the legend that was Geoff Porteous. A former paratrooper, hard as nails, Geoff had more artistic talents than I can name; he would randomly write poems and songs that were utter brilliance.

A very caring man underneath it all, he had the ability to command respect as an SSM should, without losing the human touch.

We would often have simulated battle casualties on board the ambulance, but when we didn't, we had the absolute luxury of having a full stretcher each to sleep on. There was plenty of space for us to make life comfortable when the rest of the Squadron would be sleeping on the back decks of tanks, we were lording it up.

The career progression for a tank crewman was normally clearly laid out; once class 3 crewman had been achieved, a selection was made to specialise in one of driving, gunnery or signals.

Although I wanted signals, I was sent back to Bovington to complete a class 2 driver/mechanic course at the Driving & Maintenance school.

It was a hard eight-week course, but I loved it and came away from it with the top score on the course, and a formal recommendation to return to the school to complete an instructor's course.

Italy: 15th/19th Hussars – Exercise Roman Hussar

Chapter 16. Pasta, Peaches and Wine

Northern Italy in autumn, vineyards, peaches, and the faint promise of excitement. Instead, we got fleas, boiled ravioli, and the occasional farmer with a shotgun.

In late 1988, 'D' Squadron was selected for an exchange with an Italian armoured regiment. We deployed as a unit to live and train alongside them, which sounded far more civilised than most of what the Army usually had in mind.

Exercise 'Roman Hussar' paired us with the grandly titled 15° Gruppo Squadroni Carri 'Cavalleggeri di Lodi', based in Lenta, between Milan and Biella in northern Italy, not far from Gattinara, a region better known for its wine making than its warfighting.

Geoff Porteous had been on a recce and came back with tales of what was to come. His reports were not entirely accurate!

"All the locals know we're coming and can't wait to meet us" he said. "*you'll be treated like Kings* "he said.

The Squadron took the train from Bielefeld all the way through to Milan, 120 Geordies must have been a wasp herding challenge for Geoff and the other senior NCOs, but we made it, apart from one officer; Capt. Elton Farnham, the Squadron second in command, who had to make 'alternative' arrangements. Nobody ever got a straight answer on that one.

To be blunt, the exercise was a shambles of the highest order, not because of any poor drills on our behalf, but because the Italian Regiment was made up from conscripts who really did not want to be there. The only regular soldiers were the warrant officers, and I don't think they wanted to be there either.

The accommodation resembled an open prison, no running hot water and straw mattresses so old the fleas had tenancy rights.

The food was, quite literally, boiled pasta. Every meal. Every day. For six solid weeks. Now, I don't know if you've ever met a Geordie, but they do not function without chips. You can take their beds, their pay, even their dignity, but take away chips and you're asking for civil unrest.

It all came to a head when Tom 'Paddy' O'Hara finally snapped. He stood in the cookhouse queue, plate in hand, his boiling rage matching the pasta, and bellowed, *"Where's the fuckin' proppa scran, man?!"*

The Italian cooks, more plumbers than chefs if we're honest, looking genuinely terrified. They huddled, conferred, and someone legged it to the 'Commissariato', the Italian equivalent of the NAAFI. They came back with what they clearly thought was a peace offering: a tray of Cornetto's.

And that's how a near mutiny was quelled. Not with steak and chips, but with mass-distributed novelty ice cream.

The Italian Leopard 1 tanks were crammed into a former rice warehouse and of nearly fifty tanks, fewer than ten would even start. Their only driving area was about 1km long by 500m wide.

During sports competitions, most of their teams didn't even bother to turn up, and if they did, they'd only have one trainer and a flip flop between them.

We took part in a map reading exercise, on foot of course. We were issued maps that looked like they'd been hand drawn by 'Giuseppe' aged 3 on the back of a pizza box. The route took us through a lot of orchards of varying types, apple, citrus fruits and peaches.

These peaches were huge, the size of a cricket ball, but of most importance they were ripe and ready for eating. So, the Geordies did what hungry Geordies do; went through them like a plague of locusts.

The farmers, however, didn't think it was OK; in fact, they took downright offence to their precious crops being mauled. They did what farmers do and chased at least half a dozen lads with shotguns. One of the lads later quipped it was the first time in recorded history an Italian had been seen on the attack.

Geoff did what he could to keep spirits up. At one point, he took to parading around camp in a tee shirt from the movie Alien, complete with a floppy rubber face hugger erupting from his chest. He wore it like it was standard issue, and after a few days, no one batted an eyelid

We were taken into Milan to go shopping a couple of times as a distraction, and I managed to scrape enough cash together to buy Emily some Milanese couture.

Trying to describe what size I needed for this beautiful underwear; the lady assistant gave up on words altogether and just cupped the air in front of her chest like she was trying to smuggle two footballs. '*Ow massive are ze lady's teats?*' she said, honking away. I nearly pissed myself; Milanese couture sold with Benny Hill hand signals.

The unmitigated disaster of Lenta left me with a love for northern Italy and even Italian food, although not the Army type. Some of the wines from the vineyards of Gattinara were just superb and I still look for them.

We headed back to Detmold very pleased to be leaving and wondering how we would explain the contradiction of the suntans against having had a really bad experience, evidenced by us all being ten kilos lighter.

Chapter 17. Driving Miss Daisy

We had just had a change of Squadron leader, with Maj Simon Edwards replacing Maj Moonie. Simon was tall, blond, already balding, with an infectious smile. He was an outstanding officer and was very highly respected by the whole Squadron.

He made some immediate changes; all the Squadrons in the 15/19H had nicknames that went back years. Simon realised that 'D' sqn, being so new didn't have one. In our short existence we had always played the enemy of every exercise we had taken part in.

We had our nickname; ***'The Enemy'.***

By coincidence, his tank driver was posted out on promotion, so callsign 0B needed a new driver. This position, Sqn Ldr's driver, was normally reserved for the more senior drivers, but Simon, on seeing my results from the D&M school, adopted me into his crew along with L/Cpl Ivan 'Skinny' Holmes, the gunner and Sgt Blackman, the loader/operator.

Simon had me paint the nickname of the Squadron, 'The Enemy' on both sides of the tank where all other tanks also had their names. The tank names were at the discretion of the crew, as long as it started with the letter 'D'.

We still needed a name and so ØB was christened 'Daisy', which was painted on the front left mudguard.

Simon and Geoff took a good Squadron and made it great through leadership and management that I looked at and knew was a model for success.

The highlight of the year for us was annual ranges. We would pack up everything and move on tank trains up to the range area. The Bergen-Hohne Ranges on Germany's Lüneburg Heath was NATO's busiest live-fire complex.

The tank trains in Germany were a lesson in patience. It could take up to two days for them to reach a destination that would've taken just a few hours by road. As drivers, we'd accompany our own tanks in a single carriage that trailed behind the enormous, clattering convoy. While the journey was long, dull, and cold, we never travelled unprepared. Every trip included multiple bright yellow cardboard crates of 'Herforder' Pils, universally known and loved as the “orange handbags.”

These packs of lager, brewed in the town of Herford, were a cultural institution among the British forces. You could spot the orange carriers tucked under nearly every seat or jammed into side bins on any military vehicle. *'Herfy'* was crisp, drinkable, and possessed just enough bite to take the edge off a freezing night in a sidings yard or the third hour spent watching a railway signal blink red.

If you saw someone turn up without an orange handbag, you knew they were either new or had made a grave logistical error. They were morale wrapped in cardboard, always shared, never wasted.

For weeks prior to a range period, we would be preparing the tanks mechanically and 'exercising' the Tank Laser Sights (TLS). The Chieftain's fire-control system used a very basic laser that measured time-of-flight.

It was prone to failure if not used for extended periods of time, so we would fire it into sheets of photographic paper to check that it was working correctly. This process was overseen by our REME team, Cpl Pete Carre was the man on the scene, helping our turret crews get the gun kit to the required standard.

A range day started at 03:00, breakfast was smashed down before being at the armoury at 04:00 to collect the co-axial machine guns, the commanders machine gun and the keys for the TLS.

We would be on the firing point at about 05:00 to start the days preparation; for the drivers that would mean a mechanical first parade ensuring the failure-prone engines should at least start and provide power. The gunners would start their own schedules culminating in a process known as bore sighting.

A telescope like device would be inserted into the muzzle of the barrel. The loader, precariously balanced on an old oil drum, would shout instructions down the barrel to the gunner, who would traverse or change the elevation of the gun, so that a small circle in the bore sight would be trained on a matching circle at exactly 1 km down range. Once exactly on the target, the gunner would adjust his sight to the same down range target and lock it in. This ensured that the TLS was pointing at the same thing the gun was.

Once all this preparation was complete, everyone would come under control of the SSM for the ritual of '*ammo bashing*'. Chieftain ammunition came in three pieces and was individually boxed. We had to unpack hundreds of boxes, lay the projectile, or bag charge on top of it and insert a chit, individually signed by the SSM to declare the container was now free from explosives.

By now, it would be about 07:30 and time for the gunnery instructors daily brief on the day's activities. The first round down the range would be fired at exactly, to the second, 08:00. The firing serials would continue until around 17:00 at which point we would invariably change ranges ready for the next day.

Lunch was always delivered to the range points, and it was always a stew. '*Range stew*' was the one-pot wonder that turned every live-fire day into a gastronomic game of Russian roulette. It was ladled out of industrial, green, 18-litre insulated 'Norwegian' food containers ('Norgies') that kept the mystery meat volcanic for hours.

This was accompanied by Army issue bread in waxed paper with a block of semi-frozen margarine to grease it up; which aided transition from your mouth to your arse, via the gullet. The final part of lunch was a mug of '*cofftea*', which also tasted of the range stew from the previous day.

On arrival at the new range, the tanks would be lined up ready for the morning and put to bed. The guns would be pulled through and scrubbed, rogue spent 7.62 cases from the co-axially mounted machine guns would be retrieved and the turrets cleaned and tidied. The machine guns were dismounted and taken back to the transit camp armoury.

After a terrible evening meal, we were into cleaning the machine guns, the barrels of which had seen many 1000's of rounds of tracer go through them.

People tried everything from vinegar to sandpaper and chisels to remove the chunks of carbon deposits that had built up over the day. Typically, we would hit the bunks after a cold shower at around midnight, for the whole thing to start again.

'D' Sqn was still proving itself, and I think Simon was very aware that we needed to make our mark on the Regiment. On one range period we were preparing for a night shoot. Right in the centre of the range impact area was a large WW2 concrete bunker known as

'*The Hitler Hof*', allegedly the man himself had used it to observe panzer troops training in the late 30's and early 40's.

This bunker was a highly prized target to all gunners but was officially off limits.

During the range brief Simon elected to order every tank in the Squadron to lay on to the Hitler Hof and fix the coordinates. As the firing period commenced, he proudly gave the order *"D Sqn, Fire"*. Salvos are not normal. But when he gave that command, 14 x 120mm guns let rip at the exact same moment. The sight was electrifying, 14 streaks of red converging on the same point.

We also spent a lot of time on exercise. These would normally last between 6 – 8 weeks. As a driver I spent most of my time keeping Daisy on the road and operational. The Chieftain's power pack was notoriously unreliable, but a good driver would learn multiple mechanical tricks to keep the vehicle going.

We spent most of our time hanging upside down in the engine compartment, covered in oil and grease, avoiding contact with red hot components, trying to coax a component back into life or changing consumable parts.

Every British tank comes equipped with a Boiling Vessel, the BV; a boxy miracle that kept the crew going on tea, coffee, and what passed for haute cuisine. Forget the main gun or thermal imaging sights, the BV was the most important bit of kit on board.

It didn't just heat water, it fed souls.

Tucked directly behind the breech of the main gun it could boil enough water for a full round of brews and warm up a couple of tins of compo at the same time.

The truly dedicated loaders had their own BV pantry: bottles of ketchup and brown sauce, curry and chilli powders, maybe even the holy grail, that bottle of flavour. Tabasco sauce.

A tin of beef burgers mashed into a greasy mince with a fork, blended with 'mixed vegetables', ketchup and a pile of chilli powder made a reasonable chilli con carne.

But the pièce de résistance was always the egg banjo: two slabs of Army bread slathered in margarine, with a slimy fried egg, just runny enough to guarantee the yolk ended up all over your coveralls.

A proper banjo wasn't complete without an oily thumbprint pressed into the crust.

After hours of freezing your arse off in some godforsaken training area, there was no better morale boost than hearing the faint gurgle of the BV. It meant you were minutes from a mug of sweet, strong tea and a banjo that tasted like heaven.

We could've fought the Cold War on tea alone, and thanks to the BV, we damn near did.

26 Painting reproduced courtesy of the artist, Mick Graham, showing Daisy on Patrol

27 Me and Daisy

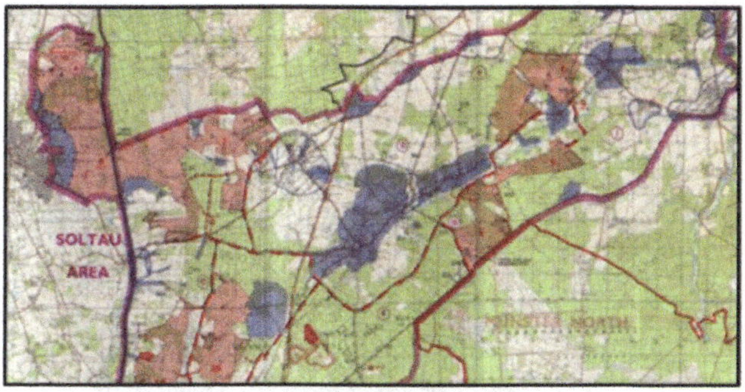

28 Map of Soltau Luneburg Training Area (SLTA)

29 Loading on OC

30 Major Simon Edwards

31 WO2 Geoff Porteous

32 Ammo Bashing on Ranges

Chapter 18. Tank Suits and Toilets

When in a '*battle position*', while the turret crew was working hard, drivers would take a chance to grab some sleep. On one occasion, whilst the turret crew were controlling the Squadron, I had the 3 day urge to take a dump arrive, it meant business and gave me 20 seconds to comply.

Normally this would mean waiting until dark, taking a shovel off the tank and going into the woods to dig a small hole next to a tree and enjoy 5 minutes of peace. More than 5 minutes and suspicions were aroused....

I let the crew know over the intercom that I was jumping out for a dump. Quickly dragging my coveralls down around my ankles and knowing I didn't have much time, I elected to squat against one of the road wheels and crap onto the track, all evidence would rapidly be mashed into the mud and sand, leaving no trace.

Army compo rations were designed to block your system for a few days, so that when it came time to give birth to the foul-smelling, concrete cylinder of regret, it was a bit of an effort to jettison the load and considerable straining ensued.

So, there I was, bent over double mustering as much squeezing power as I could, crapping on my own tank, only to look up and see a gang of German kids on their push bikes staring at me like something out of The Goonies. I don't know who was more traumatised.

The worst part came afterwards; the cleanup. The Army-issue toilet paper, a pastel pink abomination, had all the softness of a roofing tile and the absorbency of cling film. More abrasive than helpful, it felt like trying to polish porcelain with 60-grit sandpaper. And to think, I once thought doing a strip-o-gram in leather chaps at a Melton dive bar with Banksie was my personal low. This beat it by a mile and left more scars.

The weekends during an exercise were always spent *'leaguered'* up in a wood somewhere. Tank movements were prohibited over the weekends. At weekends, the four fighting troops and FHQ would leaguer separately.

During the day heavy maintenance would be carried out, but at night it was common for a radio operator to start a bonfire. The troops would naturally gravitate to the fire and pop open the cans of McEwan's Red or Blue, the favoured beers at the time, that were stored in the cooled bag charge bins.

In the days before anything vaguely resembling proper gortex waterproofs, tank crews were issued a monstrosity known as a tank suit. Essentially a green plastic bag shaped vaguely like a man, it offered the breathability of a bin liner and all the comfort of a damp coffin.

The only time it was remotely useful was during a weekend leaguer, when you were parked up, sitting still, and preferably already resigned to misery. It had to be cold, raining heavily, and you had to be utterly motionless. Any attempt at movement, or God forbid, actual work, and you'd find yourself soaked from the inside out, sweating like a glassblower's arse crack and wetter than if you'd worn nothing at all.

At one such bonfire, our troop was perched on upturned oil drums and logs, swapping tall tales and feeding the flames with anything remotely flammable; wood, plastic, the occasional can of XG279 grease. It was freezing, pissing down, and perfect tank suit weather.

Standard entertainment involved someone lobbing a compo tin of sugar onto the fire without telling anyone. As the sugar caramelised, pressure built inside the tin until; boom, it exploded like a tiny napalm grenade. The result? Flaming caramel sprayed in all directions, leaving its sticky, scalding mark. More than one Geordie ended up with second-degree burns.

One individual, who must remain nameless for reasons of dignity and potential PTSD, announced he needed a dump. *"Reet, ahm away fer a shovel recce"* he announced as he shuffled off into the trees in his rustling green tank suit, sounding like a family-size bag of Monster Munch.

After the obligatory 5 minutes, he returned, slotted the shovel back into its bracket, tucked the pink toilet sandpaper into the bin, and took his seat back at the fire like nothing had happened and the craic continued. A few minutes later, another trooper said,

"Can you smell shite?" lots of air sniffing and

"Yeah, I can…",

The individual instantly became defensive, checking the soles of his boots for signs that he may have stepped in his own shit.

"It's not me ya bastards".

Then, he took down the hood from his tank suit. There it was, in glorious, technicolour horror; perched right on top of his head, a perfectly formed Mister Whippy, decorated with a rosette of pink Army-issue toilet paper like some macabre cupcake. In the dark and the rain, he'd unwittingly dropped his coveralls and laid cable directly into the hood of his own tank suit, only to flip it up on the walk back.

No one spoke. There was nothing to say. The smell said it all.

Back in Detmold, despite living the best life driving Daisy, I was still keen to go down the signals route, I saw this as a way of getting a turret role and saw the technical aspects of signals as more challenging than driving or gunnery.

W. Germany, West Berlin: 15th/19th Hussars

Chapter 19. History in the Making

At the beginning of 1989, 'D' sqn was notified that we would be the Relief Berlin Armoured Squadron for 4 months. The role of this deployment was to free the resident Berlin tank Squadron to move down the inner German corridor to undergo a period of training that included an extensive range period.

The resident Berlin Squadron, 14/20H, was famous for its urban camouflage. For our part, we had to transport our standard green and black tanks up to Berlin; arriving as the 14/20H departed with their grey, black and brown vehicles.

Just before we deployed, Geoff Porteous was posted and another former paratrooper, Jim Lyon took his position as SSM. Known affectionately as 'The Goat', Jim had little of his predecessor's charisma and he struggled to win the sqn over.

The cold war at this point, was still very much in play, none of us knew what awaited a matter of months later.

Early in the morning, a huge convoy of tank transporters arrived at Lothian Barracks and filled the tank park. Being the driver of the lead tank, I was first up to park Daisy on the trailer of the monster Scammel Commander truck, and therefore all eyes were on me and just waiting for me to fuck it up.

The loading ramps on the trailer were very short, which meant that when they were down in the loading position, the initial lift was very steep. I edged Daisy slowly towards the ramps and felt her make contact. My eyes were glued to the transporter driver, now standing on top of his cab, arms straight up directing me to move forward.

As I eased the power on, Daisy started to climb slowly and steadily into the air, to the point where I was in my cab looking at blue sky, unable to see the hand signals from the driver. The key was a slow, steady and smooth continuous movement to prevent dangerous rocking. Many a transporter driver was rocked off the roof through poor tank driving skills.

Daisy very slowly started to nose over having reached the balance point on the ramps. Too much power, or too fast and we would slam down hard into the trailer bed, too slow and we'd either slip backwards or stop and start see-sawing, neither was a good option.

The nose came down gently on the trailer bed and we levelled out. I edged her up to the furthest point and ratcheted on the parking brake. Daisy was chocked at the front and rear and cross chained down securely.

All the hatches were closed and we were looking good. I hadn't fucked up loading a tank onto a transporter in front of the Regiment, happy days.

Eventually all the tanks and support vehicles were loaded up on about 16 transporters. With spare units the whole convoy contained at least 20 vehicles with police escorts. We set off for Berlin, 2 transporter crew members and a tank driver in each Scammel. Sitting in the lead truck, just behind the escorts I looked out and saw the most impressive convoy strung out in formation behind us.

After a day's driving we arrived at Checkpoint 'A' at Helmstedt, the crossing point from West to East Germany. Multiple Detours were required due to bridge weight limits. We spent the night on the vehicles, quite exhilarated at the days unique experience.

In the morning, we were up early and ready to roll. Our entry timing to the corridor that would take us from checkpoint 'A' through to 'B' was critical. The East German police would time us and were known to cause problems if it was deemed you may have strayed off the approved route. Checkpoint 'B' was the point from East Germany into West Berlin.

The RMP escorts kept us true to the timings and on arrival at checkpoint 'B', the West Berlin Police escorted us through the city to Smuts Barracks in Spandau. The journey from checkpoint 'B' to Smuts was something I'll never forget, all traffic was stopped for us, and the public were waving and cheering us, we all felt like liberating warriors to a besieged island; which wasn't far from the truth.

We settled into our rooms in what looked like a cheap hotel from the 1970's, I was sharing a room with Mick Harper, the Squadron Sergeant Majors (SSMs) scout car driver. Mick was, is and always will be, one of the funniest men I've ever had the pleasure to call a friend. He had joined 'D' sqn recently having spent time in the military jail in Colchester.

He'd previously gone AWOL, disappeared to Spain, and somehow blagged his way back across Europe before handing himself in.

He did his time in the glasshouse at Colchester like a true convict, and, as was so often the case with these loveable rogues, returned to the Regiment a reformed character and an outstanding soldier, but he still had to complete the JNCO cadre course.

The JNCO cadre was the Regiment's proving ground for promotion. It tested you on every aspect of soldiering and was run by SNCOs and the PTI Sgt. The course ended with a five-day exercise in the Harz mountains.

Night navigation was done in a short-wheelbase Land Rover, driven by the PTI along the dark forest tracks. When our turn came, George Carter was up front, with Mick and me in the back. One instruction George didn't give to the PTI was *"take a left turn that leads straight off a cliff."*

The Landie duly launched off the edge and careered a hundred metres down the mountain before wrapping itself around a tree.

In the back of the Land Rover was a radio rack; a matrix of steel runners and electrical connections that mounted 3 or 4 radios. Mick crashed sideways into the radio rack, breaking a rib. A nano-second later and I smashed into Mick, breaking 2 more. George and the PTI were miraculously unhurt in the front.

The 4 of us staggered out of the steaming wreck of the Rover; none of us students daring to call the PTI Sgt a '*complete fucking twat*'; that would have been rude and won us a beasting. But, feeling like a '*complete fucking twat*', he decided to put his running skills to good use and climbed back up to the track we were supposed to be on, and legged it back to the camp area.

About 45 minutes later an ambulance rocked up with a vehicle recovery section. Mick was taken to local hospital, jeered on by me and George taking the piss about him being a better crash test dummy than Trooper.

Mick was laid up in the med centre for 4 days and didn't get to complete the cadre, which was lucky for the bloke that went on to get 'Best Student' and an automatic promotion.

Life in Berlin swung from chaos to comfort. Before text messages and emails when we were away from home, we stayed in touch using free air mail letters known as '*blueys*'. Getting one was always a highlight, they were read and re-read, folded away and kept safe.

I stashed mine from Emily and the kids in my steel locker. Some were sweet, some were funny, and occasionally… they were blue in more than just colour.

One day I was walking to the tank park NAAFI at Smuts, when I started noticing something odd. People were walking past me, muttering things that sounded familiar; *very* familiar and then giggling to themselves.

A few comments were word for word what Emily had written in one of her more 'interesting' blueys.

FUCKING HARPER.

The shithead had been rifling through my stash, cherry-picking the juiciest bits and sharing them around the Squadron like saucy bedtime stories.

I never told Emily.

I don't think she'd have taken it well and frankly, I was embarrassed. But Harper's smirk every time he saw me told me everything I needed to know.

In Berlin, alongside my duties as a tank driver, I had the bonus job of being staff car driver for Simon. It was a great gig, not just cushy work, but a chance to get to know him better. We'd drive around the city and talk. One day, while heading across town, Simon confided something that clearly weighed on him.

He told me the Squadron salvo he'd ordered on the Hohne ranges had backfired, politically, not physically. It had landed him squarely in the CO's crosshairs.

The CO at the time was apparently furious and refused even to visit the Sqn in Berlin afterwards.

That single decision, that one order, had clipped Simon's career wings. I was gutted for him. He was a natural leader, sharp, confident, and the kind of officer you wanted to follow. But in that moment, I saw how fragile the whole system was. One misstep, even a calculated one, could end it all.

And for Simon, it did.

On one exercise, we were in a battle position overlooking the Fulda Gap. Simon came on the intercom and simply said *"this, gentlemen, is our actual battle location should 3rd shock Army come crashing over"*.

I would have been comfortable in battle with that crew and Simon as the boss.

Chapter 20. Checkpoint Charlie

About halfway through our tour, during a warm Berlin summer, Simon, Elton and a couple of other officers decided to go shopping in East Berlin. The Deutschmark–Ostmark exchange rate was about 10:1 in our favour, which made it an excellent day out.

The Potsdam Agreement of 1945 allowed us to cross the border freely if we were in uniform and crossed via checkpoint Charlie. As allied forces, we were not allowed to even acknowledge the presence of the East German police; If an incident occurred, we carried laminated flashcards that compelled the East Germans to summon a Soviet officer.

They didn't like this and were always keen to mess us around as much as they could.

Having had a great day wandering around East Berlin the shoppers and I loaded ourselves into the staff car to head back west side. On approaching the queue of west bound traffic, the East German police were clearly having a bad day and taking ages to process each car.

We couldn't wind the windows down and there was no AC. I was sweating buckets and stretched out over the steering wheel.

Big mistake.

I accidentally hit the horn which the Easties took as a direct challenge to their lethargy and laziness.

We couldn't apologise or explain, as the Potsdam Agreement forbade it and were marshalled into a holding area where we sat for about 2 more hours.

The only thing missing was birch twigs, 5 adults in a small car with no air flow was insanely like being in a Turkish steam room.

We laughed about it later. At the time, less so.

Chapter 21. Bare Naked Ladies

One Saturday, Mick and I grabbed a carrier bag of beer from the NAAFI and headed to one of the many parks with swimming lakes, along with a couple of the other blokes.

As we walked into the park in tee shirts and union jack shorts, we were more than slightly surprised to see an old lady walk past us totally naked apart from a big floppy hat and some sandals.

Stunned and inquisitive, we decided this required further investigation.

Walking out into the lush parkland grass, it was immediately obvious we'd stumbled into a nudist park. Being good soldiers, we knew that time spent on reconnaissance is rarely wasted, so we picked a spot not too far from a trio of naked young ladies, close enough to observe, far enough not to seem pervy.

Which, in hindsight, is exactly what perverts say.

We laid out our Union Jack beach towels. Someone had brought a battered cassette player, which was soon belting out '*Like a Prayer*' by Madonna, followed by a crackly version of '*Back to Life*' by Soul II Soul.

We stripped… and chickened out. Naked bravery, it turned out, wasn't part of our core training objectives.

A can of 'Schultheiss' lager later, and an old German man planted himself directly between us and the young ladies and proceeded to strip off. Completely. This was not pleasant, he must have been 70 years old, at least.

As this gentleman's pendulous ball bag swung free in the breeze, he attempted to sit down with all the grace of a geriatric walrus dismounting a Lilo. Mick, who had a vivid turn of phrase at the best of times, completely lost it, unleashing a stream of northern invective that Sid the Sexist, Tyneside's silver-tongued cavalier would have been proud of.

The rest of us joined in, adding our own flavour of Geordie outrage, hurling insults with all the subtlety of a flying ashtray.

Then the old boy looked up, cocked his head and, in perfectly enunciated Queen's English, said, *"Am I to take it that you gentlemen are from the Northeast of England?"*

He wasn't German. He was a senior diplomat from the British Embassy.

Obligingly, and with a certain dignity considering the situation, he adjusted his position to spare us the view. By some miracle, we never heard another word about it.

Maybe he understood that the trauma had been mutual.

Just over halfway through the tour, we were granted a week of leave, and families were finally allowed to join us. Emily and the kids were issued the correct paperwork and made the surreal drive from Detmold, through the heavily controlled corridor and into West Berlin.

I remember watching our Mitsubishi Lancer estate car pull in, covered in dust and full of snacks, the kids half-asleep in the back, and feeling a weight I didn't know I was carrying lift clean off my shoulders.

We'd rented a caravan at RAF Gatow, a strange mix of military efficiency and campsite chaos and tried to switch off. Days were spent with Chris and Anna Elliott exploring East Berlin, crossing Checkpoint Charlie with the other tourists.

The East German supermarkets were like walking into a post-apocalypse film set: cavernous concrete buildings, near-empty shelves and sullen staff.

What little that was available seemed to be reserved for Soviet officers, the good coffee, decent sausages, and anything resembling colour or flavour. The rest of it looked like it was left over from the Berlin airlift.

Still, it had its odd charms. For reasons I'll never understand, the East Germans made exceptional model train sets. Me and Matty chose a sprawling jungle of tracks and tiny stations that filled the car boot and, once home, never quite found enough space to stretch its legs.

But the real joy was in the ordinary: bedtime stories in the caravan, watching Chris set fire to the small BBQ and chasing the kids around the RAF rugby field. It was a short week, but a perfect one. For a few days, the Cold War receded, and I remembered what normal life felt like, messy, noisy, warm. It was a glimpse of what we were doing all this for.

The nightlife in Berlin was on another level; spontaneous, electric, and fuelled by that sense of freedom unique to a city still basking in post-Wall, 60's hedonism. One of the more infamous venues was a club called Mon Cherie's.

Willing, or occasionally ambushed, volunteers were escorted backstage and invited to step into the tub where they'd be joined by two very enthusiastic young ladies. As the bath rose up through the stage, the crowd would erupt in cheers and laughter, watching the participants subjected to a 'performance' that straddled the line between burlesque and bath-time wrestling, the only thing missing was a bottle of 'Matey'.

The girls were gifted, the show was… energetic, and the poor sod in the tub usually emerged grinning, dazed, and short a sock.

Soon enough, the world beyond the wire came back into focus, and 'D' Squadron remembered why we were there. Apart from one tank exercise, a simulated battle across the far banks of the Havel, complete with combat bridges to fight our way back, most of our military activity revolved around sport and infantry-style training in built-up areas.

Chapter 22. Running with a Brown Dog

Midway through the tour, Simon decided the Squadron needed a charity event and, naturally, it couldn't just be a bake sale. Elton was tasked to organise a sponsored run for the Royal Victoria Infirmary in Newcastle, a cause the Regiment regularly supported.

His plan?

We'd run the entire length of the Berlin Wall, from north to south, in teams. Simple enough, until he added the twist.

Somehow, he'd persuaded the Newcastle Brown Ale brewery to fly out a six-foot-high replica bottle of their finest. Made from fibreglass, it was hollow but still awkward to carry, like running with a stubborn child who insisted on being carried by six adults at once. The thing quickly became known as the "Brown Dog" and much like a real dog, it was something you either carried with pride or avoided eye contact with altogether.

The route was split into legs, each with six to eight people on the Dog at a time. Some stretches went through quiet suburbs where Berliners waved from balconies, bemused at the sight of soldiers in running kit carrying a giant beer bottle. Other legs weren't quite so genteel.

Kreuzberg, for example.

In the late 80s, Kreuzberg was a gritty, graffiti-covered slice of West Berlin: punk squats, smoky bars, political murals, and a permanent sense of *'don't hang about if you don't belong here'*. And here we were, a bunch of British squaddies, running through with a giant beer bottle. It was going to draw attention.

The section after Kreuzberg wasn't much easier physically, but it had the money shot: running past the Brandenburg Gate, cameras flashing, crowds cheering.

The catch?

That leg had The Goat leading it. The Goat didn't understand the concept of a "fun run." For him, maximum possible speed was the only acceptable pace, knees high, lungs burning, the kind of running that made you regret all your life choices.

Faced with the choice between Kreuzberg and The Goat, I took

Kreuzberg. At least there, the locals might hand you a beer.

Our leg started well enough, a few cheers from street cafés, kids pointing and laughing, someone shouting *"Prost!"* But then we got into the rougher part. Tight streets, high tenement blocks, shop shutters down. Faces peered out from upper windows. A couple of lads on a street corner muttered something in German that probably wasn't "welcome."

We pressed on, heads down, the Brown Dog bobbing above us.

Every so often, through gaps in the buildings, the Berlin Wall loomed into view, a long, grey slab topped with the familiar rounded pipe to make climbing harder. On the other side, East German watchtowers stood like silent sentries, their narrow windows hiding the eyes of border guards who, in theory, had orders to shoot anyone attempting to cross.

Carrying the Dog took coordination, four men on the sides, two behind to stop it tipping, swapping positions without losing stride. The fibreglass creaked with every shift. By the time we reached the end of our leg, my shoulders felt like I'd been carrying it since the Cold War started.

We handed it over to The Goat's team. In a blur of legs and grim determination, they tore off toward the Brandenburg Gate, the Dog bouncing like it was trying to escape. A BFBS reporter popped up with a microphone, clearly expecting something quotable for the broadcast.

"Where've you come from and where are you going?" he asked.

Still trying to get my breath back, I replied:

"From the start to the finish."

He gave me a look somewhere between pity and amusement, made a note, and wandered off to talk to someone who wasn't an idiot. Unsurprisingly, I never made the broadcast. But I didn't have to run with The Goat and in my book, that was a win.

Chapter 23. History is Made

Simon organised a Squadron photograph to be taken at the Brandenburg Gate. With a police escort, I drove Daisy, Skinny commanding, while Chris Elliott's tank followed close behind down the Kurfürstendamm to the gate.

We executed neutral turns, where the tank pivots and grates around its own axis, metal and rubber on stone echoing off the wall, to face back down the Ku 'Damm.

The Squadron formed up in front of the Gate, the last armoured unit ever to do so.

Later, in 2003 I was at the same spot and noticed the graze marks on the stone kerbs we'd made with the tank tracks during that neutral turn were still evident.

I'd literally made my mark on Berlin for posterity.

Our time in Berlin was up all too quickly. We returned to Detmold on the Berlin Military Train. It was novel, but after years of troop trains to Hohne, it lacked the drama of the transporters.

Watching the events of 9 November 1989 unfold felt unreal. We'd spent years training for a war that never came, parked our tanks in places designed to stop one, and suddenly the thing we'd been built to contain was simply… gone.

The Regiment still keeps a full section of the Berlin Wall wherever it's based. Most of us took a fragment of painted concrete home. But the real souvenirs were the memories, retold at every reunion, of a city changing while we were standing guard over it.

As for Mick Harper… more on him later.

34 Left to right: Lt Tam Stockwell, REME Sgt, Me, Ivan 'Skinny' Holmes, Sgt Blackman

35 The legendary 'Mon Cherie' Berlin nightclub

36 In Front of the Berlin Wall with 'The Brown Dog'

37 BFBS Reporter looking for a quote

38 View from Daisy's drivers cab on the way to the Havel

39 Daisy crossing the Havel on engineering kit

40 Handing over to the Goats team. Capt. Farnham on the bike, me in green shorts

41 D Sqn Fitter Section being interviewed

42 Exiting Kreuzburg with the Brown Dog, me on the left

43 Fourth Troop crossing the Havel

44 D Squadron on parade in front of the Brandenburg Gate. Note the Stasi Police on the Gate watching the goings on

Germany, Detmold: 15th/19th Hussars

Chapter 24. Signalling the Future

Now I had found my groove. Back in Detmold I was placed on a control signallers' course, this was unusual as normally only a single class 2 course was approved. A difficult course that dived into the depths of signal propagation and multiple other areas. I managed to ace the course and was again recommended for an instructors' course.

In 1990 I was sent again back to Bovington to complete the 3-month Regimental Signal Instructors Course. Although there were various iterations of this course across the Army, the RAC version was widely considered the most prestigious. There were attendees from nearly all tank Regiments as well as one lad, Cal Salter from the AAC. Me and Cal shared a room and got on well, supporting each other through the course.

I introduced Cal to '*The Venue*' club at Tower Park. Back then, if you were stationed anywhere near Poole and had a half-decent shirt and a few quid, you made the pilgrimage. It was billed as a "superclub" and, to be fair, it lived up to the hype.

The place was thick with strobes, dry ice, and a choking fog of Lynx Africa, sweat and cheap lager; Fosters if you were skint, 'Holsten Pils' if you were flush.

The DJ dropped *Killer* by Adamski, and the whole place pulsed, then flipped into *U Can't Touch This*. Even the hardest squaddie forgot the squaddie two-step, hammering out shapes with the confidence of ten pints and the determination to pull a lass called Alison.

You didn't go for conversation; you went to sweat, drink, and if the gods of cab fare and aftershave aligned, maybe score a number scribbled on the back of a beer mat. It was glorious, trashy freedom.

About halfway through the course Saddam Hussein invaded Kuwait. The writing was on the wall that eventually we would be involved. A couple of lads from The Royal Hussars were recalled back to Germany just before the final exams to start preparations for deployment to the Gulf region, I was very jealous.

I passed the course with good grades, but this time with a 'schools recommend' grade, meaning that I was eligible to apply to become a school instructor back in Bovington. This didn't appeal; all I could think of was getting out to the Gulf and cracking some skulls.

Initially, the British Army dispatched one armoured brigade to the Gulf equipped with the then-new Challenger 1 tanks; marking the beginning of Operation Granby. As the threat escalated, this brigade expanded into a full armoured division. During this buildup, the 15/19th Hussars were earmarked to reinforce, a move dubbed 'Op Granby 1.5.' Though it was jargon, it reflected the urgency and scale of the build-up.

Preparations began but lasted only a couple of days before we were stood down. The only members of the Regiment to deploy were unbelievably the Regimental band, who had secondary employment as stretcher bearers and porters for field hospitals.

A couple of blokes managed to find convenient postings, Elton Farnham was attached to the Scots DG and a friend of mine, Mike 'Arthur' Avery, managed to blag a job escorting armour in the transport ships from Bremerhaven. I tried that number but got nowhere.

Instead, we became the Battle Casualty Replacement (BCR) training centre at Sennelager, we had the soul-destroying job of training other units that were deploying.

My friend and neighbour, Ken did deploy with 4 Regiment AAC. I had always had my grandads' badges and accoutrements from his time with the 8th Army and the Desert Rats. I lent his original Desert Rat flash to Ken for good luck; he wore it with pride on the deployment and gave it back on his return. That badge had seen more action than I ever had at that time.

I felt I was being left behind by events that were what we trained to do.

Having had many a conversation over a beer with Ken, and seeing his lifestyle as a pilot compared to mine, I started the long process of applying for helicopter pilot training.

In the meantime, I had been promoted to L/Cpl and was actually paid this time around. On promotion I was moved from being a driver to the loader/operator position in Elton Farnham's tank. Call sign Zero Charlie (0C).

0C was a unique tank in the sqn. It was fitted with a dozer blade designed for rapid placement of scrapes for tanks to position themselves in. In the turret the radio fit was extensive and included a long-range High Frequency (HF) radio to complement the pair of standard VHF tactical radios. For this reason, the crew position was reserved for a signal's instructor.

The main role of the loader on exercise is more of housekeeper as we never carried ammunition except when on the ranges. I was cook, tea maker and sandwich king all in one. On the ranges we became so proficient in loading, that on a long-range shoot with slower velocity HESH rounds we could attain 3 rounds in the air at the same time. Being crewed as loader/operator was easily my favourite role on a tank crew

I had my eyes very firmly on my next challenge, to get assigned as a troop corporal within 'D' sqn as I had been further selected for promotion to Corporal, but commanding my own tank.

I really wanted to go to 4th Troop. I knew the Troop Sergeant, Gary Black, well and held him in high esteem. But by then I was too useful where I was. The courses, the qualifications, the radio fit on 0C, all of it meant I stayed put. I became a victim of my own success before it could turn into command.

Canada, BATUS: 15th/19th Hussars

Chapter 25. Sultans and the Sin Bin

The 15/19H battle group was the first to be assigned one of seven annual training programmes at the British Army Training Unit Suffield (BATUS) in Alberta, Canada. This was a large-scale series of exercises that brought together the full range of arms and units that combine to create a battle group following the end of hostilities in the Gulf.

'A' and 'B' sqns plus 'C' and 'D' companies of the Royal Irish Rangers formed the 15/19H BG under command of our new CO, Lt Col Braithwaite. The CO was keen to impress and decided to re-staff the Regimental command troop.

Along with another Cpl, I was posted to Command troop, 'HQ' sqn, sadly leaving 'D' sqn behind.

Command troop is the brain of the BG, equipped with specialist armoured vehicles configured for communication and control known as Sultans. The Sultan was an armoured box on tracks, loaded with multiple communication systems. Two or more could be joined together using pull out tents to create a command complex that could be rapidly moved.

Callsigns 0A, 0B and 0D were the 'main' command centre and were always moved together. The CO would command the BG from main; if he was in his tank, the Regimental second in command would assume control.

Call sign 0C was known as 'step-up' and was the sole command vehicle when 'main' was moving between locations, this meant BG command always had a foot on the ground.

I was assigned as the commander of 0C, which I was very happy with. I had a driver, Jamie Young and a radio operator, L/Cpl George Carter from my JNCO cadre. We also had the Regimental Signals Officer, Capt. Freddie Champion. As a crew we operated exceptionally well together, we would arrive in a location and be up and running in a matter of minutes.

I liked Freddie Champion, he was a pleasant officer, but his appearance was a tad comical. With fiery ginger hair and a stereotypical complexion to match, he was unfortunately an easy target for practical jokes.

Freddie was a keen wildlife spotter, and Alberta's plains offered plenty to keep him twitching. He was especially fixated on seeing a honey badger. One day, someone treated the rubber eyepieces of the vehicle's binoculars with a generous smear of black boot polish, then held them close to their face and shouted, *"Honey badger! eleven o'clock!"* Freddie launched himself from the back of the command vehicle, grabbed the binoculars without hesitation, and pressed them tightly to his eyes.

After a few moments of intense scanning, he lowered them. No honey badger in sight, only to reveal himself now resembling a ginger panda. At that exact moment, the RSM arrived in his scout car with his driver, Paddy O'Hara, and spotted the entire crew doubled over with laughter.

"What the fuck is going on here?" he barked.

Still unable to speak, I simply pointed at Freddie. The RSM clocked his boot-polish eyes and nearly fell off his scout car, wheezing with laughter himself. It was one of those timeless squaddie moments; stupid, perfectly timed, and utterly unforgettable.

The training programme saw us completing multiple individual exercises with time back in the camp at Ralston in between. Once the vehicles had been stripped, cleaned and any maintenance carried out we had the chance to enjoy the delights of the nearest town, Medicine Hat.

Medicine Hat, or '*The Dut*' in fluent Geordie, was legendary. A neon oasis after the desolation of the prairie, it offered every kind of bar, club, and questionable establishment a squaddie could dream of. At its heart stood the infamous Assiniboia Inn, known far and wide, and affectionately, as 'The Sin Bin'.

The Sin Bin operated under its own laws of physics. We turned up, 15 blokes strong, and booked into a single room, each one of us using a name from the current England football team. No ID, no questions, no raised eyebrows. Just keys and a nod.

Every place in town seemed to offer some form of '*gentlemen's entertainment.*' The Sin Bin's answer to the arts was a resident performer known as Stacey Stacks ('possibly' not her real name).

Stacey had a rare talent: floor paintings. Using a strategic palette of vibrant colours and even more strategic parts of her anatomy, she'd produce masterpieces on a roll-out canvas with flair, energy and…well, gusto.

Other bars and clubs were more traditional in their approach, encouraging minor contributions in either single dollar bills, or one-dollar coins known as '*loonies*' which had recently replaced the paper money, in return for exclusive views of a range of party tricks.

The Army laid on adventure training activities; white water rafting, skydiving, kayaking, all available for a token contribution. It was a rare moment of generosity, and not to be sniffed at. Between that and the local entertainment in 'The Dut,' the off-duty distractions were almost as memorable as the military ones.

The Medicine Man exercises culminated in full battlegroup, combined-arms training. Even though we were operating as a light unit, the final scenario was genuinely impressive. Tanks roared across the prairie, artillery batteries thumped out live fire missions, and infantry did what infantry does best: run around looking angry, stabbing things, and blaming everything on the blokes in vehicles.

The grunts favourite sport was shooting up anything stowed externally on the wagons, especially sleeping bags, which were routinely shredded in the name of battlefield realism.

When it was all over, the CO was debriefed by the Brigadier in charge of the exercise. Against all odds, we'd done very well as a battlegroup. It didn't make up for missing the Gulf, but it was some consolation to know we'd proved ourselves. Live rounds, real tactics, but no enemy. We'd made it work.

45 Command Troop, 15/19H

46 Me on stag in a Sultan

47 BG Preparing to hit the Prairie

48 Track Bashing

49 The command troop team getting an erection

50 The 'Gag & Puke', Ralston

51 The 15/19H Battle Group

Germany, Detmold: 15th/19th Hussars

Chapter 26. Selection and Separation

On return to Detmold, we had a period of leave to enjoy time with our long-suffering families. Having previously applied for pilot selection, my application had progressed to an interview with the CO. Being only 22 at the time, some concern was raised about my age, but Simon Edwards must have advocated for me, because despite the concerns, the CO approved my application.

In early 1991 I was sent to the Royal Air Force Officer and Aircrew Selection Centre (OASC) at Biggin Hill, the famous WW2 fighter station near London.

The tests were mostly computerised and designed to overload you deliberately. Artificial horizons were rotated into unfamiliar attitudes, and you were expected to identify the aircraft's position instantly. Other tests involved complex diagrams of gears and pulleys, where a single direction of rotation at one point required you to predict the movement elsewhere.

The point wasn't the answers themselves, but how quickly and accurately you could process unfamiliar information under pressure.

Finally, there was another interview, assessing suitability for aircrew roles, leadership potential, and motivation under scrutiny.

If successful at Biggin Hill, the selection progressed to an appointment at the military hospital in Woolwich, just down the road, where candidates were subjected to an EEG. Electrodes are glued to your scalp while strobes fire at various frequencies…

We also gave what felt like litres of blood to prove we were medically sound.

With giant blobs of dried super-glue still in our hair, we were sent back to unit in Germany to await the results of OASC.

I knew operational requirements would unquestionably take priority over my hopes and dreams.

Shortly after returning home following OASC, the Regiment was informed that we would be deploying to Cyprus just before Christmas of 1991 to become the UKs contribution to UNFICYP, the United Nations operations still ongoing to man the green zone separating Northern, Turkish Cyprus from the southern, Greek Cyprus.

Whilst this was exciting news, it potentially threw a spanner in my plans, as deployment could have ended the selection process entirely.

I learned that I had been successful at OASC, and my blood was considered up to the standard, the next phase of selection awaited.

The next stage of selection was Army specific and took place at the Army Air Corps Centre in Middle Wallop. This stage consisted of several written exams that covered military skills such as map reading, recognition of enemy vehicles and signals.

I found these exams straightforward. On successful completion a further interview with a retired senior AAC officer took place, but the results of the Middle Wallop stage of selection aren't released for a little while, so it was back to Detmold once again.

I discovered I had been successful in the second stage and would now progress to the third and final stage, flying grading. Flying grading was thirteen hours in the ubiquitous DeHavilland Chipmunk T10, where talk stopped and flying began.

Chapter 27. Chipping Away

I reported to Flying Wing at the end of September 1991 with eight other hopefuls. We were issued flying suits, a handful of navigational tools and fitted for flying helmets. Each student was assigned a flying instructor from a pool of retired, mostly former RAF, pilots.

My first instructor was Mr Pat Bushby, a jovial and very pleasant man. Pat would describe a flying exercise, demonstrate it to me and I would then follow him on the flying controls to learn the correct technique for each exercise. Once grasped, I would have to fly the exercise to demonstrate back that I had learned the correct manoeuvre.

In the first flying hour I encountered a bit of an issue that I wasn't expecting, I was chronically and violently airsick. The smell of the Chipmunk and the unfamiliar movements were just too much. I returned from each sortie with at least one full sick bag to present to the rest of the course.

Despite the near continual puking I was grasping the basics of flying and demonstrating to Pat my rate of learning. For the second 6 hours of grading, we all had an instructor change, I was now paired with Mr Pete Jennings, the opposite of Pat Bushby.

Pete Jennings was a cantankerous ex-RAF pilot who firmly believed that pilots should be commissioned officers. Sitting behind me in the tandem cockpit, the berating started almost immediately. The airsickness continued but was subsiding as I became accustomed to the motion and smell of the aircraft.

The main challenge in learning to master the Chippy was the landing. Being a tail dragging aircraft, unless the touchdown was perfectly executed so that all three wheels touched down at the same time, a bounce was inevitable. The severity of the bounce depended on the touchdown but bounce it did.

And every bounce felt like another silent mark against you.

Pete Jennings gave zero encouragement or feedback on progress, I had no idea whatsoever if I was making the grade or not, no idea of any areas requiring improvement or what my chances were. I was struggling to remain polite with this man who made his negative opinions very clear.

The only thing stopping me answering back was the knowledge that he held my future in his hands.

The final, thirteenth hour of flying grading was classified as a final handling check and was done by a third, independent instructor.

After climbing out of the old Chippy for the final time, walking back to the hangar to hand my parachute back and consult with the rest of course, I had no idea of my performance. None of us did; feedback was forbidden. We handed back our flying clothing and prepared for the final interview of the whole selection process and assembled in the flying grading crew room for the final time, second guessing our own performance.

Had we been successful or was this the end of the road?

We each individually marched into the Chief Flying Instructors (CFI) office, came smartly to attention and saluted. With the CFI was the Head of Chipmunk Squadron, Mr Fred Trowern. Invited to remove our headdress and to take a seat, then the results came.

I had been successful. I was beaming from ear to ear.

Then came the caveat.

I was classified as a 'training risk' due to chronic airsickness. This meant that unless I got over it within 10 flying hours, I would be terminated from further training.

Of the 8 initial candidates that assembled for flying grading, five went on to become pilots. I returned home for the final time to await loading onto an Army Pilots Course.

Two years later, long after flying grading was behind me, Pat Bushby was tragically killed when his aircraft was involved in a collision with a landing Chipmunk. It was a terrible loss to the basic fixed wing team, and he is remembered with deep respect.

Back in Detmold, with my flying future still hanging in the balance, the Regiment was now well into its preparations for deployment to Cyprus and command troop was no different. We now knew that we would be deployed to St Davids Camp just outside of Nicosia with the rest of 'HQ' sqn, whilst the two fighting Squadrons would become rural and urban Squadrons respectively.

52 Flying Grading Course 13/91

Cyprus, Nicosia: 15th/19th Hussars

Chapter 28. Amalgamation and Division

Our role was to man the sector control room for BRITCON. In early November 1991 we were bussed to RAF Gutersloh and flown out to Cyprus.

Just prior to deployment we were gathered as a Regiment and told that our 350-year-old Regiment would be amalgamated as part of the 'Options for Change' review.

We knew that we would be amalgamated with another cavalry unit, and rumours were rife as to which that might be. The wise money had us amalgamating with the 14/20H, our equal in seniority and another northern Regiment. No one was particularly excited by the idea.

We were instead told we would amalgamate with the 13/18H, which came as a genuine relief; they had a solid reputation and were also northern with a recruiting area in South Yorkshire.

We would become the 'The Light Dragoons'.

This new name proved to be a masterstroke; it honoured the forebears of all four Hussar Regiments and offered a fresh start for soldiers on both sides of the amalgamation. After some initial hesitation, both Regiments embraced the change with cautious optimism and a genuine sense of renewal.

On arrival in St David's Camp, my body was in Cyprus, but my head was already in Middle Wallop. Whenever I was on duty, which was often, I was reading papers from previous courses and preparing for ground school.

I was reasonably confident, but formal academics had never been my strength, and I knew ground school would test that.

Chapter 29. Doors and Departures

Keen to give myself a head start, I contacted the UN Army Air Corps detachment in Nicosia, 16 Flight, home to a couple of Gazelle helicopters. I got through to the flight QHI, SSgt Geoff Sutcliffe, explained I was stuck at St David's Camp with the UN waiting for an Army Pilot Course, and cheekily asked if I could pop over for a familiarisation.

Geoff couldn't have been more helpful and on my next day off invited me over for a brew and chat. After meeting the flight and chatting for a while, Geoff casually mentioned that I'd be going flying with him.

After a thorough pre-flight brief, we walked out to the helicopter, I was nervous, having never been in a Gazelle before but knowing this would hopefully be my life. I climbed into the left-hand seat and Geoff into the right.

He gently talked me through the pre-flight check sequence and then, just before starting the engine, asked me to close my door,

"You know how to close it, right?"

"Yeah. No problem."

I was lying. Unfortunately, I grabbed the wrong handle, the door jettison one; the one that was black and yellow, gave it a yank and before Geoff could say *"not that one wanker"*, the door fell off the helicopter!

I was mortified as Geoff called the Sqn technicians on the radio to request they come out and refit the door. Fortunately, the door was easily refitted and a further safety brief conducted to reaffirm my knowledge of what to touch, and importantly what not to touch.

The coup de grâce came when Geoff called for departure clearance,

"Tower good morning, this is Army Air 123, request start and departure to the north from current position"

"Army Air 123 good morning, clear start and when ready, depart to the north, oh, and don't forget all your doors!"

I just wanted to melt into the seat and disappear, not the start to my flying career I had hoped for!

Geoff demonstrated some low-level flying, and all humiliation rapidly evaporated; it was everything I had hoped for, exhilarating, exciting and addictive.

Shortly afterwards, my joining instructions arrived for Army Pilot Course 342, starting in March 1992. It was now very real, it was happening, and I tried to get back to Germany before Christmas so that we could pack up the house and prepare for the move back to the UK. I'd been away from home so much, I really needed time with the kids.

The request was denied. Apparently, the operations team would be undermanned without me. Eventually, I left Cyprus in mid-February 1992, which gave me only a matter of a couple of weeks to pack up the quarter, move to the UK and report for the most difficult and important course of my life.

So far, my only flight in a helicopter was mired in humiliation, my 13 hours of flying the Chipmunk, now 8 months previous, was mired in chunder and the threat of being *'chopped'* unless I conquered the motion sickness. I had hated my instructor and now faced with 6 weeks of ground school and another 40 flying hours in a Chippy. Reality hit, had I bitten off more than I could chew?

53 Christmas 91 with Red Cross Parcels

54 Building the 'Echo Beach' Swimming Pool

55 The SDC Ops Room

56 SDC Ops Room 2

57 HQ Sqn Christmas Party

58 Command Troop Vehicles in SDC

Part 4: Level Flight to Steady Climb – Life in the Air

Middle Wallop: Army Pilots Course 342 – The Fuglies

Chapter 30. Ground School Phase

The Army Pilots Course (APC) was broken down into three distinct phases with additional training requirements squeezed in between. It was made very clear to the fifteen of us on day one that people failed or were 'chopped'.

Sometimes for not flying well enough. Sometimes for reasons that weren't explained at all, everything counted; attitude, drive, motivation and commitment. You could be fine on a Monday and escorted off the premises by Friday.

The message was simple enough: stay sharp or be Returned to Unit.

Dress code for day 1 was service dress, or No. 2's as it was more commonly called. This was for the course photograph, a feature of all courses at flying wing.

Tradition dictated that any student leaving the course, for any reason, would have their faces unceremoniously coloured in, or a cross put through it.

Our formal course photo was pinned on the board in the flying wing, where Course 341 would decide on our nickname, another tradition for incoming courses. As we were all squinting into the low spring sun when the picture was taken, none of us were at our gorgeous best, and we 'earned' the handle "*The Fuglies*", as in 'fucking ugly'.

Making up the Fuglies were 6 officers, 2 senior NCOs and 7 junior NCOs. One very notable member of the course was Cpl Pimbahadur 'Pim' Gurung, the very first Gurkha soldier to be selected for pilot training. Pim was a clerk by trade, and his father had been a notable Gurkha soldier before him. Pim's grasp of the English language was perfect, or so we thought.

That afternoon, the very first test was upon us, the requirement for our annual Battle Fitness Test, or BFT. Failure would result in instant RTU. Everyone passed.

The next 6 weeks were spent in the classroom learning all aspects of dead reckoning navigation, aviation meteorology, use of navigational aids, basic aircraft technical aspects to include principles of hydraulics, piston and jet engines, principles of flight and airmanship. My Chieftain driver mech course served me amazingly well for a lot of the technical aspects.

The classroom was unfortunately located in the wings of one of the former WW2 hangars.

Watching students on courses ahead of us walking out to their aircraft and listening to engines firing up and propellers biting into the air was one hell of a distraction. As we approached the examination phase, we learned who our basic fixed wing (BFW) phase instructors would be, I'd heard a rumour that we would be assigned the second instructor from flying grading for some degree of continuity. If true, this would mean the return of Pete Jennings into my life.

It was true.

Talking to Sgt Bob Skeldale of the Royal Engineers, from the course ahead, who was Pete Jennings current student on BFW, I learned that Bob was living the same nightmare I had suffered on grading.

Throughout the ground school phase I became quite close to one of the officers on the course. Capt. Jonathan Slade was a doctor from the Royal Army Medical Corps and was in the process of qualifying as a Specialist in Aviation Medicine. Jon was not a stereotypical officer, he was very down to earth, very funny and very willing to help me through some of ground school subjects.

We were to become close for the duration of the APC, at one point though he was summoned for a *'friendly chat'* from another more senior officer, who advised him against fraternising with a junior NCO on a flying course. I suspect it was a short conversation…

As we started the exams for the ground school subjects, inexplicably, under exam pressure, Pim's grasp of English seemed to vanish, but with some assistance he did eventually pass all the exams.

Although I would consider my relationship with Simon Edwards as close, my relationship with Jon was a more conventional friendship with a meeting of minds, senses of humour and outlook in life. It was still unusual though for a Captain and a Corporal to strike such a friendship, but this was an unusual situation.

59 APC 342 "The Fuglies"

Back Row: Cpl S Norris| Cpl S Hicks| Cpl P Gurung| Cpl M Tacon| *Cpl D Crang*| L/Sgt J White| Cpl J Termini| *Sgt N Hawley*
Front Row: Lt J Facer| Lt M Edwards| Capt. S Tully| Capt. M Baxter| Capt. J Slade| *Lt E Power* | *Sgt K Dean*

Chapter 31. Basic Fixed Wing Phase

My reunion with Pete Jennings was, predictably, joyless. He hadn't changed: still grumpy, cantankerous, and looking like someone had pissed in his tea.

We now had five flying hours to 'go solo'. Fail, and the clock would keep ticking without you.

This wasn't so much of a problem for some, they had only just come off grading and been loaded straight onto an APC, a few of us however were quite rusty. Failure to go solo quickly enough meant review time, and after that, RTU.

The course photograph had its first face coloured in at this point. A Sergeant from the Royal Marines, already a helicopter aircrewman, failed to go solo and was RTU'd. It shocked everyone to the core, if he could get binned with his extensive flying experience, any one of us could.

My fear of motion sickness was justified. I was once again a vomit dispenser of notable efficiency, but I managed to go solo within the allowed five hours, which at least bought me some breathing space. Instead of RTU'ing me on the spot, the Army decided I was a training risk worth retaining.

I was summoned to see Fred Trowern. He was polite but clinical, explaining that my progress would now be monitored formally for signs of improvement. If none was forthcoming, I would be referred to the RAF Desensitisation Unit at Farnborough.

When I found out what the desensitisation programme involved, the fear of God entered stage left. The process, known affectionately as 'modern torture,' involved being strapped into a two-seat Hawker Hunter and flown aggressively until your inner ear surrendered. In other words: puke until you stop puking.

Whether it was the looming threat of the Hunter or natural desensitisation, I'll never know. But either way, I never went to Farnborough. And for that small mercy, I remain eternally grateful.

Having gone solo, the fixed wing sorties increased in complexity, we progressed from flying circuits and practiced forced landings through navigational exercises and aerobatic manoeuvres. Pete Jennings continued to be himself.

One day Pete gave me a basic navigational route to plan and then present for evaluation before flying it. This sortie was a deep low in my experience with him; he rejected my planning at least 3 times for spurious reasons and then, when we got airborne, entered a state of relentless haranguing.

Everything I did was seriously wrong, or so he had me believe, at one point he started hitting the back of my helmet with his navigation ruler. It carried on into the post-flight debrief and I was nearing my breaking point. I walked back into the crew room sweating and boiling with anger. I had passed the sortie, but only just. In frustration I launched my flying helmet across the room.

I sat at my desk, head in my hands, silent. Jon collected my helmet and brought it over, offering a listening ear and some witty comments in his own, gently theatrical manner. I cracked, came to my senses and took my helmet down to the RAF flying equipment bay for a new visor, I told them I'd accidentally dropped it.

I often spoke to Bob Skeldale about Jennings, just to clear my head and to take in that it wasn't personal, he was the same with everyone.

We were now trusted to fly solo on longer navigational exercises, normally large loops around southern England, practicing not only dead reckoning, but stretching our airmanship skills and contact with ATC. We were given diversions to deal with and simulated engine failures at random moments in flight.

On one long solo nav, Steve Hicks and I briefly abandoned professionalism and met south of Salisbury for a short, possibly illegal, dogfight. We survived, laughed, and never spoke of it again.

As we approached the allocated 40 hours of flying in BFW, the Final Handling Test (FHT) approached. Again, failure to pass would mean the possibility of being placed on review. Passing meant moving up to what we had all gone through selection for, flying helicopters, the collective goal for all our individual reasons.

I guess my own reasons included the continuing need to prove to myself, and the Army, that I was better than the 'me' that had been let loose on the RAVC and the stables troop.

At the time, that seemed like a reasonable ambition.

"Train hard, fight easy," they say. I'd been trained hard, and I passed

the FHT with no real issues, which suggested at least some of it had stuck.

In my final briefing with Pete Jennings, he said something that stayed with me. To this day. After congratulating me and shaking my hand, he looked me in the eye and said, *"Remember, the only person who can make you a better pilot is you. Good luck. I know you'll be great."*

After everything he'd put me through, I'd written him off as a bastard.

Turns out, he was a useful one.

I took his final words at face value and used them as fuel.

He was right. The only person who could make me a better pilot was me.

Chapter 32. Survival and Aeromedical Training

Between passing BFW and starting Basic Rotary Wing (BRW), we were sent on a series of survival training courses. First up was the Fleet Air Arm's aeromedical wing in Portsmouth, where we sat through lectures on oxygen deprivation, or hypoxia.

Jon was in his element, translating the medical lectures into something the rest of us could vaguely understand. His five-minute briefing on reverse osmosis was outstanding, mainly because he stopped at five minutes.

After the theory came the practical: a session in the hyperbaric chamber.

We were issued canvas flying helmets and oxygen masks, and once fully kitted out, we looked like a squadron of WW2 Japanese fighter pilots. Pim really leaned into the kamikaze look, far more enthusiastically than the situation required.

Hicks, helpfully, was muttering "Tora! Tora! Tora!"

We were sealed inside the hyperbaric chamber with a doctor who looked far too cheerful, but then he knew what was coming our way. In front of us was a surreal spread of children's toys, shape sorters, crayons, and paper pads. It looked like playtime at a psychiatric ward.

"Number 3, take off your mask."

Off came the oxygen, and fighter pilot 3 was handed a simple task: *"Place the shapes into the ball."* At first, they'd manage fine, then gradually, and hilariously, coordination and logic deserted them. The basic task became impossible, and the subject oblivious. Once back on oxygen, they'd recover almost instantly, with no memory of what had just happened.

Steve and I were seated as Numbers 7 and 8. *"Numbers 7 and 8, masks off,"* was the last thing I recall.

Apparently, we were handed crayons and instructed to play a game of noughts and crosses. What followed, according to eyewitnesses, was a heated exchange. I accused Steve of cheating. He took umbrage.

A full-blown row broke out between us, two grown men arguing over a children's game while slowly starving ourselves of oxygen, so animated the staff nearly intervened.

Back on oxygen, we had no memory of it at all.

I still maintain he was cheating.

Staying with the Royal Navy, we made our way to the Royal Naval Air Station in Yeovilton, home to the dreaded Helicopter Underwater Escape Training Centre, better known as '*The Dunker.*' It was a simulated helicopter fuselage suspended over a deep pool.

We were allocated cockpit seats inside the fuselage and strapped in as if preparing for a real flight.

"Drop 1," called the Chief Petty Officer running the exercise.

Drop 1 was straightforward. Drop 2 inverted us, leaving us hanging in our harnesses. Drop 3 added multiple rotations before inverting, lungs burning as we bomb-burst out through anything resembling an exit.

Then came Drop 4 - complete darkness. All lights were turned off, and the fuselage rotated twice after being submerged before stopping. Eyes and lungs straining, we searched for any trace of light or a slight contrast that might hint at an opening. When we spotted a faint grey patch in the blackness, we surged toward it, surfacing like breathless Polaris missiles.

From the Navy, we moved on to the RAF Search and Rescue Centre, where we were given a crash course, literally, in sea survival. Military logic at its finest: the Navy gave us the high-level challenges, the RAF the water-based nightmares.

The Fuglies quickly developed sea legs as we were bundled onto a small vessel and taken out into the unforgiving English Channel.

Once we were well offshore, the skipper opened the throttles and carved a hefty wake behind the boat. We lined up, each of us mentally preparing for a dramatic leap. The direction came: *"Go!"* Off we went, one after the other. I performed my best impression of a back somersault with pike into the churning swell.

Surfacing, I spotted the last of the Fuglies bellyflopping into the water in the distance.

I gathered my salty thoughts and initiated my 'stole', a horseshoe-shaped lifejacket integrated into the Load Carrying Jacket (LCJ). It exploded to life, snapping me to the surface like a cork. Then came the next trick: pulling the cord to deploy the life raft.

It burst open in a flurry of tangled plastic and poor ergonomics, and I flailed my way into it with all the grace of a pissed squaddie climbing out of the bath in Mon Cherie's.

Once inside, under the dubious shelter of its sagging roof, I began bailing it out. The raft leaked badly and came with a sponge the size of a digestive biscuit. To my amazement, I wasn't seasick, perhaps I really was desensitised at last.

We bobbed violently in the chop with only a small sea anchor for stability. I spotted Jon and Steve in their own rafts nearby and decided to paddle over for moral support and a chinwag. That's when I noticed their rafts were much firmer than mine. Yep, mine had an 'effing' puncture. I was slowly but surely sinking.

Did they help? Did they bollocks! They sat there laughing their tits off while I did an impression of a man slowly drowning in a bright orange crisp packet.

Eventually the RAF launch returned. The instructor bellowed at us to separate for recovery, not by rescue helicopter as promised, but by the boat crew. I was knackered by this point, having spent most of my energy bailing and keeping what was left of the raft upright.

By the time they hauled me out, I was more Wilson than Tom Hanks.

After a night drying out in a hotel, we returned to Middle Wallop, where the real excitement was about to begin: BRW.

This was the moment everything so far had built towards: months of selection, setbacks, sickness and stubbornness. I was finally about to take the controls of an Army helicopter again, for the first time since 'Doorgate.'

At the same time, Emily and I had our own update. She was pregnant again, due at the end of December.

Which, as it turned out, was exactly when I was meant to be getting my Army flying wings.

Assuming I made the grade of course.

Chapter 33. Basic Rotary Wing Phase

Located in a different hangar at Middle Wallop, Basic Rotary Wing was manned by civilian flying instructors, all former military pilots. I was paired with Mr Bob Weston, a former Royal Navy pilot and all-round top bloke.

Bob was gentle, compassionate and a superb instructor who spoke with a very slight lisp; I knew I was lucky to have him. After my experiences in BFW, I could finally focus on my flying skills, without the added stresses that come with an 'old school' instructor.

BRW was about unlearning fixed-wing habits. Helicopters punished over-control, and every input bled into the others. It demanded finesse, not force.

My first sortie with Bob was just a familiarisation of the Gazelles capabilities with a running commentary of what he was thinking and doing to control the aircraft. After this, every time we climbed into the aircraft it would be me, with a time limit of 15 hours flying before going solo.

For the first few sorties of BRW, we never left the airfield. Bob would hover-taxi us to one of several hover squares, white-painted hover squares marked with a cross in the centre. They were basic in appearance but brutal in purpose. This was where student Gazelle pilots learned to hold a hover with precision, building the muscle memory needed to control the aircraft.

The Gazelle was light, twitchy, and brutally honest. Bob introduced the controls one at a time, then in pairs, then all three together. The aim was simple: stop chasing the aircraft and let it settle.

As confidence grew, so did the tempo. Soon, we were flying full circuits, complete patterns around the airfield designed to simulate approaches and departures from real-world landing zones. A perfectly flown rotary circuit in the Gazelle began with a smooth transition from hover to forward flight, climbing out at 60 knots.

Sharing the circuit with multiple Gazelles and Chipmunks was controlled chaos.

Downwind was all about rhythm: checks complete, spacing and speed right, turn base at forty-five degrees.

Final approach meant adjusting the collective to keep the descent shallow, maintaining 60 knots, then bleeding off speed and height as we reached the threshold. At 15 - 20 feet, the trick was transitioning smoothly into the hover, raising collective, applying pedal, and arresting any forward momentum before settling onto the touchdown point like a leaf landing on water.

And this wasn't done in isolation. There could be six or more Gazelles in the same pattern and just as many Chipmunks at higher levels. The circuit was alive with radio chatter: *"Gazelle Five-Three, lifting for left-hand circuit." "Chipmunk Two-Nine, final, touch and go."* You were expected to maintain spacing by sight, listen for every call, and know who was where.

Occasionally, the controller would step in, but for the most part, it was down to student judgement and instructor nerves. Perfect flying wasn't about flair, it was about calm, confident control. No drama. No drift. Just a clean take-off, smooth turns, accurate altitude keeping, and a gentle return to the dispersal area.

Bob used to say that when circuits went well, no one noticed. *"Boring is beautiful,"* he'd grin. In a place like Middle Wallop, boring was also safe. And in the Gazelle, that was more than good enough.

Before solo, we had to demonstrate real engine-off landings.

Bob would wait until we'd completed a clean circuit, then calmly reach up and close the fuel shut-off lever on the overhead panel with the words, *"practice engine failure, GO."* The engine would spool down, the noise would fade, and everything felt… different. Quiet.

Suddenly, we weren't flying the aircraft anymore. We were managing the imminent crash.

At around 50 feet, we'd raise the collective slightly to start arresting the rate of descent. At about 15 feet, we'd gently flare, easing the cyclic back to lift the nose and bleed off the vertical energy.

As the flare decayed, we'd ease the nose forward again at around five feet, level the skids, and use what collective we had left to cushion the landing.

With a bit of forward speed, the Gazelle would slide onto the ground, and once the skids were down, we'd lower the collective fully and come to a stop.

Get it right, and it felt smooth, almost controlled. Get it wrong, and it very quickly wasn't.

Learning to do it myself was a different story. Just like hover training, Bob handed over control gradually, first the descent, then the flare, then the whole sequence. At first, every move felt too early or too late, and the margin for error was razor thin. We weren't allowed to get it wrong at 500 feet; every practice was done to the ground.

The thought of a misjudged flare or a heavy landing was never far from my mind. But repetition breeds familiarity, and soon what had felt like a leap of faith became something closer to instinct.

Autorotation stopped being terrifying and became just another part of flying helicopters. Still dramatic, but now controllable, if not quite beautiful yet, like Bob's.

Steve was the first to crack solo, followed gradually by the rest of the course, except one. Dave Crang, who I had been on flying grading with, struggled through hover work and was eventually failed. Another Fuglie had bitten the dust.

My turn came on 08th July 1992, I flew a couple of warm-up circuits to the required standard, followed by a practice forced landing. On the ground, Bob brought the engine back to full power, eased the throttle forward, and turned to me.

"Right," he said, unstrapping himself.

"I'm getting out. Let's see a couple of nice circuits."

He paused at the door. "*She's going to feel light and right skid low, be ready for it.*"

Ever thoughtful, Bob took some photographs of my first rotary solo on his way out.

I hover taxied to the departure point, transitioned into forward flight, and climbed into the circuit. As I reached the crosswind turn, I scanned for traffic, then reality slapped me. Bob wasn't there. It was all on me.

They say every pilot remembers their first solo. I don't. But this, this was my second solo, and it's the one I remember. As I made my radio calls, I had to swallow the lump in my throat. It was a strange, unexpected surge of emotion, an unfiltered moment of pride and disbelief.

This was the most important moment of my flying career to that point, and one I wish everyone I love in this world could experience at least once in their life. I landed after the second circuit and Bob climbed back into the aircraft. He shook my hand and congratulated me.

During BRW, we also had theory sessions on helicopter systems and servicing, because once operational, we'd be expected to handle some basic maintenance tasks ourselves.

These were run by Warrant Officer Class 1 Cassidy, a towering and sharply intelligent REME aircraft technician who had taught countless pilots before us. He had a calm, matter-of-fact teaching style and a disarming presence.

In the first session, he asked Steve a question about hydraulics. Steve, being Steve, casually dropped a random swear word into his answer.

Without missing a beat, Mr Cassidy stopped the class and announced, *"As a Christian, I don't appreciate swearing."*

Jon, Steve, and I took this less as guidance and more as a direct challenge to our comedic genius.

From then on, we tried to slip in as many swear words, especially the blasphemous ones, as subtly as possible, just to see how far we could push it without getting caught. It was going brilliantly, with the three of us barely containing our silent laughter each time one of us succeeded. But then Jon went too far.

He overcooked it and got caught mid-sentence.

The look Mr Cassidy gave him wasn't anger, it was disappointment. The sort of disappointment reserved especially for officers. A Captain, no less. Cpl Steve 'Nozza' Norris couldn't contain himself and exploded in laughter. Jon melted into his chair.

And with that, the game was over, that lesson endeth forever.

Back in the cockpit, we moved beyond the circuit and into navigational exercises, starting with daytime sorties and eventually pushing into night flying. Each of us was briefed and dispatched solo by our instructors to put our new skills to the test.

On one such occasion, the routing took us south from Middle

Wallop toward the Southampton area. Once on track and clear of controlled airspace, we were expected to practise various manoeuvres: steep turns, climbing and descending Rate 1 turns, and solo general handling under unfamiliar surroundings.

The only catch? If you were operating near controlled airspace, you had to coordinate with local ATC. In this case, that meant checking in with Southampton Approach for a basic flight information service.

In the excitement, Jon forgot that part.

Instead of checking in, he flew straight into the approach path for Southampton's active runway and began his manoeuvres, entirely unaware that he was now a slow-moving obstacle in the middle of controlled airspace.

He wasn't responding to calls and, allegedly, forced a Dash-8 to take avoiding action.

Not ideal.

When Jon returned to Middle Wallop, he was summoned to the CFI's office, where the errors of his ways were explained to him in no uncertain terms, and at max volume. He emerged sheepish and red-faced, and from that day forward, he was no longer Captain Slade. He was '*Centreline Slade*'.

Another core skill we had to master was low-level map reading, using the standard 1:25,000 Ordnance Survey map. In BRW we were flying down to 300 feet, which, while high by tactical standards, was the lowest we'd flown so far.

No one was more surprised than me at how much I struggled. I'd spent years as a vehicle commander; map reading was second nature.

Now, minutes into a nav sortie, I was disoriented.

Five minutes in, I was properly lost. And the panic was setting in.

Bob spotted it straight away. It wasn't the map; it was the speed. I was used to crawling across the ground at 20 miles per hour in a tank, not streaking along at 120 knots.

The visual flow was overwhelming. I was so far behind the aircraft I could almost see myself in the distance.

I was struggling.

But Bob had a plan.

Rather than hammer me with more traditional nav legs, he broke things down. He started giving me short, structured routes, bite-sized tasks using large, obvious visual cues. I didn't need to follow the line on the map at first, just get from A to B, then from B to C, using standout features like bends in roads, rivers, or villages.

It helped me reframe the flight, not as a test of where I was, but as a series of navigational stepping stones to where we were going.

Gradually, they spread further apart, and I began to relax.

Eventually, I wasn't lost. I was ahead of the aircraft.

Bob broke the bogey, and I finally got it.

Just like that, the curse lifted, and with it the threat of review.

Around the halfway point of BRW, we all faced the Intermediate Handling Test (IHT). It was the usual milestone. We all passed with few issues. We now progressed to night flying, more circuit work and navigational exercises flown at 300 feet above the ground.

About this time, Pim started to struggle too, but for different reasons. The instructors recognised he had talent, and so instead of RTUing him, he was removed from APC 342 and 'back-coursed' to APC 343 where he would eventually complete training and be awarded his wings.

By now, our map reading sorties were taking us over vast swathes of the countryside. We were introduced to a new and humbling challenge: confined areas; small clearings in dense woods and forests scattered around Hampshire.

Each one had a name and a difficulty rating, based on the size of the space, tree height, and internal obstacles. The most infamous of them all was Harewood 6. It was brutally tight, so tight, in fact, that you couldn't land directly. You had to fly a high approach to the top of the tree line, hover, and then descend vertically into the space.

It was technically challenging, but the real hazard wasn't the trees, it was the pheasants. Confined area landings tended to freak out the local wildlife. On more than one occasion, a startled pheasant shot up from cover straight into the rotor disc.

One unlucky bird made it all the way through.

No one forgot the mess.

Least of all the pheasant.

The Final Handling Test, or FHT, was the make-or-break point of BRW. It was the last major flying assessment before we moved on to the Tactical or Advanced phase, and it wasn't just a box-ticking exercise.

Pass the FHT, and you were through. The next time you flew, it would be tactical, nap-of-the-earth, low-level, hedge hopping sorties, the kind that felt more like combat than training. Fail it, and you might be given review time. Or you might not.

Either way, the FHT was the gate. And everyone felt it.

Emily's pregnancy was ticking along nicely, and the kids were growing up faster than I could keep up with. On the rare weekends we went up north, I was running on fumes. One Friday in Stockton we arrived late in pouring rain, the kids half-asleep, bags everywhere. I trailed muddy prints across Bobby's kitchen floor, and she went off like I'd slaughtered a pig on her tiles.

I snapped back, something smart about only being trained to hover, not levitate. It was petty, but it wasn't really about the mud.

That weekend I used revision as a hiding place. It kept me out of arguments, but it also meant I was tired, tight, and carrying home into the cockpit. I could still fly. The question was whether I could keep flying with everything else pulling at me.

The Final Handling Test (FHT) was now upon us. The course had gone quiet in that way it does when everyone knows what's coming, but none more so than for Sergeant Ned Hawley of the Royal Corps of Transport.

In the weeks leading up to it, Ned had begun to unravel. A run of poor sorties had knocked his confidence, and he was pushing too hard, trying to force it back.

He was clearly overreaching to claw his way back and was granted a stay of execution. After a period of review he unfortunately failed his FHT with no prospect of further flying hours. Ned was RTU'd.

We were already four Fuglies down, and we hadn't even reached the hard part yet.

Even while on course, we were still subject to the usual barrack duties. For us Junior NCOs, that meant stints as guard commander, and for the officers, it was orderly officer. These duties were routine, but during the bi-annual Middle Wallop Air Show and Heli meet, things were switched up a notch.

I drew the shitty end of the stick and was rostered as guard commander for the Survivors' Party, the post-show event where military formality gave way to extreme excess. Dress was standard Regimental Guard Dress. In the 15/19H, we wore a dark blue rollneck instead of a standard shirt. It looked smarter and set us apart.

The Orderly Officer, WO2 Paul Connor, a QHI from Advanced Rotary disagreed.

One look at my rollneck and I was dismissed on the spot, sent home in the rain to change into the approved, but boring shirt. Dignity lost, lesson learned.

The barrack guard was reinforced for the air show. Pim was my second-in-command, which was good to see, and we'd been augmented by a small Home Guard detachment, polite, keen, and utterly unsuited to crowd control or life without a cup of tea in hand.

By the time I got back to the guardroom, reports were coming in that things were kicking off in the gym, where the party was. A Home Guard chap offered me a brew and said it was "getting lively."

Outside, we heard chairs whistling through the air and a lot of shouting. Royal Marines. Very drunk Royal Marines.

I grabbed Pim and went in fast and loud. No easing in. No negotiation.

Incredibly, it worked. The hall went quiet, apologies followed, and the situation deflated almost instantly.

Paul Connor arrived moments later with his geriatric reinforcements, zimmer frames made ready, only to find it already under control. He didn't say much, but I could tell he was quietly satisfied, even if he'd never admit it.

I'd survived my first near-riot, lost a minor battle over a shirt, and somehow come out ahead.

60 Moments before first rotary solo

61 First Rotary solo.

Chapter 34. Advanced Rotary Wing (ARW) Phase

We progressed to Advanced Rotary Wing (ARW). This wasn't just the next chapter; it was a full step-change in tempo. We thought BRW was demanding, we were rudely awakened.

ARW was where they separated the pilots from the passengers.

All the instructors were serving AAC Qualified Helicopter Instructors (QHIs), and their personal reputations were on the line. Each one would be tied to our name for the rest of our careers if we passed. No QHI wanted to be known as the one who signed off an idiot and sent them to an operational Squadron.

ARW was commanded by Capt. Dominic Mayall, a respected young officer, pilot, and leader; sharp, composed, and very clear about what was expected of us. We'd proven we could fly, but could we fight the Gazelle, not just fly it? Could we operate low, fast, and unseen, by day, by night, and on instruments, one of which was a tatty piece of red wool tied to the nose.

That instrument rating alone was no joke. The Gazelle had no ILS, minimal instrumentation, a basic Doppler navigation aid, and very limited capability in icing conditions. But we'd be expected to fly instrument departures, arrivals, and procedural approaches with total precision. On top of that, we'd learn to fly in mountainous terrain, and master the art of nap-of-the-earth flying, staying as low and undetectable as possible, while navigating at speed.

We knew one thing going in: from here onwards, the failure rate only increased.

On our first day, we reported to the tiered lecture room at ARW. The full QHI team was waiting. We were introduced to our instructors, and sure enough, like clockwork, I was paired with Paul Connor. I should've seen that one coming. Paul gave nothing away, as usual. Meeting up with him I said, *"I hope the roll neck isn't going to cause any more issues?" "Only if you want it to"* he replied.

Then Dom Mayall addressed us. He laid it out in stark terms. There would be no quarter given. If we struggled, we might be granted review hours, but only after serious consideration. We were expected to meet the standard first time, full stop.

Every sortie would be formally graded: Blue for above average,

Green for average (meeting the standard), Brown for below average (not acceptable), Red for a failure.

He paused, then added, *"Given enough time, a monkey can be taught to fly. But we don't have the time. If you can't keep up, you won't stay."*

No one said a word. We didn't need to. We were all thinking the same thing. It was game on.

There were no surprises in the system. Only consequences.

After a couple of ease-you-in type sorties, more for the instructors to assess our baseline skills, we were thrown straight into Instrument Flying.

To simulate being in cloud, or Instrument Meteorological Conditions (IMC), a black peak with side panels was clipped to our helmets, blocking all peripheral vision except for the instrument panel directly in front of us. A black Styrofoam screen was wedged between the doorframe and the console, cutting off any chance of glimpsing the outside world. Try to cheat, and all you saw was your own soul in a void of black.

We learned to fly solely on instruments, responding to precise instructions from the instructor. *"Fly heading 270, maintain altitude 3,000 feet, speed 100 knots."*

Our scan of the instruments had to be continuous; the scan became a rhythm you didn't question, only obeyed. A miniscule correction on one instrument would affect the others. Scan, adjust, repeat.

Once we could hold straight and level flight, the instructions ramped up. *"Turn right, heading 293, climb to altitude 4,500 feet."*

Now you were expected to fly a Rate 1 turn, a constant, coordinated turn at exactly 3° per second, meaning a full 360° turn would take precisely two minutes. This brought the turn and slip indicator into your scan and demanded enough power to hold a 500 feet-per-minute climb, while maintaining 100 knots.

Precision and discipline were everything.

The exercises became more complex by the day. *"Enter a climbing turn to 6,000 feet,"* Paul instructed one day, *"and at every 500-foot interval, call out a girl's name in alphabetical order. Roll out on this heading."* It was a test of mental agility: juggling flight parameters, verbal tracking, and situational awareness, all while keeping your scan going. If you nailed it, you'd reach 6,000 feet and roll out perfectly on heading.

At the end of most IF sorties, we'd contact a nearby RAF station and carry out an instrument approach, either a Precision Approach Radar (PAR) or a Surveillance Radar Approach (SRA). In a PAR, the controller talked you down a 3° glide slope while providing heading adjustments in real time.

"Approaching the glide path, start your descent now. Turn left two degrees. Make your heading 268. Above glide. Adjust rate. On centreline. On glide."

Minute corrections, calmly delivered, guiding us to Decision Height, where the instructor would either take over for landing or execute a missed approach to bring us back to Wallop visually.

Strangely, I loved Instrument Flying. Where others dreaded it, I thrived. It just made sense to me, and I was good at it. Maybe too good in my own mind.

After one particularly sortie, where I'd been bang-on the numbers from start to finish, Paul asked during the debrief, *"So, how do you think that went?"* Cockily, I replied, *"Easily a blue sortie."*

Wrong answer.

He failed me. Graded it red and made me repeat the sortie.

The message was clear: stay humble. Confidence was fine. Overconfidence would be punished.

I was gutted, but the lesson stuck. The Instrument Rating Tests (IRT) came, and we all got through.

Sorties were now coming thick and fast, most with a tactical overlay. We moved into mutual sorties, two students sharing the cockpit, one flying, one navigating, both being judged.

My first mutual was with Capt. Paul 'Hagar' Hatchard. Hagar had been back-coursed for medical reasons to the Fuglies from the course ahead and, coincidentally, was from Bolton Abbey, the village just beyond the farm where I'd worked with Dinny during my gamekeeping years. We hit it off immediately. Hagar would later become central to one of our more memorable course stories, but more on that later.

Following the Instrument Rating Test (IRT), I completed a handful of additional sorties with Paul. As we neared the Intermediate Handling Test (IHT), the instructors began to rotate. The aim was to expose us to different styles, and different views of our performance, ahead of final assessment.

I respected Paul a great deal. After Pete Jennings, he was almost a teddy bear, despite the no-nonsense approach, which I'd come to appreciate.

My ARW IHT was flown with Major Dwayne Stevens, a senior flying standards officer. A straight-talking Yorkshireman from the 13/18H, he'd come up through the ranks and brought a calm, competent air to the cockpit, and, just as importantly, a proper accent.

Once I'd demonstrated all the required flying exercises to a satisfactory standard, advanced quick stops, practice forced landings, simulated emergencies, we spent much of the return leg to Wallop chatting about the imminent amalgamation of our respective Regiments to form The Light Dragoons.

With the IHT now behind us, everything shifted to tactics. The focus was now on how to fight with the Gazelle, not just fly it. We began sessions in the Invertron, an archaic artillery fire mission and support simulator, where we learned how to direct both light and heavy artillery batteries with precision. Training soon progressed to the live ranges on Salisbury Plain, where we directed and adjusted live artillery fire.

Jon and I were paired for this phase, which naturally meant the devil's horns came out.

As part of our fire mission orders, we were required to give a brief description of each target, typically something like "*tanks in the open.*" But having absorbed a whole new vocabulary of three-letter abbreviations, we couldn't resist pushing the envelope.

Our targets became *"3 × CWP in the open*". It was harmless fun, until the crusty old Royal Artillery Major in charge caught wind of it.

He didn't see the humour.

Convinced we were taking the piss out of his gun crews; he threatened to knock us both out. Fair to say, the message was received.

The daily routine settled into a familiar rhythm: get briefed, plan the sortie, back brief the plan, fly the mission, debrief. Rinse and repeat. We burned through dozens of maps. Tactics dominated everything now, day and night, and our night flying had progressed to using NVGs to operate low-level and in and out of confined areas.

We were flying in pairs, providing mutual support, and without even realising it, we were becoming competent Army aviators.

Around this time, I was assigned a new instructor, Sgt Chris Keane. He'd just completed his QHI course at RAF Shawbury and was, quite simply, the best instructor I'd ever flown with. Happy, jovial and slightly rotund, Chris had started life as an infantryman but in the cockpit he was patient, articulate and had an uncanny ability to explain exactly what was needed.

Chris could have taught the Pope to pray and had him doing it better than before.

He was my instructor for a long nav. We were given some say in the route and the crew pairing, so Hagar and I put ourselves together and planned a route up to Ilkley. The idea was to overfly his family farm at Bolton Abbey, then drop into the Cow and Calf Hotel on Ilkley Moor, which conveniently had a registered landing site.

Chris agreed. Hagar flew first, with me in the back. We passed over his family farm and the nearby woodland where I'd worked with Dinny years ago, before continuing to Ilkley for lunch and a student changeover.

We'd arranged the HLS in advance but hadn't thought to warn the landlord to keep quiet about it. Why would we? Well… maybe we should have. He took it upon himself to inform the local paper, The Ilkley Gazette, about our visit. When we touched down, a reporter/photographer was already waiting to capture the moment.

The article appeared the following week with the punchy headline:

"Pilots Drop in for a Drink"

Completely inaccurate, of course, but when has that ever stopped a newspaper? My Mom sent me the clipping. I handed it straight to Chris so he could get ahead of any fallout with the chain of command. Fortunately, no drama followed.

After the long nav's came Mountain Week; five days based at RAF

Llanbedr in North Wales. We flew multiple sorties through the Snowdonia National Park, where the scenery was as breathtaking as it was nerve-wracking.

Towering ridgelines, narrow valleys, sudden down draughts and freak gusts, one moment you were flying through a postcard, the next you were fighting the controls, trying to balance an optical illusion of a lake that appeared to be sloping! The weather had a mind of its own and didn't tolerate complacency.

As a way of repaying the national park for hosting us, we lifted underslung loads of gravel up the mountains to support repairs to walking tracks. It was practical aid for the park and excellent training for us.

It was shortly after returning from Wales that we suffered the final casualty of the course. Capt. Eoghann Power had been struggling with low-level navigation. The issue wasn't just the map reading; it was also the lightweight Doppler navigation system known as TANS. While most of us had come to grips with it, Eoghann never quite got the hang of it. To be fair, it was a piece of garbage.

He was placed on review and assigned Paul for his final assessment. Eoghann went into the sortie confident he'd be fine.

He wasn't.

When they returned early, which was an omen, his face was purple with rage, and the tension was visible. Word quickly spread about what had happened. Apparently, not long into the flight, the TANS had played up. Instead of managing the problem, Eoghann snapped.

"RIGHT! THAT'S IT!" he'd shouted at Paul.

"Let's go home!"

Paul, famously unflappable but not known for tolerating that kind of behaviour, made the call. That was it. Eoghann was gone.

The final hurdle was Exercise Woodlark, a week-long war game where APC 342 deployed to a faraway war zone that looked suspiciously like Devon. Ground crews played the live enemy and we were sent out on belt-fed tactical sorties that brought together every element of our training from the past year.

Day and night, we used reconnaissance skills to locate the enemy, and the kill them by calling in fire missions and fast jet attacks. We were exhausted, running on fumes, and all eyes were fixed on the wings parade. Just one or two sorties left, and a final FHT to get through.

Having been victorious in Devon, we returned to Wallop for tea and medals, and after a few revision sorties, the Fuglies were declared ready for Final FHTs.

Word circulated that an officer on the previous course had passed his F/FHT but failed his final NVG sortie the night before his wings parade. Despite having passed the final flight test, he was binned. Another reminder not to get complacent.

62 Me with Hagar and the Cow & Calf Pub Landlord

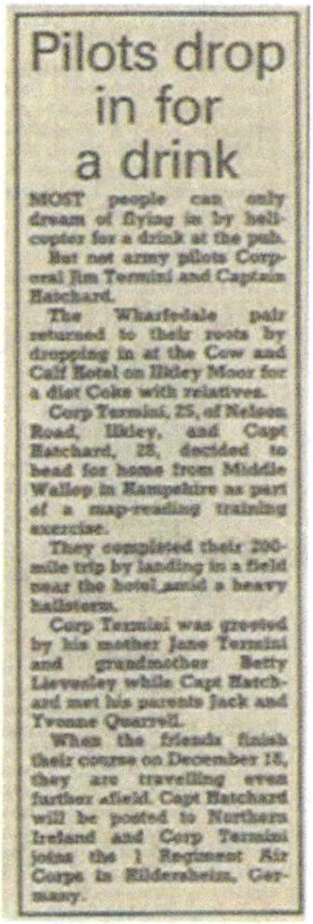

Pilots drop in for a drink

MOST people can only dream of flying in by helicopter for a drink at the pub.

But not army pilots Corporal Jim Termini and Captain Hatchard.

The Wharfedale pair returned to their roots by dropping in at the Cow and Calf Hotel on Ilkley Moor for a diet Coke with relatives.

Corp Termini, 25, of Nelson Road, Ilkley, and Capt Hatchard, 25, decided to head for home from Middle Wallop in Hampshire as part of a map-reading training exercise.

They completed their 200-mile trip by landing in a field near the hotel amid a heavy hailstorm.

Corp Termini was greeted by his mother Jane Termini and grandmother Betty Lievesley while Capt Hatchard met his parents Jack and Yvonne Quarrell.

When the friends finish their course on December 18, they are travelling even further afield. Capt Hatchard will be posted to Northern Ireland and Corp Termini joins the 1 Regiment Air Corps in Hildesheim, Germany.

63 The Ilkley Gazette Article

Chapter 35. Brickbats and Bouquets

Emily was due to give birth at any moment. With that in mind, we were given a list of possible postings to operational squadrons and the chance to request one. With four children, transport wasn't a luxury, it was a tactical requirement. The only viable option was a Volvo 740 estate, mainly because it came with a third row of seats.

So, I requested 651 Squadron, part of 1 Regiment AAC based in Hildesheim, citing the need for a tax-free car.

I joked with Jon that the hierarchy would take one look at my request and bin it. *"A Corporal? Four kids? Must be a welfare case."*

Along with three students who had been back-coursed to us, ten Fuglies passed the FFHT. Our wings parade took place on Friday 18th December 1992 in the museum of Army flying. At the start of the course, I had been a 15/19H soldier, but with the amalgamation complete, I was now a Light Dragoon.

Using a set of new uniform badges sent by the RSM, I became the first Light Dragoon to be awarded the Army Flying Badge, or the last 15/19H, whichever way you look at it.

Maj Gen. Toyne-Sewell, the presenting officer, announced the course awards. Steve Hicks took the Bob Bowles Trophy for Best Student. Martin 'Fishy' Baxter won Best Fixed Wing. Marc Edwards topped the exams.

I was awarded the Westland Trophy for Most Improved Student.

Jon immediately dubbed it "The Knackers Award" for starting off as a complete liability and somehow becoming acceptable.

Both Steve and I were graded as high-average pilots in our final reports, a solid endorsement of our performance throughout training. We were also recommended for conversion to the Lynx, which was exactly the outcome I had hoped for.

After the obligatory photographs with family and friends, I noticed someone queuing to shake my hand. Pete Jennings. The only fixed wing instructor to attend, he'd made the effort to come to the parade. He shook my hand and said, "I knew you'd be great. Stay safe."

I'd been wrong about him. About his motives, his intentions. His methods might have been extreme, but they worked.

Mr Jennings, thank you.

From there, attention shifted sharply to domestic life. The plan was to hand back the keys to the house on Monday 21st December, then drive up to Teesside to spend Christmas with Emily's parents before heading to Germany.

My posting request had been granted and we'd already picked up our brand-new car.

On the Monday morning, I handed the house back. Five minutes later, Emily went into labour.

We drove straight to Andover General in a car packed to the roof, and on it, with our entire life looking like the Clampetts. Technically homeless, our third daughter, Francesca Lois, arrived at around 11:00 a.m.

By early evening, with a seven-hour-old baby in the back, we set off for Teesside.

When we arrived, Bobby was a different person. We had a stream of visitors coming to meet Fran, and every time someone walked through the door, Bobby would beam and say, *"Jim's a pilot, you know."*

I hated it. I was still the same person, but now I was suddenly worthy. The change in her attitude made me like her even less.

There were plenty of friends from 15/19H home for Christmas, and we had a great time catching up, Kenny and Joanne, Chris and Anna. My report date for 651 (Premier) Squadron was 10th January 1993. It didn't come as a surprise.

What did the future hold? How would I be received by my new Squadron colleagues?

Everything ahead was new. I was excited to the max and nervous all the same. There'd been moments I wasn't sure I'd make it past grading, let alone earn my wings. New baby. New car. New status. Same me. Just better trained and very happy.

Figure 64 The Fuglies Crew Room Presentation Photograph – Me rear row, second from the right

65 Maj Gen Toyne-Sewel and Mr Peter Jennings

66 Emily, Peter Jennings, Chris Keane and the kids

67 Presentation of The Westland Trophy

Germany, Hildesheim: 1 Regt AAC, 651 Sqn, Gaz Flight

Chapter 36. A Brave New World

In the New Year of 1993, we arrived in Hildesheim to join the AAC as the most junior of junior pilots. It was a grimmer, more industrial place than Detmold, and it felt like a long way from anything familiar. 1 Regiment shared the barracks with 1 RTR, and the sight of Chieftains nearby was a small, but welcome comfort.

My sqn, 651 was nicknamed 'Premier', as numerically it was the first of the AAC Squadrons. We wore the 'Maid of Warsaw' flash on our sleeves, a WW2 battle honour earned supporting the Polish 1st Division.

I reported to the Gazelle Flight offices, commanded by Capt. Richard Felton, a young captain from the Gloucestershire Regiment. Capt. Felton was a classic young thruster, ambitious and clearly going places. The flight second-in-command was SSgt Keith Anderson. My first task was to meet the boss, Maj Colin Baulf.

Colin 'Baulfers' Baulf was a Falklands veteran. He had been attached to 3 Commando Brigade Air Squadron of the Royal Marines and had earned the right to wear the coveted Commando dagger on his sleeve; a rare badge of honour in the Army Air Corps. He'd flown extensively during the war, often alongside Sgt Mac McDaniel, another Gazelle pilot who had recently transferred from the Marines into 651 Squadron.

Mac was larger than life. A towering figure with black wavy hair and a moustache Saddam Hussein would have respected, he could be intimidating one moment and hilarious the next. He was the kind of bloke who filled a room before he even walked into it; and you never forgot he was there.

I took an instant liking to Baulfers. He was one of those Officers who struck the perfect balance of confidence and competence, and it was clear from the outset that he could be trusted. He also had me figured out in no time. After a few questions about my hobbies and interests, he leaned back and quipped, *"So, not a gladiator then!"* I had to laugh, fair call.

He then told me I'd been loaded onto the next available Lynx conversion course, which, typically, wasn't until July.

Whenever I was promoted previously, I'd spend the first week catching myself glancing down at the new stripes, making sure they were still there. Now I found myself doing the same with my wings. For a few days I felt quietly proud every time my eyes caught them. Then, like every stripe before, they just became normal. They stopped being something I'd earned and settled into being something I was.

I settled into Squadron life quickly. The first series of sorties were with the Sqn QHI, SSgt Dan Sanderson. Dan's job was to firstly assess my flying ability and whether I would be safe with a line aircraft commander, and secondly take me through Role Qualification, the process that would qualify me as a Gazelle reconnaissance pilot. Early on I also had sorties with Richard Felton and Keith, again, so they could assess my skills as a line pilot.

I was pleasantly surprised to meet an old friend from my time in 15/19H, Sgt Pete Carre, the REME electronics technician. I hadn't realised he had completed pilot selection ahead of me. We were authorised a mutual flight together, a simple run up to Hohne and back, made memorable purely by flying with Pete.

In early February I deployed on my first tactical exercise with the AAC as a junior pilot, eager to see how different life in the field would be like without sixty tonnes of steel beneath me.

Chapter 37. The Para, the Gamekeeper and a Psycho

The Squadron assembled at the obligatory stupid-o'clock start time. Ground crews fussed over vehicles while the aircrew, bleary-eyed and coffee-starved, shuffled into the morning met brief.

The weather forecast was predictably grim. The forecaster, known as 'stuttering Sid Cirrus', delivered the bad news with the enthusiasm of someone who'd already accepted his weekend was ruined: for the first two to three days, flying conditions in the Harz would be officially crap and below limits.

Then Baulfers delivered the bad news: a snap 'escape and evasion' exercise in Himmelstur, a local training area wrapped in near freezing fog with a fine drizzle, guaranteed to ruin any party.

We were muscled up, bundled into Land Rovers and driven out into the murk. I was with Lt Bob Scarlet, a former paratrooper turned AAC pilot. He had the quiet confidence and enthusiasm of someone who'd survived worse ideas than this one.

Bob and I hatched a cunning plan: use the fog to leg it somewhere remote, then dig in and keep our heads down.

It worked, almost too well.

We found a hollow under a thick canopy of trees, lined with dense brush and soft ground. Flat from the outside, it gave just enough shelter for us to disappear. We belly-crawled in, hollowed out a spot, lined it with branches, and hunkered down in our sleeping bags.

Warm (ish). Dry (ish). Invisible.

And then... nothing. For over 48 hours.

We had no idea the exercise had already moved on without us.

Eventually, we heard the iconic music of a Lynx overhead, circling.

Bob volunteered to check it out. Turns out the entire Squadron had been *'captured'* hours earlier and taken back. We were the only idiots still in the field, a former paratrooper and a former gamekeeper gone native.

Apparently, there was genuine concern. No comms. No sign of movement and mutterings about pacific islands and Japanese snipers.

Back in Squadron lines, dry and in fresh kit, the weather finally lifted enough for us to deploy.

I was flying with Sgt Connor "Psycho" Judd, a quiet bloke with a passion for massive handguns. Entirely legal in Germany, of course. His pride and joy was a Desert Eagle, completely impractical, but completely Connor.

The flight to the gliding site in the Hartz took just over an hour. It was chucking down, but visibility was workable. Just. As we came into land, I noticed something odd, no ground crew.

Not a single person came out to help us camouflage the aircraft.

"Right," I thought. *"This is weird."*

In a tank Squadron, if one crew arrived late, everyone mucked in. You didn't wait for orders, you just squared the problem away. That culture wasn't universal it appeared.

It shouldn't be about orders; it was just what professional soldiers did. The unspoken rule was simple: we all get squared away, or none of us do.

Connor and I set up the aircraft camouflage nets and then headed over to where a few tents were pitched near the glider club buildings.

When I heard we'd be getting proper aircrew tents, I was cautiously optimistic. That lasted right up until I discovered the tents were still in their storage bags

In the rain. After three days.

They hadn't even been looked at, let alone pitched. Turns out, optimism is just naivety in denial.

Sodden canvas weighs a ton, stinks, and is a bastard to pitch. And while the groundcrew had apparently been here for days, they'd achieved precisely nothing, except drinking beer in the clubhouse.

Another new experience: central catering.

On Chieftains, we had, of course, lived on 4-man ration packs, usually cooked by the loader and as described, was surprisingly good.

This stuff though? Inedible. Some tragic stew that had been boiled to death, cremated for good measure, then abandoned like a Romanian orphan. It tasted like it had given up halfway through cooking, just slumped in the pot and accepted its fate.

No spice. No texture. No joy.

After "dinner", the SSM, Nigel Hunter, gave us a rapid-fire admin brief, including naming the poor sods selected for pan-bashing, cleaning the scorched wreckage that the 'cooks' had left behind.

When my name was called, I saw red. I wasn't about to scrub burned sludge off catering tins while half the Air Troopers hadn't lifted a finger in days.

There were dozens of Air Troopers in the bar, and I was scrubbing burnt slop out of catering tins? Not a chance.

I made my displeasure known. I muttered something about the Army cadets being more professional.

Baulfers must've heard I was fuming. He wandered over, beer in hand, wearing a grin that suggested this was about to get worse.

"What's up, then?"

Still seething, I snapped:

"Sir, I may be a junior pilot, but I'm still a corporal in the cavalry!"

Baulfers doubled over laughing. Mac had set me up beautifully.

From that day on, and every time Baulfers and I crossed paths after, I was forever "*the corporal in the cavalry*".

After a couple of tactical sorties squeezed into a dwindling window of decent weather, we returned to Squadron lines in Hildesheim, but the fiasco wasn't over.

Unsurprisingly, the aircraft arrived back well before the ground crews, and that was where the real trouble started.

To move the Gazelles, a wheel had to be clamped onto the 'heel' of each skid. A long steel pole was then slotted into the jack and levered over an overlocking cam. Get it wrong, and the pole would whip back like a spring-loaded battering ram. Broken noses weren't unheard of.

Once the wheels were on, a tractor would hook onto the front and tow the aircraft into the hangar.

With no groundies in sight and rain still falling in sheets, we decided to do the right thing; crack on and do it ourselves.

That's when it went pear-shaped.

One of the 'Westland wheels' was unserviceable. The rubber and steel pad beneath the locking clamp was missing, meaning the tightening spigot sat directly on the aluminium skid. When we jacked it over, nobody noticed the spigot boring into the metal. By the time we had the aircraft rolling, the damage was done.

Each of four helicopters had a skid punctured and written off. Not a bad day's work for a single dodgy wheel.

To make matters worse, we hadn't considered that none of us had an airfield driving permit. And yes, driving the tractor without one was strictly verboten.

So: unauthorised driving, written-off aircraft components, and a total breakdown in procedure.

The shit hit the fan.

Richard Felton went ballistic, and boy could he!

Luckily for me, I was still just a corporal, albeit in the cavalry. The seniors caught the full force of the blast. I just got some light freckles from the spattering.

The week after, the whole flight was knee-deep in cleaning aircraft and every bit of equipment that wasn't bolted down and needed drying out. By Wednesday afternoon, tradition dictated an early knock-off, the sacred 'sports afternoon.'

Given we'd spent the previous day's being hunted through the Himmelstur during escape and evasion training, it seemed a fair and reasonable request.

The only problem? No one wanted to be the poor sod to ask Richard Felton for permission. His reputation for humourlessness was well established.

Eventually, Mac volunteered.

Ever the showman, he donned a helmet and body armour and kitten-crawled, yes, crawled, into Richard's tiny office, hoping a bit of absurdity might soften the blow.

The rest of us were loitering outside like a pack of schoolkids outside the headmaster's office, giggling as we listened in. *"Sir! may I respectfully request a sports afternoon for Gazelle Flight, please SIR?"* came Mac's overly formal, muffled voice.

Unfortunately, Richard didn't find it nearly as amusing as we did. He reacted about as well as you'd expect someone allergic to joy to react. Suffice to say, we didn't get the early knock-off. But credit where it's due, at least Mac tried.

In May, Richard Felton was posted out, presumably to somewhere that didn't require a sense of humour and was replaced by Captain Callum Claypole of the Worcestershire and Sherwood Foresters.

Despite also being an infantry officer, Callum was a very different breed: approachable, good-humoured, and generally less inclined to make you feel like you were on trial every time you spoke.

Richard went on to command an Apache Regiment in Afghanistan, where his intensity no doubt came in handy. He eventually reached the rank of Lieutenant General and collected a healthy stack of medals and honours along the way, all thoroughly deserved. Still, none of us were terribly surprised.

You don't forget a man who could make asking for a sports afternoon feel like negotiating a ceasefire.

The transition from armour to aviation brought another unexpected culture shock. In a tank crew, we lived in each other's pockets. Everyone knew their job, trusted each other completely, and usually stayed with the same Regiment. That stability bred tight, instinctive teamwork, there was room for character, but not ego, just competence and mutual respect.

In contrast, the AAC was split along clear lines: techs, groundcrew, and aircrew. While the pilots worked exceptionally well together, many of them former cavalry, infantry or artillery officers and other ranks, with strong professional standards, that sense of tight-knit cohesion didn't always extend across the AAC Regiment.

The divides were structural, not personal, but they shaped everything from daily routines to how problems were solved. Teamwork existed, but mainly within these silos. It was hard not to miss the clarity and unity of a tank troop, where everyone pulled in the same direction, because you always had to.

Germany, Gutersloh: 1 Regt AAC, 651 Sqn, Gaz Flight

Chapter 38. New Barracks, Old Ways

In June 1993, 1 Regiment packed up and departed Hildesheim as part of the broader drawdown in Germany. The RAF, ever gracious, vacated RAF Gutersloh, previously home to Harriers, Pumas, and Chinooks, and handed the whole operation over to the Army.

Naturally, the Army responded in the only way it knew how: by systematically dismantling everything that had worked. The place was promptly rebranded as The Princess Royal Barracks.

Alongside us came three Regiments of the newly formed Royal Logistic Corps (RLC), who not only joined the garrison, but assumed command of the entire site.

The new Commanding Officer, a Brigadier from the RLC, immediately made the most baffling decision seen since the British Empire decided woollen uniforms were a good idea for tropical climates and red tunics provided excellent camouflage.

He was a big, lumbering man with the air of someone who had misplaced his ironing board twenty years earlier. His uniform never quite sat right. Fitness?

He looked like he'd struggle to pass a cholesterol test, never mind a BFT.

But it was the glasses that really set him apart; a battered pair of hornrims, one arm lashed together with what looked suspiciously like sticking plaster. The effect was pure Jack Duckworth from Coronation Street. Once the resemblance was spotted, it could never be unseen.

From that day forward, no rank, no title, no gravitas: to us he was simply "Duckworth."

First on his list was commandeering the Air Traffic Control tower, apparently for the views over his new kingdom. The RAF ATC detachment wasn't finished yet and fought him off, retaining control of the tower and, with it, a shred of sanity.

Then came the brilliant reshuffle. The hangars, designed and built for helicopters, were assigned to the Regiments of supply trucks. The helicopters of 651 were shoved into Hardened Aircraft Shelters

(HAS), originally built for Harriers. 652 and 661 were both crammed into a single Chinook hangar like an aviation-themed game of Tetris.

There was, however, one small issue, the rotor diameter of the Lynx meant they couldn't fit in the HAS. The solution? Erect temporary RUBB tents at the entrances.

The madness didn't stop there.

A new perimeter fence was installed, well, partially. Apparently, the budget ran out midway through construction, so rather than loop around sensibly, the fence was redirected straight across the main runway, neatly bisecting the airfield in two. Aircraft were now parked outside the secure area, in tents.

The trucks? In hangars, inside the wire.

Of course, there was poetic justice.

The trucks promptly cracked the hangar floors and were moved outside. The hangars were then repurposed for storing blankets and water bottle tops. To summarise: aircraft in tents, trucks outside, and aircraft hangars full of miscellaneous kit.

A masterclass in military logistics.

Amid all this madness, my own time on floppies (Gazelles) was coming to an end, my last flight on the Gazelle before Lynx conversion was a mutual with Corporal Phil "He Man" Brundle, a lad from Leeds, built like a pipe cleaner and courtesy of the Royal Signals. Naturally, I couldn't have picked a better co-pilot for a send-off.

Being my final sortie, I wanted to go out with a bit of flair, not just another loop around the patch. With some creative justification, Callum authorised a night mutual to recreate the original Dambusters route.

We launched and headed south, tracing the original profile flown by the Lancasters of 617 Squadron. First stop: the Möhne Dam. Then across to the Sorpe, and finally to the Eder. We didn't have bouncing bombs or a bomb aimer yelling "*steady... steady...*" but I did have 'He Man' calling out altitudes and making sound effects. No Richard Todd, but close enough.

Flying that route at night, imagining what it might have felt like to be part of the RAF's most daring raid, was everything I'd hoped for, absurd, brilliant, and just serious enough to feel like a proper farewell. A fitting way to end my time with the wonderfully twitchy Gazelle AH1: flying low, laughing hard and grinning like an idiot the whole way home.

Next time I flew, it would be in something much bigger, heavier, and far less forgiving.

Middle Wallop, AACC: Lynx Conversion, Jul – Sep 1993

Chapter 39. Promotions and Missiles

On arrival in Gutersloh, I was promoted to acting Sergeant. Because I was still a Light Dragoon, my long-term future technically sat with them. Fortunately, the Light Dragoons Commanding Officer had flown with 1 Regiment himself and understood exactly what was at stake.

When my three-year flying tour ended, I'd be given a choice: return to the Light Dragoons or transfer permanently to the AAC.

No contest.

The Lynx conversion unit at Middle Wallop was much more laid back and less stressful. By the time most people went through conversion, they had spent some time in a line Squadron, so we were treated with a tad more respect, plus, we didn't have to jump through quite as many hoops.

The first part of the course focused on getting to know the aircraft. The Lynx felt like a monster after the Gazelle, and now there were two engines to keep happy instead of one.

Steve Hicks was on the same conversion course, and old habits die hard. Also on the course was Mac, who had previously converted, so for him, this was a refresher.

The course included elements from the ARW phase, such as mountain flying, underslung loads, NVG work, and tactics. But the big new feature was our own weapon system, the TOW missile. The Lynx could carry up to eight TOW missiles, and we were trained to operate the system from both seats, flying as the pilot and as the TOW system operator.

We were each given one missile to fire on Salisbury Plain. TOW missiles had a reputation for going rogue. And if it went rogue? That counted as a miss. No one wanted that, it was a total waste of time, effort and money, besides which, firing a missile was a high point on the course.

We also learned Forward Arming and Refuelling Point (FARP) drills; basically, a helicopter version of an F1 pit stop, except instead of slick tyres, we were getting re-armed in mere minutes. Once reloaded, we'd head to a firing point to acquire, track, and shoot a target.

"Target acquired, come into constraints," the TOW operator would call. The pilot would then align the aircraft with the sighting system using a small instrument on his coaming. *"IN CONSTRAINTS."* Then from the operator, *"Firing in 3, 2, 1...firing now."* A whoosh of smoke, a slight bang, and the missile would depart the tube.

My missile 'gathered' perfectly into the sight, and using the joystick controller, I steered it straight into the heart of the target. A direct hit! The missile dragged kilometres of guidance wire behind it. I then hit the wire-cut switch twice, just to be sure. A small explosive charge detonated and severed the wires.

We turned away from the firing point, ready for a crew change and to discuss our efforts with the Helicopter Weapons Instructor (HWI). Naturally, we pretended to be cool about it, but inside we were all secretly fist-pumping.

The worst part of the missile range was collecting the wire afterwards. It was extremely strong and difficult to cut; each missile left two wires. The whole course gathered to collect it, mindful of protecting wildlife and anyone walking through it.

Nearby was an old aircraft fuselage used as a hard target. I snapped off some parts of it and handed them out for people to use as a bobbin to wrap the wire around.

One Major, about to take command of a Squadron, knew better, *"no thankyou Sgt, I'll just use what god gave me"* electing to wrap this high tensile steel wire around the crook between his thumb and pointing finger and his elbow, as if it were a garden hose.

An hour later, we all met up to dump our conveniently wrapped spools of wire into a large barrel. All except the Major, who strolled in about ten minutes later, looking as if he'd just lost a wrestling match with a bear. His wrist at a comically painful 90-degree angle to his forearm, and his thumb had begun to turn a worrying shade of purple and black.

We had wire cutters with us, but it was an absolute nightmare trying to free him from his own stubbornness.

I fell in love with the Lynx, the mark 7 was sleek, powerful and very agile. I loved climbing up into the cockpit, rather than getting down into it as with the Gazelle. The sound of the Rolls Royce Gem 42's, combined with the BERP blades was pure music.

She wasn't perfect, the engines were prone to failure, but the aircraft itself was robust, except when spares were in short supply, which was most of the time. If she bit, it tended to be without mercy.

Whilst on conversion, I missed both Alex's and Jessica's birthdays, again. It was becoming a pattern, and one I hated more each time it repeated.

Emily was, by now, resigned to me not being around for important family occasions. We always did our best to make up for it when I was home, but life as an Army family is brutally difficult. You can't plan anything. When the Army wants you, it means now, and there's bugger all you can do about it.

My promotion to Sergeant came with an overdue upgrade: a real house with a garden. After years in German flats, typically two apartments awkwardly knocked through to create enough space for a family of our size, the change was massive.

With four kids, space wasn't a luxury, it was survival. Matty and Fran got their own rooms; Alex and Jess shared, which worked well given their shared obsession with ponies and anything equestrian.

The flats had a bleak, uniform charm. Beige carpets that never cleaned properly, magnolia walls unless you chose one of the approved pastels, and a block-job rota that ignored rank, exhaustion, or the number of kids you had. When your name came up, you cleaned. No excuses.

The toilets were unforgettable, the infamous German *'inspection shelf'*, which ensured your latest effort would sit proudly on display until flushed. A grim curiosity that never felt normal.

To top it off, each block of flats came with a block job rota, dictating when it was your turn to clean the communal stairwells, landings, and bin areas. The Army made no distinction between rank or number of kids: once your name was up, out came the mop and bucket.

It didn't matter that you'd just come off a late engine pack change or had four kids climbing the walls, the hallway wasn't going to clean itself. And woe betide anyone who skipped their turn; the unofficial neighbourhood watch of squaddie wives ensured full compliance.

Still, it was ours. And no matter how sterile or impractical the Army housing might have been, we made it feel like a home, beige carpet, pink walls, inspection shelf, Army stick furniture and all.

Emily was put in touch with a British civilian who ran a tax-free Toyota dealership in Bielefeld, not far from our apartment. He offered her a job selling cars to squaddies who couldn't really afford them, which she threw herself into and genuinely enjoyed.

Sadly, it didn't last long. The owner ran into health issues and let her go. One minute she was shifting RAV-4's to lads on instalments that would outlive their postings, the next she was back at home wondering what had just happened. It was a shame. For the first time in a long while, she'd found something that was hers alone.

Looking back, it was one of the first signs we were drifting apart, even if neither of us recognised it yet.

Germany, Gutersloh: 1 Regt AAC, 651 Sqn, Lynx Flight

Chapter 40. Hunting and Heroes

Coming back to the Squadron lines, there had been some personality changes. Mac and I arrived in Lynx flight together. The Flight Commander was Capt. Dan Perren, an Australian Army aviator on exchange. In typical Aussie fashion, he was relaxed about most things and enthusiastic about Friday beers, a combination that worked surprisingly well.

The flight second in command was CSgt Darren 'George' Hunt, a Royal Marine on exchange. George was, is and always will be a legend and a personal hero. Kind, funny, strict and frighteningly fit, he was the epitome of a warrior. George had been awarded the Military Medal in the Falklands war and would only recount his actions after a lot of bullying and coercing.

We had a fantastic flight, a real blend of old-school professionalism that 'got the job done'. Around the same time Baulfers was posted out and we welcomed his replacement, Maj Norman Hooper. A very kind and thoughtful man, he brought a very different style of leadership; one that reminded me more of a parish vicar than a field officer. The contrast with Baulfers' more hands-on style was striking.

The SSMs had changed too, with Nigel Hunter moving on and a young Karl 'Ripper' Miller stepping up. Ripper was ambitious and keen, perhaps promoted earlier than most would have expected, but he certainly brought energy to the role.

Despite the changes, Lynx flight settled quickly. With George's steadying influence and Dan's enthusiasm, we were ready to tackle anything that came our way. Soon enough, the new dynamics took hold, and we found ourselves consolidating our skills, both in the cockpit and on the ground.

The Light Dragoons had deployed to Bosnia just six months after amalgamation, and I was missing out. Again. But the UK hadn't deployed any aviation assets, so there was little I could do.

The next few months were spent consolidating flying skills across a wide range of scenarios. One day we might be doing general handling, the next could be tasking in support of units conducting training, but we always had an eye on the situation in Bosnia.

We built our NVG flying in stages. Cat 1 was a single aircraft down to 200 feet. Cat 2 introduced multi-ship work. Cat 3 was the full lunatic menu: multiple aircraft in close formation, under wires and bridges, separated by no more than a rotor span.

To make it possible, the aircraft were fitted with infra-red station keeping lights, invisible to the naked eye but glowing under NVGs. In line astern, the lead aircraft looked like a floating smiley face: exhausts as eyes, lights as a grin.

Formation flying at night, NVG-only, felt like something between a ballet and a knife fight.

Inside the cockpit we used electroluminescent panels that cast a pale blue hue over the instruments, and self-purchased microphone boom-mounted lip lights, crude but ingenious. A quick press of the mic with your lower lip gave just enough light to check a map without destroying night vision. Primitive, effective, and utterly essential.

The QHIs developed our skills steadily until we were all confident on NVG Cat 3, from that point on, we would only fly without NVG at night to the minimum required. I loved flying on NVG, it was exhilarating and difficult, but highly rewarding.

Flying so close to other aircraft at night that you could see rivet lines, anticipating the other aircraft around you, with our lives in each other's hands brought about a bonding that is difficult to describe.

We put these skills to good use on an exercise in March of 1994. Exercise Pilger Reise (Pilgrims Journey), was all of Lynx flight deploying to a German Air Force base in Mendig, Bavaria. We were supporting a hunter force who were tracking students on the escape and evasion part of a NATO Special Forces course at the Long-Range Reconnaissance Patrol (LRRP) School in Weingarten.

Fittingly, the motto of 651 Squadron is '*Let Us Prey*', and that's exactly what we did.

We were given free rein to hunt.

I was crewed with Keith, who had also recently crossed over to Lynx flight, for the full week. Mac was on a short posting to Northern Ireland. Dan came up with a plan that saw us hunting in pairs so that we had complete coverage during the periods we knew the students would be moving.

Using the Thermal Imaging (TI) sights on the TOW system, we would get airborne about an hour before first light and patrol on a grid pattern. The tactic was simple: make the runners think helicopters were everywhere, but one of the aircraft would be static, scanning wood lines and likely laying up areas for signs of fires or even body heat.

We were highly successful in this stalking tactic, locating the runners and then talking ground forces on to them.

In the post course debriefs nearly all the runners claimed that the sound of helicopters was constant, it wasn't, that was just their perception.

On the long transit back to Gutersloh, we had about 1.5 hours to kill. We called up the German air defence system and let them know that if there were any fighters in the air, we would be happy for them to try and shoot us down.

"Lone Ship, Lone Ship this is Army Air 842, a formation of 4 lynx, 21 miles south of Kassel, open for trade".

In a deep, guttural German accent; *"Army Air 842 here ist Lone Ship guten morgen sir, standby for trade"*

A few moments later he came back: *"Army Air 842 here ist Lone Ship, callsigns hammer Von und hammer zoo, a pair of F4s in your 6 o'clock vectored for trade, game on; viel glück"*

We could see the Phantoms from a long way out, their telltale smoky exhausts giving them away.

"Lone Ship, Army Air 842 visual the bogeys; game on"

In real life, our only hope was to get below tree top height, as fast as possible. In that environment, below the fighter's radar, we were pretty much safe. The fighters would always claim the kill; we knew for a fact it was bullshit; we were watching them circle in the distance looking for us.

I was taken back to my childhood at the Schenectady airshow, watching the F4 Phantoms of the Blue Angels and Thunderbirds perform, now I was fighting them.

Just after we returned to Gutersloh, the mood quickly shifted from the thrill of the chase to something far darker.

The news hit like a gut punch: a Lynx had crashed in Northern Ireland after an uncontrolled runaway down on the No. 1 engine. Apparently, both engines had started showing varying power indications, and an emergency landing was initiated.

There was a loud bang from the No. 1 engine, but despite that, the crew managed to land in a boggy field three miles north of Portadown, Armagh, with minimal damage to the aircraft. But as they evacuated, a fire erupted and quickly consumed the entire helicopter.

The commander was Mac, and the pilot, Capt. Sean Tully, a good friend, and a Fuglie.

Just weeks before, a Lynx had been shot down by the IRA in Crossmaglen, South Armagh. The crew survived that, despite losing the tail rotor and enduring a severe fire. Allegedly, they had been delivering stock for the Naafi on the base, including an underslung load of strawberry milk!

How the hell did they survive that? I thought, and then, in the very next breath, I thought about the crew in Portadown; how lucky they'd been to walk away, how quickly fortune can change.

In May 1994 I was sent to Middle Wallop for the AAC Guidon Parade. The guidon is the Army Air Corps and cavalry equivalent of the infantry's Colours and it was being presented by Prince Charles.

Before I left, George handed me some 651 Lynx Flight 'zap' stickers to pass on to a new pilot finishing conversion.

That pilot was Lance Sergeant Paul Mason of the Coldstream Guards. I handed them over with a *"see you in a few weeks"*. He barely looked up. He was a stereo typical Guardsman, upright and a buzz cut you could polish a lump of coal with.

In July 1994, 651 were invited by the German Airforce to attend an airshow in Laupheim, an airbase in Bavaria. The Germans had invited Squadrons from across NATO, including 2nd Squadron of the 6th Cavalry Regiment of the US Army. 2/6 Cav were equipped with the Apache and were based in Illesheim.

We got on well with these guys, Paul and I became good friends with CW2 Vince Stella, an Italian New Yorker with the same sense of humour as us. As an event it was boring; nothing like the Thunderbirds or Red Arrows, but we made some good contacts and had a very memorable hangar party.

Figure 68 Lynx Flight 651 Sqn, Gutersloh

Chapter 41. The Hunters become the Hunted

The time came soon afterwards where it was decided that 651 would take part in a full E&E exercise. This wasn't a spur-of-the-moment fuck-around. This was the full monty. 5 days on the run followed by full Resistance to Interrogation (R2I) training.

We were comprehensively briefed on survival, evasion, and what to do when capture came. The only one missing was Marcus Bell from Lynx Flight, who was away on a course.

On the first morning we were all dressed in 'flying combats', the most ridiculous, not-fit-for-purpose clothing ever issued to aircrew. We were briefed and given a scenario along with a couple of nuggets of intelligence to remember.

After being blindfolded we were taken to individual drop off points in pairs, simulating escaping from a downed aircraft. We were permitted to carry only what we would have in our flight bags, a tobacco tin sized container with survival kit inside. The drop off areas were dotted around the Harz mountains area not far from Kassel.

Dan and I had been paired. We were dumped out of the back of the truck and got our bearings. We had map fragments and basic instructions to get us to a checkpoint. We set off in good spirits; the weather was cold but wasn't wet for a change.

The first thing we did was find a laying up point to lay low until it got dark. We identified a suitable area and headed quickly to it. Surprise surprise, we found another crew already there, Capt. Ian Marlow and Cpl Mark Hostler from Gazelle Flight had the same idea.

We pooled plans. Dan had smuggled in a stash of Deutschmarks, concealed in a long, thin plastic cigar tube that was still a little too slippery for comfort…

The new plan was simple. Firstly, we needed to eat, we'd seen a small hamlet not far away. With funds secured, Dan went shopping. He returned an hour later with four bratwurst, pommes mit rot/weiß sauce (ketchup and mayo) and we ate like kings.

The next morning, a taxi 'mysteriously' rocked up. We piled in and took it to about a mile from the first checkpoint. We knew the hunter force were around, but even if we were caught, they'd need us hungry, tired and demoralised for what came next.

The hunter force was drawn from the Queens Dragoon Guards (QDG), a Welsh Cavalry Regiment, and we knew they would enjoy putting a boot in whenever possible.

After 72 hours of going from checkpoint to checkpoint, notionally under 'resistance' control we eventually arrived at a wooded area. Our very own '*Rene Artois of the resistance*' told us to hide in a specific area. It was all too obvious that this would be the point of capture. About twenty of us tried to hide in a tiny area. It was more Easter egg hunt than life-or-death evasion.

A couple climbed up trees as high as they could, trying to hide in the little amount of foliage there was, others found bushes and shrubbery. I found a hollow under a large tree and scraped all the dead leaves away, lay down in it and got one of the tree dwellers to bury me in the leaf litter.

I heard the QDG rock up, their Welsh accents giving the gig up. I heard them shouting at the guys high in the canopy to *"get the fuck down"*, lots of replies along the lines of *"Bollocks, come and get me if you can boyo"* echoed across the woods. Eventually, the exercise Directing Staff (DS) had to intervene and order them down.

At the same time as a boot slammed hard into my head, I heard *"I think I've got one!"* I stayed silent, another boot crashed into my head; they had me.

I was dragged out, hooded, cuffed and thrown into the back of a truck. I could see under the hood a little, most of the Squadron were already there.

A short while later the truck drove off; I felt it contact tarmac and speed up. The training had told us to escape as early and as fast as possible; it would be much easier than waiting. Under the hood I could see that there was only one guard in the back, right next to the tail gate. I took my hood off, my hands still bound, but in front of me.

I reached for my aircrew survival knife that was permanently attached to our flight suits or flying combats, about 6 inches long with a curved blade. I cut the plasti-cuffs off. Keith was sitting next to me, I gave him a nudge; I was free, there was only one guard. *"Let's wait for the truck to stop, overpower the sheep-shagger and go out over the tailgate,"* I suggested.

At that point, the guard busted me and Keith, saw me with a large knife. He panicked. Keith took pity and told me to give him my knife and sit back down, I was gutted, I really wanted to go over the tailgate.

After what felt like hours, we arrived at the enemy facility. My knife was handed over by the guard with a *"that fucker there, it was him"* pointing at me. Thanks mate, that'll make my life easier.

A dog handler from the RAVC was there, his dog was a vicious bastard and even though I knew the rules of engagement, it scared the shit out of me when it got really fucking close.

We were taken to a holding cell. A concrete bunker that had been sprayed liberally with a hose leaving pools of near freezing water on every surface. Individually we were taken for 'initial processing'.

A female 'tactical questioner', a true professional in her field, made us strip naked.

Standing there, bollock naked, in front of an attractive female whilst freezing cold was a deliberate ploy to humiliate. It worked.

"Run your fingers through your hair" she barked,

"Lift your arms", "now turn around, bend over and spread your cheeks." She was checking for contraband.

Then came the killer instruction:

"Peel back your foreskin."

This wasn't an option for me.

"I <u>cannot</u> answer that question sir" I said, pointing at the area of interest.

After processing, we were given an old WW2 style greatcoat and a pair of ill-fitting boots with no laces, taken to holding cells, placed in stress positions and subjected to deafening, spiralling jazz music.

The stress positions, the bone splitting cold did its work. Occasionally a Welsh accent would scream in your ear about some unapproved twitch or other movement.

At intervals over the next 36 hours, we were subjected to every trick we'd been warned about and deployed resistance methods until roughly 24 hours had passed, and then slow release of the information we had been primed with back in Gutersloh.

In one 'interview' session, I recognised the 'gentleman' asking questions, he had been a member of the REME attached to 15/19H. He should never have been in the same room as me.

Eventually our CO, Lt Col Arthur Gibson came into a room after the final interview. *"Do you recognise me?"* he asked, *"yes sir, Col Gibson, the CO", "very good, the exercise is complete. The scenarios are over, do you understand?"* he said in a very matter of fact tone.

We then each had a thorough, personal debrief from both the exercise DS and the team that had interviewed us. Me recognising one of their team didn't go down well.

The escape portion was a bit of a hoot. I enjoyed it. The resistance to interrogation though was a different story. I went into it knowing that no one was going to smash my knees with a 2 x 4 or waterboard me. I was confident, but the truth is that these guys are so professional that they don't need to. It was a sobering experience that I don't recommend to anyone.

Meeting up with the rest of the flight to see how they'd survived, it turned out that Marcus had been brought down to the exercise on his arrival back from his course. He didn't have a clue what was going on and so during the interviews, he just had conversations with the interviewers like they were his long-lost best mate.

He told them everything they wanted to know. In some detail too.

Marcus had earned his callsign, from now on, he was *'Squealer'*.

As we prepared to depart, the CO held a brief parade to wrap up the exercise. He offered a summary and shared some parting words from the female questioners, who had apparently convened and reached a unanimous verdict: Mac was voted man *"most unlikely to satisfy."*

It was a fitting end to an exercise that had tested every aspect of us; physically, emotionally, and disturbingly intimately. To this day, the mere hint of spiralling jazz makes my eyelid twitch and brings back the smell of damp concrete and piss.

Chapter 42. Callsigns

At this point I should take an aside and explain the personal callsign system. Callsigns were an American import, introduced via exchange officers. The Royal Marines always have a US Marine exchange pilot, we had one on Lynx conversion, callsign 'bad finger'. Then, the Royals on exchange with the Army, brought the system with them to us. The idea was simple: you could refer to someone over the radio without using their real name.

Callsigns were always based on a fuck-up, a dumb incident, or something you'd never live down. You never chose your own.

George was the assigner-in-chief of callsigns in Lynx Flight. His own callsign was 'Mutley', a nod to his Military Medal. Mac was 'Dogs Knob' due to his (allegedly) unfeasibly large scrotum! Paul Mason became 'Badger', courtesy of a streak of grey in an otherwise immaculate crewcut.

Dan was obviously 'Skippy'. Keith became 'Dumper' after landing in a field for reasons of digestive urgency.

My callsign? Well, I was, and forever will be, 'Vortex'. It sounded quite cool. It wasn't. It referred to my desk, which functioned as an administrative black hole.

Middle Wallop, AACC: Command Course, Sep - Dec 1994

Chapter 43. Command and Loss

By mid-1994, I was considered ready for a commander's course and prepared hard for the pre-command assessment with the Regimental QHI.

After ground and flying assessments, I was cleared to attend the course in September that year, back to Wallop again. Another year missing the girls' birthdays.

The command course lasted a full month. From the start, we were paired, Lynx with Gazelle, operating as a tactical unit. My Gazelle partner was Sgt Birty Cross, a talented Royal Marines pilot. Living together in the Sergeants' Mess meant we learned each other quickly. We were both naturally aggressive in our tactics and, more often than not, we achieved what we set out to do.

I was due my annual aircrew medical during the course, so I booked in with the doctor at Middle Wallop, a certain, now Maj, Jon Slade. 'Centreline' was doing my medical and, with what I can only assume was malice aforethought, prescribed me a pair of aircrew spectacles that were as ugly as they were unusable. I could see him trying not to laugh as I got dressed.

Needless to say, they never saw the light of day.

Sgt Martin 'Ozzy' Osbourne from 14/20H was also on the course. A fellow cavalryman, he struggled at times but got through on determination alone. We spent many evenings in the Mess playing snooker and talking nonsense.

On 22 September, I was deep in preparation for my command FHT, scheduled two days later, when the news came in from Germany. A Lynx from 652 Squadron had crashed on exercise near Mendig, the same place we had deployed to for Pilgers Reise.

Then the worst was confirmed. All on board XZ650 had been killed.

Those of us from 1 Regiment gathered in the Mess, trying to piece together what had happened, all of us in shock.

The crew were Sgt Les Berrisford and Cpl Andy Beck, both highly respected aviators. They had just dropped off a visiting General at Mendig and departed over the perimeter fence, heading back to the

Squadron location.

At around 600 feet, a critical component in the rotor system failed.

A blade separated from the aircraft.

What followed was catastrophic. The aircraft instantaneously broke up in the air and came down over a wide area. The main wreckage crashed into a field near Welling and burnt out. Lynx fleets across multiple nations were grounded.

The cause was later traced to a manufacturing fault in the tie bar, the component that secured the blade to the rotor head.

We later heard what had happened on the ground. Friends and colleagues had to guard what remained of them until the investigators arrived.

Life on the course carried on, as it always did.

Whilst on any flying course at Middle Wallop, we flew regardless of the time of day, which meant meals were snatched whenever a gap appeared in the programme. Usually, that meant the 'aircrew feeder', a mess that was permanently open and always good for something quick.

Near the end of the course, I was grabbing a sandwich with Birty, when I overheard a group of student pilots on the next table. One of them was loudly claiming to be the first from the Light Dragoons to get onto a pilot's course.

I let him finish his speech.

Then I dropped my light blue beret, complete with Light Dragoons cap badge, onto the table in front of him.

"Not quite."

I picked it up and left.

With the fleet grounded, it wasn't certain my FHT would go ahead at all. It did, but both the examiner and I were subdued.

1994 had been a brutal year for Lynx crews. Three aircraft lost. Good men killed.

Birty and I passed our final tactical and flying assessments, but returning to Gütersloh felt different. The usual sense of achievement had gone. The Regiment felt quieter, as if something had been taken out of it.

Passing the course mattered, professionally and financially, but it didn't feel like a win.

The Army did nothing to support us. Not even the basics, clear communication, was provided. A few of us submitted confidential ACORN reports, hoping our concerns might be acknowledged. Each of us received a handwritten reply from a Human Factors specialist.

It reassured no one. Nothing changed.

Emily and I talked about Les and Andy, but not in the way we should have. Not properly. Not honestly.

Should I have a will? Would the Army look after my family if the worst happened?

With the fleet grounded, we didn't fly again until after Christmas.

Figure 69 Memorial to Les and Andy. Welling, Bavaria

Chapter 44. Flying from a Desk

Although there was no flying, it didn't mean we weren't kept busy. The US forces in Germany were hosting a multinational computer-assisted exercise (CAX) in Grafenwoehr, southern Germany, Exercise Atlantic Resolve.

Callum Claypole and I were 'volunteered' to attend to represent the AAC component of 4 UK Division. As the RAF couldn't provide any personnel, or more likely they found some excuse not to, I was further volunteered to represent 18 Sqn RAF, who operated the Chinook Helicopter.

I had a full day to work with 18 Sqn to try and understand their SOPs so that I could reasonably programme the computers to simulate Chinook operations.

Callum bagged the day shift; I was helicopter operations for night shift and on a reverse cycle. My role was little more than drinking strong Folgers coffee to stay awake, and programming the computers to simulate aviation patrols, anti-tank operations and anything else that the AAC could be tasked for.

One of the quirks of programming the patrols was that, being American, the system required a callsign to be associated with each patrol. Normally we would use the callsign, 'Army Air' for anything other than tactical communications where we would use a secure, alphanumeric Battle Code (BATCO) callsign.

I used all the '*Warhammers*', '*Dragons*' and '*Talons*' I could think of.

At 03:28 I didn't think anyone would notice the callsign '*Gunfucks 35*'.

I was wrong.

A senior American officer spotted it and, unsurprisingly, wasn't impressed. Within minutes, a purple-faced yank burst into my cabin, relaying the diplomatic crisis I'd apparently triggered.

In the same cabin was a detachment from the Coldstream Guards, Paul's old mates, they had some lads from the Welsh Guards tagging along. I turned to one of the Welsh guys and asked for callsign inspiration. Without missing a beat, he offered:

"Cymraeg 35."

Perfect. It sounded official, and we knew full well no American controller would be able to pronounce it.

Problem solved and revenge served in one helping.

On return to Gutersloh, Dan gathered us together and told us his time with us was up and we'd soon be getting a replacement Flight Commander, another Aussie. This was a blow to morale. Everyone in the flight liked and respected Dan, the new bloke might be completely different.

And he was, apart from the Aussie accent that is.

Dan had arranged for the flight to meet Robert 'Robbie' Everett, his replacement. Gutersloh was famous for its '*Weihnachtsmarkt*', or Christmas Market, so a few Glühwein's in the market would be the perfect location to break the ice.

There we were, huddled around a couple of oak barrels, 3 sheets to the wind on gluhwein, freezing cold, when he appeared late. He'd got lost. Red Flag 1. As an Aussie I could understand he wasn't used to the German mid-winter cold, but turning up in a Paddington Bear duffle coat didn't help. Red Flag 2.

It took George at least 20 seconds to issue his callsign.

'Paddington' had arrived.

Mac, being Mac, went to town on Robbie, just ripping into every aspect of this Captain's life, not giving a fuck if he was being offensive. The Australian Army only has commissioned officers as pilots, and that's what Robbie was used to. He was clearly very uncomfortable being spoken to like this by a SNCO.

Germany, Gutersloh: 1 Regt AAC, 651 Sqn, Lynx Flight, Jan – Mar 1995

Chapter 45. Transferring Traditions to Habits

The start of 1995 felt odd. The Lynx fleet was beginning to come out of grounding, with aircraft undergoing a complete strip-down of the rotor head. Each tie bar was removed, tested, and replaced.

A handful of ours were not far from failure. We all knew how close we'd been to being in Les and Andy's shoes, and that realisation hit home. Hard.

It was unsettling to see that XZ650 wasn't the only aircraft with significantly degraded components.

Now a newly minted aircraft commander, I had to go through the Combat Ready (CR) check with Regimental standards, so I was given limited priority for flying hours, while others barely flew at all.

Around the same time, I had to return to Hohne for one last act as a Light Dragoon, an interview with the CO to finalise my transfer to the AAC and to approve my promotion to substantive Sergeant. George and I flew up, and I'd cleared it with the RSM to land on the parade square in front of RHQ.

The RSM was now Gary Black, once 4th Troop Sergeant in my 'D' Squadron days. He welcomed us, offered me his office to change, and took George straight to the WOs' and Sergeants' Mess. After my interview I joined them for lunch.

The Sergeant's Mess of The Light Dragoons is not just a dining room; it's a place steeped in centuries of tradition. The moment you walked in, you were met with the Regimental Guidon, housed by, and protected here in the mess.

This was an atmosphere thick with history, portraits of long-dead cavalrymen and battle honours stretching back to the 18th century.

Regimental silver gleams in display cabinets, drums and sabres from past campaigns stood proudly on show, the walls were adorned with relics from the 13th/18th and 15th/19th Hussars.

At its heart stood the Emsdorf Helmet, the sole surviving relic from a 1760 battle that earned the first formal cavalry battle honour in British Army history. That battered helmet, preserved in glass, was a reminder that this Regiment didn't just inherit tradition, it helped invent it

It's a living, breathing tradition, where every mess dinner, every toast, and every footstep across the polished floor carries the weight of those who came before.

The Sergeant's Mess of the AAC, by comparison, tells a more modern story, one born not of centuries-old cavalry charges but of rotor blades, machines and engines. Barely 50 years old, the AAC mess lacked the deep-rooted lineage of the Regiments from which many of its members were originally drawn.

Aside from the extraordinary wartime legacy of the Glider Pilot Regiment, its walls display more framed photos than battle honours, more prints of helicopters than relics. Rituals here often felt like habits inherited from other Regiments, rather than traditions grown from AAC history.

Where cavalry messes carry the weight of generations, the AAC messes attempt to build cohesion through social routine. Which is no bad thing. The port may be passed correctly, but few remember why.

To those raised in older Regiments, it could feel like history was something the AAC read about rather than inherited.

For all their differences in age, formality, and historic weight, both cavalry and AAC messes share something deeper, a reverence for those who didn't make it back. Raising a toast to absent friends is a ritual that cuts through time and type, a moment of silence where the rank fades and the loss lingers.

Equally enduring is the unspoken tradition that when a mess member passes, their bar bill is not just written off but ceremonially maxed out first; free drinks in their name, on the mess they helped shape. It's a final gesture of inclusion, respect, and remembrance.

I was equally proud of both messes, but for different reasons. The cavalry mess stood as a living museum of Regimental history, steeped in tradition, identity and memory. The AAC mess, newer and more austere, lacked the weight of centuries but still embodied service, sacrifice, and camaraderie. One honoured the past. The other focused on the present. Both mattered.

A few weeks later, Robbie had advised the flight that an opportunity had arisen to attend a mountain flying course attached to the Austrian Army, I threw my hat in the ring and, to my surprise, got the slot.

Austria, Aigen-im-Ennstal: Hubshraubergeschgader 2, Alouette 3, Mar - Apr 95

Chapter 46. Where Eagles Dare

Dan Sanderson and I arrived in Aigen-im-Ennstal as the two exchange pilots selected for Austria. If you remember, Dan had been the Squadron QHI when I first joined 651 back in Hildesheim, so we already knew each other well.

We were met at the Squadron HQ and taken to a local Gasthof where we would be staying, the Austrian Army didn't have mess facilities.

We were introduced to the pilots and crews of Hubshraubergeschgader 2, or Helicopter Squadron 2, a jovial mix of young and old. The home of the Squadron had been used as the location for the airfield scenes in the 60's movie, *Where Eagles Dare*, an iconic airfield.

The first objective was to get qualified on the venerable Aloutte 3 (AL3), a light utility helicopter used extensively for mountain rescue. Capable, practical, and, to my eye, utterly fugly.

With a powerful, single engine, it can carry out high-altitude rescues, transport injured individuals and perform aerial surveillance in rugged terrain.

Unusually, it sat 3 crew abreast in the front, with seats for 4 more people in the back. The third crew member was the 'engineer/crewman/everything else'.

I was expecting a full and intense conversion course like Wallop. Wrong. A couple of handling sorties, engine-off landings, some nav, and we were signed off on type.

The Austrians requested that Dan and I fly some low-level routes, an environment we were very used to, they weren't and wanted to learn some of our techniques. We planned a low-level route and briefed the crews; Dan would take the lead, and I'd be No. 2. When we got back to Aigen after some exhilarating low-level flying, the Austrians climbed out with eyes on stalks.

Horses for courses, where we were very comfortable, and weren't that low, it was new for these guys, as much as flying in their environment was new to us.

One evening Dan and I were eating in the Gasthof and talking about the days flying. In the background we kept hearing a familiar name being mentioned in the low chatter of Austrians talking in German. *"Blah blah blah 'Dick Kalinski' blah blah",* I looked at Dan, *"did he just say Dick Kalinski?"*

Dick Kalinski was an AAC legend. Known lovingly as Klink, Dick was a former paratrooper who'd flown everything, done everything, and left a trail wherever he went, Dick was one of those people that you want to know and be associated with.

When we weren't at the Gasthof, the Austrians were very hospitable, taking us to their homes to meet their families and to eat and drink traditional Austrian fare with them. The Austrians loved their 'Grappa'.

At most officers' homes we were plied with both wine and grappa, I hated the grappa, it tasted like petrol laced with battery acid. While the officers' homes were filled with hospitality, there were moments that left me uncomfortable. Among the superb wine and food, I couldn't help but notice the black-and-white photos of their fathers and grandfathers, in uniforms from a time long past – the best place for them.

Having been plied with it before, Dan and I were ready to politely decline what we anticipated would be another round of the same potent liquor. But our latest host had other plans, he wanted to toast the newest AL3 qualified pilots in Austria with a "special grappa".

With a sly grin, he assured us this was something we'd *"love."* I should have known better. I didn't.

"Special grappa" and *"love"* should never appear in the same sentence.

When our host finally produced the bottle, it was as if he'd unearthed a relic from a forgotten time, dusty and shrouded in mystery. The moment he popped the cork, a pungent aroma flooded the room, sharp and overpowering, something like pine disinfectant mixed with old boots and a hint of medicinal herbs!

"There's no way I'm drinking that shit" I said, instantly wishing I'd been more polite.

Crestfallen, our host explained this was no ordinary grappa, but a spirit born from a berry that grows only in the highest pastures and flowers once a millennium, or something like that. Instead, I'd just insulted a thousand years of Alpine folklore in one ungrateful gulp.

Now qualified on type, we progressed to high-altitude flying from Aigen. Gradually, we pushed higher, working up to the tallest peaks in Austria: the Kleinglockner and its bigger brother, the Grossglockner.

The Kleinglockner was only about 20 metres lower than the big lad, at 3,770 metres.

At 3,700 metres the aircraft felt as sick as I did. Controls went mushy, engines starved for air, and nothing responded quickly.

On qualification we departed Aigen and joined the rest of the Austrian Squadrons in Lienz, a small town south of Salzburg and not far from the Italian and, at the time, Yugoslavian borders. The Hubshraubergeschgader 1 operated the UH 1 Huey and its twin-engined sister, the Bell 212.

These machines were genuine Vietnam veterans purchased from the US and there were a couple of US Army helicopter pilots attached.

The attachment now moved on to rescue operations by day and night. The ground school was delivered in German, with a translator doing her best to keep up, but the message came through well enough: don't crash and don't make the headlines. Simple stuff.

Then we met the US pilots.

One of them was a total, 100%, card-carrying, 24-carat idiot who should not have been allowed anywhere near a simulator, let alone an actual aircraft. I'll call him 'Buck'. Buck could not fly. Fact. Even his colleague hated him.

On our first night in the hotel in Lienz, Buck got so shit-faced he could barely walk, let alone fly the next morning. He went outside to get some fresh air and promptly puked all over the stone steps outside reception.

Buck, being Buck, decided he'd better clean it up before anyone noticed. So, he found a bucket of water and threw it down the steps, thinking it would wash the pavement pizza away.

Austrian mountains are not known for being warm in March. The water froze instantly, turning the steps into a perfect ice chute, which someone immediately demonstrated by going arse over tit down it.

Buck returned home to Vicenza the next day, mired in disgrace.

On the operational phase, although we weren't allowed to fly solo, we were given some very basic rescues to take part in, largely, in fact definitely, the ones where the person was already dead and the weather was near perfect. It wasn't glamorous, but it was real mountain flying, and it taught me exactly how unforgiving that environment could be.

Chapter 47. Rescuing Dead Cows

The most memorable rescue in Austria didn't involve a person at all. It involved a cow.

It's common for cows in the high pastures to be struck by lightning during a storm and obviously get barbequed. The Austrians are very environmentally conscious and don't allow carcasses to be buried up there, they must be brought down for disposal.

My mission was to bring the carcass of a dead cow down from a high pasture and deliver it to a truck on the nearest road.

This was a big cow. A really very large beast.

We landed next to it to assess the weight. The procedure was to sling it by the head using a slipknot, which would tighten under the jaw. It was the safest way to hook them up, stopping them from swinging once airborne.

The engineer eyeballed it, decided we'd survive without a chainsaw or a second trip, and hooked it up.

We lifted, using every ounce of power the aircraft had, then slid gently downhill to the road a couple of miles away.

The road had been closed to allow us to place the carcass on the ground near the ramp of the truck and so had attracted an audience.

I set the cow down within winching range of the truck, slipped to the right a little and landed, shutting the engine down. The commander went over the road to talk to the tourists who were gasping in abject horror at the sight of a cow, underslung from a helicopter by the neck.

I could tell they were not happy travellers. The crying was the biggest clue.

Col. Melser, my commander, came back to the helicopter stifling a giggle. The tourists were appalled that we had moved a living animal in such a manner, Melser looked over his shoulder towards the carcass, and feigning surprise that the animal was dead, said *"Well! It was OK when we took off!"*

Having seen the UH1s flying around with their iconic sound of freedom thumping across the valley, I was desperate to have a go. I managed to talk one of the First Squadron instructors into giving me an hour. Sitting in a Huey for the first time felt like stepping into every Vietnam book I'd ever read. Louder, bigger, and rattling far more than seemed healthy.

The whole aircraft had that famous thump thump bounce, like it was happily battering the air into submission rather than flying through it. The cockpit felt vast, the cabin behind me felt big enough to host a small wedding, and the view out the chin bubbles was straight out of a film.

As we lifted, I had Robert Mason's 'Chickenhawk' and the standard Vietnam soundtrack of *'Fortunate Son'* rattling round my head, half expecting jungle to appear below and someone to shout for a medic. It was a simple flight, nothing clever, but it stamped itself on me. An hour in a Huey is all it takes to understand why so many pilots fall in love with them.

An hour wasn't enough. I ended up logging a total of nearly 3.5 hours of flight, including some high mountain time.

At the end of our time in Austria, we had made good friends, eaten a lot of good food, drank a lot of good wines and suffered the grappa experience. I learned how to escape downdraughts, ride updraughts, Lessons that stayed with me.

Personally, I'd experienced a lot of small teaching points, and learned from them, I still had Pete Jennings words about being a better pilot bouncing around my subconscious.

The mountain rescue pilots of Hubshraubergeschgader 2 were extraordinary in many ways and my time with them, as short as it was, stayed with me all my flying career.

It's normal when on exchange, if qualified under another nations flying programme to wear the wings of that country.

The Austrian Army doesn't award wings as such; they have a version that is worn by everyone in aviation. Not being a badge man, I wasn't bothered, but Dan wore it. The next time I saw him at Wallop he was still wearing it. I didn't take the piss. Much.

Germany, Gutersloh: 1 Regt AAC, 651 Sqn, Lynx Flight, Jan – Jul 95

Chapter 48. More Pasta, Peaches and Wine

Back in Gutersloh, I had to requalify on the Lynx after time flying a different type. Thankfully, the Regimental standards check went smoothly, I hadn't forgotten everything.

Toward the end of April, Robbie told us we'd been "selected" to participate in Exercise Roman Flight, an exchange with the Italian Army. It would be held at a military airfield between Venice and Trieste, where they operated the A129 Mangusta.

I was being rostered increasingly as aircraft commander. Just before our deployment to Italy, I flew a low-level map reading sortie with Robbie, joined by another crew for pairs work. Crossing under high-tension pylons, the standard procedure was to fly roughly 10 metres from a pylon at about 4 feet above ground, then spot-turn to observe the second aircraft passing through.

I was 'Pilot Handling' (PH) duties, Robbie was navigating. After clearing the wires and executing a spot turn, Robbie suddenly yelled, *"WHAT THE FUCK ARE YOU DOING? Watch your fucking tail rotor!"*

I stayed calm. *"I am, and there's nothing wrong. What's up?"*

Robbie, having flown Kiowas, wasn't used to the Lynx's tail rotor. The Lynx could spin 360 degrees in six seconds. Robbie never quite trusted that, and I never quite convinced him.

As the junior aircraft commander, I was 'volunteered' to plan the route for the squadron, 4 Lynx and 4 Gazelles, from Germany to Italy.

Switzerland was neutral, Austria had too much terrain, no one knew that better than me. So, the route went southwest into France via Saarbrücken, fuel at Dijon, overnight in Lyon, then on to Nice and across to Italy. Simple enough; in theory.

France wasn't a NATO participant at the time, so we needed full diplomatic clearance and civilian fuel arrangements. When we tried to enter French airspace at Saarbrucken, we were delayed by ATC for 30 minutes, likely too busy dealing with their Gauloises and espresso.

Fuel became an issue.

I quickly found a nearby French military airfield near Dijon and called ahead. Miraculously, they answered and let us land at Longvic, refuelled us, and provided last-minute accommodation and food.

Dinner was a surprise though: some kind of boiled purple mash with bread. *"Monsieur, this is lentils,"* said our host. 'Squealer' Bell, my pilot looked horrified: *"LENTILS?! Lentils are for kids to make collages out of; they're not supposed to be eaten!"*

We cracked on early the next morning, the weather still marginal, but within safe limits. A quick suck of fuel in Lyon and we continued south down the rivers.

As we approached the ATC zone controlled by Orange in southern France, I spotted a chance for a shortcut. I called up Orange Tower to request a zone crossing for the formation. In my best Ilkley schoolboy French, which sounded more like Inspector Clouseau than anything useful, I went with:

"Ohranjh Tower, bonjour, British Armée de Terre huit-quatre-deux, une formation of huit helicopters."

I tried it twice.

The reply came back belt-fed: an incomprehensible, heavily accented burst of French.

"Orange tower, British Army Air 842, apologies, in English please".

I never heard from them again.

Our final refuel point before transiting into Italy was at Nice, the airport was gearing up for the Monaco Grand Prix the following week. Nice has a very active heliport, more used to dealing with VIP flights, so despite the Prior Permission Required (PPR) being completed, being French, they knew nothing about our formation coming in.

I then went to pay for the fuel for 8 helicopters using a civilian 'carnet' card.

"Non non non, monseeerrr, we do not accept zose! Zey are verrry much contre la règle!" came the sniffy reply. Well, I wasn't bloody paying for it! The only thing I had left in my folder was, ironically, a NATO travel order we'd been given for fuel in Italy. I slid it across the desk like it

was a winning poker hand.

To my amazement, the French civilian fuel company accepted a single scrap of paper as payment for many thousands of pounds' worth of aviation fuel. French bureaucracy at its finest: They wouldn't accept a credit card, but a dog-eared NATO chit was apparently as good as gold.

With memories of Lenta and Roman Hussar in my mind, we landed in a neat formation, very impressive, and were met by the Italian CO. Although better than Lenta, and by a considerable margin, it was still as rough as a badger's arse.

As with Roman Hussar it was supposed to be an exchange, in this case the Italians would fly our aircraft, and we would fly theirs.

The Mangusta, although sold as an attack helicopter, was little more than a tandem-seat Lynx without the good parts. Its only real weapon was the same TOW system we already operated, and even that felt like an afterthought.

If the Apache was a wolf, the Mangusta was a pug with asthma.

The Italian Regiment we were working with had zero serviceable aircraft. Not one. Their flight line looked like a museum of suspended ambition. Several airframes sat open in hangars with chunks of concrete on the pilot seats where conscripts had dropped them straight through the canopies.

In ten days, the only flying achieved was a couple of NVG Cat 3 sorties, and even those were restricted to pre-approved routes. No low level. No tactical work. The exchange, once again, lived up to its reputation.

Shite.

To fly back to Gutersloh, we had a total crew change so that the flying hours and experience were equally spread across the Squadron. I couldn't believe the Italians had managed to repeat the shambles of Roman Hussar, they had and in true Gucci style.

At home, Emily had moved on from the Toyota job and was now working for Volvo in Gutersloh, still selling tax-free cars to squaddies with more confidence than credit history.

The move suited her. It was closer, steadier, and came with people she got on with. She threw herself into it, and this time it felt steadier than before. She had her routine, her customers, her stories from the showroom, and she came home brighter for it.

I was glad for her. She'd found something that gave her a sense of purpose again, something that was hers and didn't orbit the Army. And without either of us saying it aloud, we'd both started living in slightly different rhythms, each doing our best to carry on.

Roman Flight turned out to be George's final exercise with us. His exchange was ending and he was heading back to the Royal Marines as a Warrant Officer.

A man like George deserved a proper send-off. Anything restrained would have been disrespectful.

So, we planned a dinner in town, followed by a full-blown bar crawl to honour the legend.

All was going well, no major incidents until Squealer managed, with one of his one liner's, to get us thrown out of a club. Spotting a striking black lady with platinum blonde hair sitting on her own, obviously waiting for someone, Marcus decided it was a good idea to engage her in conversation. His opening line of *"so, are you a natural blonde?"* was not as well received as he had hoped.

It may have been lost in translation, but lost it was, and she went ballistic.

Now outside the club, in the rain, the only place left open for a beer was the local brothel. It wasn't just a hostelry for 'that' purpose, half of it was a regular bar/club where it was normal to just have a drink or two, but given its 'alternative' purpose, it had later opening hours.

We headed over. The manager, a cheerful English woman from Teesside, seemed unfazed by a dozen drunk squaddies piling in and running up a tab. Hospitality at its finest.

As people began disappearing in taxis without paying, Paul, Keith, Mac and I executed a tactical withdrawal. Unfortunately, so did everyone else.

At the first set of traffic lights the taxi stopped at, Paul made a break for it. The rest of us were less athletic. He'd seen the blue lights.

Minutes later we were standing in the street, hands out, explaining ourselves to armed German police, desperately trying to ensure this didn't enter the system as '*absconding from a brothel*'.

We sheepishly trudged back, escorted by Herr Flick and his sidekick.

There's nothing quite like handing over handfuls of Deutschmarks to an unimpressed Teessider. It wasn't dignified. It wasn't cheap. And it wasn't forgettable.

George left the flight the next morning exactly as he'd lived in it, respected, slightly embarrassed, and still quietly in charge. We were worse off without him.

Replacing George was Sgt Steve 'Suicide' Priddle. Keith took over as Flight second in command.

Chapter 49. Grief and Loss

At the start of August '95, as we started the day, Ripper called the whole Squadron into the briefing room, where the aircrews would normally have their daily briefings. His best friend, and our Squadron Quartermaster Sergeant (SQMS) Mick 'Stumpy' Gwinnel, had been tragically killed in a motorcycle accident.

The Squadron was gutted, Stumpy was very popular. His wife asked Ripper to lead the pallbearers. Paul and I were among them. Stumpy was to have a funeral with military honours.

The only one of us that had ever received formal training in funeral drill was Paul. His old unit, The Coldstream Guards were based in Munster, not too far away, and they had a practice casket, weighted to simulate the conditions experienced in the actual funeral.

Paul and I flew up to Oxford Barracks in Münster to collect, it so that we could rehearse properly.

We soon realised on loading it in Munster, that it wouldn't fit in the cabin with the doors closed, we had to fly with the cabin doors locked open, the casket lashed to the floor.

Flying back to Gutersloh we crossed the Munster/Osnabruck airport-controlled airspace. The controller was clearly eyeballing us with his binoculars; he made a comment about having an *"interesting load"*.

Stumpy was laid to rest in Hannover Cemetery with military honours, 651 did him proud.

By the middle of 1995, the situation in Bosnia was deteriorating and the UK took the decision to deploy 24 Airmobile Brigade to the region. 3 Regiment AAC, based in Wattisham was activated for deployment, but didn't have enough crews. Lynx crews from 651 and 652 were hastily readied and dispatched. All except me. I was 0.1 of an hour uncurrent in unassisted night flying.

6 fucking minutes.

The requirement was that each pilot had to be 'fully current' across all flying disciplines. 3 Regiment would not budge, and I had to watch the rest of my flight get ready and deploy, initially to the UK, and then out to theatre.

As it happened, within a week, they were mostly back in Germany, not needed after all. Some crews did deploy with 3 Regiment to Ploce, Croatia. Only a week after we laid Stumpy to rest, we heard that a Lynx had crashed into the Adriatic.

Then the worst possible news. The commander was 'Ozzy' Osborne from my command course; the pilot was Cpl Iain MacDonald from 652. In the cabin were Air Trooper's Roger Willingdale, 652 Sqn and Graham Witherstone, 663 Sqn. All were killed, whilst a third Air Trooper Ian Brain was rescued.

In a tragic twist of events, back in Gutersloh, Stumpys' widow was at home with her best friend, Iain MacDonald's wife. They were together when the news of Iain's death was delivered.

Ozzy and Iain had been practising missile evasion techniques, how to 'break a lock'.

The AAC had deployed crews without this basic combat manoeuvring exercise being formally trained.

They had impacted the water with substantial force. The helicopter remained afloat for a short period of time, Ozzy had initially escaped but swam back to rescue his crew. In doing so, he was taken down with the sinking aircraft and drowned.

Ozzy gave his life trying to save his crew. There's no higher tribute a soldier can earn.

I wish I could say I cried when I heard the news. But I didn't. Just like I hadn't cried when we lost Les and Andy in the tie bar failure. That one hit hard, harder than I admitted at the time. No one really talked about how it felt. No pastoral care, no support or a friendly word in your ear. Nothing.

We wore our grief like we wore our flight suits, zipped up, clean lines, no creases. That's what being a pilot, especially in combat, is. It's performance. It's an ego construct that says, *"I'm indestructible. It won't happen to me, I've got this."*

And so, we did what pilots do, we flew the next sortie. We briefed the next mission. We told jokes in the crew room and gave each other shit for bad landings. Anything but face what we were carrying. When I got parked over 0.1 hours of night currency, I was furious, embarrassed, even. I didn't realise then it could have been the reason

I wasn't in that cockpit when the Lynx hit the water. Six minutes. That decimal point may have saved my life.

Ozzy and Iain's deaths cracked something open in me. Not in a loud, dramatic way, but quietly, like a slow hydraulic leak you don't notice until the systems fail. I'd always told myself that what happened to Les and Andy was a freak accident, wrong place, wrong time. But now I couldn't lie to myself anymore. The pattern was undeniable. Good men, great crews, wiped out in a flash.

But here's the thing: we're not trained to carry that weight. We're trained to fly, to lead and to fight, to complete the mission. There's no course on how to mourn someone you laughed with at breakfast, flew with, got drunk with and did dumb shit with. No SOP for grief.

So, you put it somewhere. Deep. Behind the visor, behind the jokes, behind the rank. And you keep moving. Because pausing to feel it means admitting you're not invulnerable, and for a pilot, that's the same as being unfit to fly.

It catches up with you. Quietly. In unguarded moments. In dreams. In how tight you grip the cyclic and count the blades, just to make sure they're all there. In the way you avoid flying past certain places or reading certain names in the logbook. We flew like we were invincible, but we buried friends like mortals. And the worst part? We kept going, and the system kept failing us.

70 The Crew of XZ 668 return home

71 Remembering Iain and Roger in 652 Lines

Germany, Gutersloh: 1 Regt AAC, 651 Sqn, Lynx Flight, Aug – Nov 95

Chapter 50. Back in the Air

By late summer the pace picked up. The spares situation eased and we were finally flying properly again.

The Lynx required a lot of air testing. Any work on the rotor head would normally result in a 'track & vibe' air test. A monitoring system plugged into a laptop measured vibration caused by rotor imbalance.

This meant minute adjustments of pitch on each main rotor blade until they all tracked exactly on the same plane, eliminating as much of the vibration as possible.

As well as the obvious effect of vibration on the aircraft, it punished the crews more.

Flying the Lynx came at a cost, one we all paid with our spines and eardrums. The armoured seats were essential for ballistic protection, but ergonomically, they were a disaster. Designed for survivability, not comfort, they locked you into a rigid, hunched posture. Sheepskin covers softened the surface and protected arse grapes, but did nothing to cushion the relentless whole-body vibration that pulsed through the airframe.

Every sortie slowly ground your spine down like a pestle in a mortar.

A new qualification opportunity arose. All Aircraft commanders could conduct basic air tests, but some were more complex and carried real risk.

The head of Regimental flying standards (RQHI) decided that he would personally authorise individual aircraft commanders to conduct the more complicated air tests where detailed knowledge of aircraft systems was required.

It meant I was now first in line for the dullest flying tasks on the programme.

Chapter 51. Immersion Suits and Clogs

In September, Lynx Flight launched from Gutersloh to embark on Mountain Flying training in the Grampian mountains of Scotland. This 7-day detachment involved a long flight from Germany, through Holland to RAF Manston in the UK.

Our stops in Manston were frequent. We'd call in for fuel, a sandwich and a piss, and to get in or out of our immersion suits.

These "Goon Bags" were sealed at the neck, wrists and ankles to keep you dry and vaguely alive, and were compulsory for any flight over water.

You never forgot you were wearing one.

The stop at Yeovil landed differently. We'd already seen the ruined and perished tie bars from our own aircraft, and how close they'd been to failure, but seeing dozens laid out in front of us brought it into sharp focus.

Lynx rotor heads sat stripped down on benches, tie bars removed, tested, replaced. Some were within tolerance. Some hadn't been far off the limit. The engineers worked through them methodically, measuring, inspecting, signing off each component with quiet precision.

There was nothing dramatic about it. No urgency. Just a calm, deliberate process that made it clear how seriously it was being taken.

It hit home. Not in a flash, but in that slow, uncomfortable way where you realise how much you've been depending on parts you never see.

By the time we reached Edinburgh, the weather had closed in, making any prospect of mountain flying unsafe. We parked the aircraft up where they'd be secure and adjourned to a hotel to rethink and replan.

The new plan was to offer Edinburgh University Officer Training Corps some familiarisation flights, one of our officers had gone through UOTC there.

Paul and I were crewed together for most of the trip, and we had a blast. By then we were becoming proper friends. We shared similar backgrounds, were ambitious, outcome-driven, and absolutely loved the job. But more than that, we genuinely enjoyed each other's company; the kind of easy camaraderie where the hours in the cockpit and the hours in the bar rolled seamlessly into one.

Whilst in Edinburgh, it would've been rude not to pump a bit of cash into the local hostelries. Ripper had gone ahead and lined up a minibus to get us around. Driving through the city centre, we spotted another minibus, this one full of women. A hen night.

Keith, something of a Mike Yarwood wannabe, couldn't resist. Sliding the window open, he bellowed in his best mockney accent:

"Get yer bangers aht!"

Unbelievably, the bus was full of real cockneys, who, without missing a beat, roared back:

"Nah, you get yer cock aht!"

Ripper nearly stoofed in there and then, tears streaming down his face.

Later that night, improbable as it sounds, we ran into the same hen party again. Keith dropped the Mike Read act and switched into his version of a Dubliner.

"Och nar, yer the spittin' image of yer wee ma, so y'are."

This cockney lass looked him up and down, wide-eyed, and fired back:

"Ere, Arr do you knar me muvva?"

She was totally taken in.

The flight up to Edinburgh had been a great trip, but the return south was spectacular!

On the way back, we stopped at RAF Wittering for fuel and then, as usual, Manston for assessment of the weather and the goon bag shuffle. On arrival, the Squadron QHI assessed the weather and deemed it unsuitable for a cross-channel flight.

As Paul and I walked from the aircraft to the Visiting Aircraft Section (VAS), we were greeted on a first-name basis, we'd become familiar faces by now. The VAS team booked us all into a hotel in Margate, The Ivyside; which happened to be a decent night out.

Being mindful of our *'bottle to throttle'* limits, we made sure to head back early, the SQHI wanted skids up early, having "identified a window of opportunity." The next morning, after a full formation rotor start and a fly-by to thank the VAS team, we headed toward Antwerp.

The instrument flying, and the fact that we were all NVG Cat 3, certainly helped. Thunderstorms had started to form, and things got a bit hairy. The rain was hammering down, winds were strong and unpredictable, by the time we landed in Antwerp, the unanimous decision was to stay overnight and wait it out.

No VAS this time to sort out our problems. Keith booked us a hotel using a NATO travel order. When we arrived and gathered for a detachment briefing, the SQHI had a stack of weather reports for the proposed route to Gutersloh. It didn't look promising.

This meant the usual bottle to throttle limits were off. Then Norman Hooper, the OC, made a rookie mistake, he told us his credit card would cover everything. Clearly, he had no idea who he, or it, would be dealing with.

After a dozen surf-and-turfs and more beers than anyone cared to count, Norman's bill had edged over £2,000. Me, Paul, Keith, and Steve Farmer, the SQHI, quietly excused ourselves from the larger group. We weren't fleeing, of course, just relocating to a quieter bar to conduct a thorough assessment of the local Trappist beer scene. For science.

At some point in the evening, buoyed by the effects of Belgian beer and the misplaced confidence that only a group of aviators abroad can summon, we stumbled across a karaoke bar. To say we entered with the subtlety of a mariachi band would be generous. The place was already half-full, mostly locals nursing drinks and singing earnest renditions of Europop.

Naturally, we decided to elevate the tone.

It started innocently enough, Keith did a passable *Living on a Prayer*, and Steve took a shot at *Sweet Caroline*, complete with the obligatory shout-backs. But then we hit our stride. Paul suggested Bob Dylan, which should have been a warning sign. The screen rolled up *Blowin' in the Wind*, and that's when it all went wrong.

Paul, never one to pass up a cultural exchange, led the vocals with his own reimagined chorus:

"The answer, you Cloggy bastards, is blowin' oot yer arse… it's blowin' oot yer arse!"

We belted it out with the passion of a rugby team after a home win. The rest of the bar, oddly, didn't seem to appreciate our creative take on 60s protest anthems. The barman cut the mic mid-howl, and a very short man with tight trousers pointed to the door with a scowl that needed no translation.

It was the fastest we'd moved all night.

Once outside, we agreed it was probably time to call it a night. But the story of our heartfelt tribute to Dylan's lesser-known, flatulent period lived on for months. Especially back at the flight line, where any mention of karaoke was immediately met with someone crooning, *"…blowin' oot yer arse…"*

The second night in Antwerp was understandably subdued. We didn't return to karaoke, and Norman's credit card was, shall we say, no longer in circulation.

The flight from Antwerp back to Gutersloh was, for the most part, uneventful, a high-level trundle home with the usual gentle banter over the radio and the occasional comment about needing a shit or a shower, depending on who you asked.

That was until we were passing over the heavy industrial belt near Düsseldorf.

Just as we skirted the edge of a massive steelworks complex, one of the stacks erupted with a sudden, thunderous belch of flame, a proper industrial fireball, the kind you could feel through the windscreen even at a distance.

A heartbeat later, Keith struck again. This time he came over the radio, channelling Eddie Murphy from his legendary *'Raw'* stand-up set, the famous "cook-out" sketch:

"Now THAT'S A FYER!"

Even through the static, you could hear the grin in his voice. We were in stitches.

Silence for a moment, then the radios lit up with laughter. It was childish, inappropriate, and utterly perfect. That brief flash of nonsense somehow summed up the mood: a flight of happy aviators, somewhere over Germany, finding joy in a massive steel mill burp and a well-timed Eddie Murphy impression.

After everything we'd been navigating, the long hours, the relentless maintenance cycles, the pressure to stay current, this trip gave us something we hadn't realised we were missing; each other. The laughter and shared chaos rebuilt something solid. The banter, the bad singing, the bar-hopping, and the strict bottle-to-throttle discipline; it forged a cohesion you can't fake. We weren't just a collection of aircraft flying in formation.

We were 651 Squadron. And for the first time in a long while, we knew it.

Germany, Gutersloh: 1 Regt AAC, 652 Sqn, Lynx Flight, Nov – Dec 95

Chapter 52. Leaping with Frogs

In November, 651 was unexpectedly broken up. 652 Squadron had been hit hard, after a run of losses that had hollowed the Squadron out. It felt like they just couldn't catch a break. With 1 Regiment now under the command of Lt Col Gary Coward and preparing for deployment on Operation Resolute 2 (May to December 1996), most of Lynx Flight was to transfer across to 652 to reinforce their numbers.

Where 651 was known as 'The Premier Sqn', 652 was affectionately known as 'Frog Sqn,' thanks to its long association with the nimble Westland Scout helicopter, small, agile, and prone to tactical hopping.

Robbie remained as Lynx Flight Commander, with WO2 Steve 'Cholly' Cholerton from 652 continuing as second-in-command. Our new squadron boss was Maj Chris Collett, a tall, bespectacled man with a precise pencil moustache.

Already in 652 were Capt. Marc Edwards, another Fuglie, and Lt (RN) Austin Rally, a young Naval aviator on exchange. Chris Keane, who'd been my instructor back at ARW, was now our SQHI. I was genuinely pleased to see him.

Though deployment was still months away, we had a mountain of work ahead. We needed to build new crew dynamics, improve aircraft availability, fit-out airframes for operations, and maintain full flying currency. After missing a deployment over 0.1 hours of night flying, I was not letting that happen again.

First priority: get to know each other.

Steve Cholerton earned his callsign "Rottweiler" within ten minutes, loud, intense, and slightly terrifying at first, but ultimately a great bloke, and hugely respected. Everyone warmed to him quickly.

Cpl Ian Cook became "Trojan" for reasons only his wife understands. And anyone who spent more than ten minutes with him. If he didn't tell you, he'd show you.

Chris Keane by now had grown a thick, bushy moustache, earning him the callsign "Bunny", a nod to Eddie Murphy's 'Aunt Bunny' from Raw. The humour was sharp, relentless, and exactly what you needed to bring two flights together as one.

The merger worked surprisingly well. Two flights became one, forged by professionalism and banter.

'Bunny' led a solid training programme, heavy on NVG sorties. We flipped into reverse cycle, flying at night, sleeping during the day, for week-long blocks. At the time, I was commuting on a Triumph Trident, a classic three-cylinder motorcycle. Emily had the car to commute to work, so the Trident was my only option.

Chapter 53. Riding on a High

One night, Paul and I were flying the second NVG wave, only two of our six aircraft were serviceable. Bored and restless, we had an idea.

A brilliant idea.

Why not ride the Trident, two-up, around the perimeter track... with no lights... on NVG?

It was NVG practice, we told ourselves. Good training. Excellent judgement.

NVGs thrive in low light, but hate red light, hence the blue-lit cockpits. So, there we were, me riding, Paul on the pillion, bombing around the twisting perimeter track as fast as I dared. Then the worst-case scenario exploded into reality.

The fuel low warning light, which was bright red, flared to life.

Our NVGs overloaded instantly. And shut down. I slammed the brakes on. We skidded to a stop, ending up on the verge staring directly at the canal, just metres away.

There was silence, except for some very heavy breathing, then uncontrollable laughter.

We rode back slowly, NVGs tilted back, locked out of the way, and said nothing as we re-entered the crewroom.

That, we decided, would be one of those stories best left untold.

Until now.

As Christmas approached, the mood across the Squadron settled firmly into "*work hard, play hard.*" Robbie, always a generous host, invited all the pilots and their wives over for festive drinks. The invite read: *"bring a dish."*

Emily, a great cook, decided to make a tropical fruit salad. As she chopped and prepared it, I couldn't help noticing the growing pile of discarded rinds, peels, seeds, and skins. That's when the mischief got the better of me, I couldn't help it, the horns were out.

I took the entire heap of fruit waste and blended it into a foul, pulpy sludge. The bowl looked a little empty, so I topped it up with scraps from the green waste bin, teabags, crusts, some leftover beans on toast from the kids' lunch, and a bit of plain yoghurt for texture. Back into the blender it went. Then I carefully covered the result in cling film and added a label:

"Jim's Fruit Melange."

We arrived at the Everett's home, warmly greeted with mugs of steaming Glühwein. It was winter in Germany, freezing outside, festive inside.

I found Keith and Paul and whispered, *"For fuck's sake, don't touch the shit I brought."* Keith passed the warning to Cholly, who told Cookie, and the message quietly spread around the room. The only person who hadn't been briefed was Robbie.

I was talking to Keith in a doorway leading to the living room, beer in hand, when I spotted Robbie across the room; with a bowl of the Melange.

Shit!

I was laughing so hard I couldn't make a sound, the air to my lungs was one-way only; out. Just a thin strand of drool leaking down my face as I slid down the doorframe, wheezing and gasping for breath.

Robbie, ever the polite host, had sat down next to Emily for a friendly chat. Seeing me, collapsed in laughter, he glanced at her and, perhaps not wanting to offend her, asked what was in the fruit melange. She looked confused.

"Oh, that's not mine," she said. *"I made the fruit salad."*

Robbie, in a moment of heroic obliviousness, took a generous spoonful.

I genuinely thought I was going to pass out. Keith was trying to lift me up just so I could open my lungs.

A few minutes later, Robbie casually walked up to Cholly and said, *"I don't know what's wrong with Jim,"* he said, in his Aussie accent, *"but his dish tastes like shit"*.

That was it. Keith and Paul were reduced to baggage. Cholly and Cookie followed. Soon the entire room was howling. Even Robbie, once he'd pieced it all together, saw the funny side.

1995 was a year defined by stark contrasts, moments of profound tragedy woven tightly with unfiltered humour and growing camaraderie.

And then came the laughter, the kind that only exists in military life. From bombing around the perimeter track on NVGs, to getting kicked out of karaoke bars in Antwerp and blending up fruit peels and bin scraps and passing it off as "*Jim's Fruit Melange*" at the bosses Christmas party.

These weren't just jokes.

They were pressure valves, the only way to let a bit of it out.

Because behind the humour, 1995 hurt.

The loss of Les and Andy. Then Stumpy, Ozzy, Iain and Roger. Good men. Friends. Gone in seconds. Each death left a scar we didn't have the time or tools to process.

The next time you see veterans raising a glass to 'absent friends', that is why they do it.

The break-up of 651 and the integration into 652 brought more change. Bosnia loomed. Training intensified. And yet somehow, we laughed even harder. The worse it got, the funnier we became.

That's how it works. Black humour, laced with practical jokes.

What stayed with me wasn't just the sorties or the milestones, it was the contradiction. One minute we were bent double with laughter, the next standing in silence for another crew lost.

That tension, between grief and banter, between duty and absurdity, defined the year.

1995 taught us what it meant to be part of something bigger than yourself.

Germany, Gutersloh: 1 Regt AAC, 652 Sqn, Lynx Flight, Jan – May 96

Chapter 54. Combat Frogs

January arrived in a hurry, straight into troop lifts across the German training areas, preparing the infantry and ground troops for Bosnia. At the same time, I heard a crew from The Light Dragoons had struck a large mine and all had perished.

Life and death were very real.

I had kept in touch with Vince down at Illesheim, flying down a couple of times to catch up with him and the guys in 2/6 Cav. He and I were still determined to find a way to work together.

That opportunity came when I reached out to a Royal Artillery air defence battery equipped with the new High Velocity Missile (HVM) system. I asked if they'd like the chance to scrap with Lynx and Apache's. Vince and I collaborated to create a three-day programme that pitted UK Air Defence and AAC against the Apache, which we knew we'd be flying soon enough. HQ AAC at Middle Wallop even sent observers.

We had the training areas booked, intelligence briefs done, everything lined up, until the weather intervened. 2/6 couldn't cross the Sauerland due to heavy snow, so we pivoted.

Lynx and Gazelles would fly against the HVM battery instead. The Gazelles simulated heavy artillery and fast jet threats, and we came in with TOW attacks.

Unsurprisingly, the HVM guys claimed they were untouched while we'd been wiped out. We called it a draw and had a beer.

The snow wasn't all bad, it gave us rare training experience with snow landings.

As a deploying unit, we also got to live fire up at Hohne, critical preparation, since we'd be reconfiguring the aircraft in-theatre. The right-side TOW boom would be removed, reducing missile load to four, and replaced with a GPMG doorgun. Our groundcrews were now being trained as doorgunners.

Paul, with his background in heavy support weapons, led the door gunnery training. I took charge of the TOW side. We were engaging vehicle targets out to just under 3,000m.

Meanwhile, I headed back to AACC at Wallop to complete the Helicopter Weapons Instructor (HWI) course. It was relatively straightforward; ground school sessions, instructional sorties with a QHI and HWI in the back, and finally a real-world validation with a Lynx conversion student.

My assigned student was Lt James Gwynn, who happened to be joining 652 just before we deployed, a great way to get to know him early. I liked him straight away.

Training accelerated rapidly, more NVG work, more tactical sorties. One aircraft was finally kitted out for operations, so we started learning the extra operational gear: secure 'HAVEQUICK' radios, Radar Warning Receivers, IR jammers and full NVG cockpit lighting.

Navigation was upgraded in some aircraft to a GPS based system, but individually we bought early Garmin handheld GPS units, clunky, limited, but still a gamechanger, even with a screen the size of a postage stamp. We began using them to convert grid references into waypoints and build routes.

Part of the beat-up training included a trip to the UK to refresh our dunker skills. The Army, in its usual logistical genius, sent us from Gütersloh on a bus, then onto a ferry, to Yeovilton. The orange suitcases came out for the bus trip from hell.

By the time we reached the ferry in Calais, we were already in that dangerous state of being bored, happy, and very 'hydrated'.

Things happened on that crossing that shouldn't have. We were a squadron of professional pilots. You wouldn't have known it.

At all.

A fat sweaty truck driver who took his socks and shoes off for a nap, woke to discover he'd be finishing his journey in bare feet.

One REME tech made the mistake of expressing an opinion Paul hadn't asked for. He regretted it instantly.

His first stop on English soil was an NHS dentist.

When we finally arrived at the dunker, the man running the place wandered in looking like someone had dared him to reinvent himself halfway through a mid-life crisis. He used to be a big, burly Petty Officer, a barrel of a bloke with a beard you could store tools in, now wearing full WREN kit, skirt included.

He introduced himself as "Caroline" with the confidence of a person who'd not only crossed the Rubicon but burned the bridge behind him.

During the safety brief, Caroline proudly announced that the dunker had recently been upgraded with actual Lynx armoured seats, recovered from a crash off Ploce. *"So, you boys'll be getting the real experience,"* she said, pointing at the fuselage with real pride.

The room froze. Even the air seemed to step back.

Keith's chair scraped hard enough to echo.

Robbie moved first, getting to him before he could erupt. Cholly was already on his feet as well, but for very different reasons, a scowl building across his face like bad weather.

Cholly wasn't the type to get confused by uniforms. Caroline's skirt wasn't going to save anyone if he decided to make a point.

Explaining in short, cold, unmistakable Anglo-Saxon that we wouldn't be strapping into those seats. Not today, not ever. Caroline blinked, confused. She had no idea why those seats mattered or what name hung in the air the moment she mentioned where they'd come from.

Iain Macdonald never made it out of one of those.

Caroline seemed to wilt. The seats were changed out in record time, accompanied by apologies delivered in a voice that still didn't match the skirt.

We all passed. Nobody joked about the seats again. Iain was never far from our minds.

Chris Collett proved to be an excellent boss. By now, he had a firm handle on who to trust and who would deliver. Paul and I found ourselves handed plenty of secondary tasks, jobs that kept us busy, useful, and most importantly, not bored. But it didn't go unnoticed.

Robbie seemed increasingly irritated. Whether he felt his authority slipping or just didn't like how things were playing out, he began pushing back, undermining, disrupting and creating friction. During our final pre-deployment exercises, it became obvious.

The tipping point came after a sortie near Welling, Mendig, where we had arranged a fly-in to Les and Andy's crash site to pay our respects.

Frustrated, Paul and I went to Keith and Cholly to ask what the hell was going on. They agreed something had changed and said they'd speak to Robbie.

Allegedly, Keith confronted him bluntly: *"Look, they're just Sergeants. You can fuck them about all you want. You're the Flight Commander. Just remember, in a couple of weeks, they'll have access to all the live ammunition they want."*

It wasn't meant as a threat, that wasn't how Keith operated, but Robbie apparently took it as one. Whatever was said, something shifted. His attitude softened noticeably after that. We weren't mates, but he stopped going out of his way to screw with us.

72 Mason| Priddle| Termini

73 Mason| Anderson| Termini

74 Termini| Mason

75 Termini| Bennet| Mason

76 Priddle

Pre Deployment Training Hohne and Mendig

Chapter 55. Homes on Fire

Home life had its own pressures. The kids were doing well, though Fran had evolved into something we hadn't expected, marching off to school in a pink tutu and Disney princess outfit, proudly declaring herself *"ready."* I loved that streak in her. Emily didn't.

I knew she was carrying a lot. Six months solo with four kids while I deployed would break most people, but she didn't show the strain. Not to me anyway. Looking back, that silence was the warning sign I missed.

Then, a few weeks before we were due to leave, Vince and his wife Colleen came to stay for a weekend. They were meant to arrive Friday afternoon. Emily, who should've been home from the showroom, wasn't. No message. No call.

I rang her office. No answer. Tried again. Eventually her boss, Scott, picked up.

"She'll be home soon," he said.

"Where is she?" I asked, already knowing I wouldn't like the answer.

"We've been at a hotel for a conference."

It landed like wet cement. Heavy. Cold.

Emily had been away before, for "training courses" or "marketing events." I'd never questioned it. Maybe I didn't want to. But now the shape of things shifted, and the truth walked out from behind the curtain.

Her distance. The tension. The clipped answers. The feeling that home wasn't quite home anymore. It all made sense in one sickening click.

And what did I do?

I did the one thing soldiering had taught me too well. I cracked on. Kept moving. Pretended nothing had changed. Because deployment was coming and the job didn't stop just because my world had started to tilt. It was easier to grip the routine than face the fact that the one thing I'd been working so hard to protect, my family, was already burning.

The last few sorties we flew in Germany were just currency flights, ticking the boxes, filling the logbook, giving us breathing space once we hit theatre. On paper they were normal. In the cockpit, I was normal. But every time I shut down the aircraft and went home, I knew I was stepping into something I couldn't fix with checklists, drills or a calm head.

Former Republic of Yugoslavia (FRY): 1 Regt AAC, 652 Sqn, Lynx Flight, Jun – Dec 96

Operation Resolute 2

Chapter 56. Why

All the aircraft commanders in the Squadron formed the advance party, scheduled to arrive in-theatre on 11th June. Among us was one young officer, Lt Tim Peake, a Gazelle commander, or "floppy puke" as was the customary term of affection. At the time, he was just another junior officer with a bad haircut and a quiet confidence. None of us could have imagined just how far Tim would go, from low-level recces to actual low Earth orbit.

The day before departure followed the familiar rhythm of pre-deployment admin. We paraded through the orderly room to be issued our IFOR and BRITFOR ID cards, plus a pair of metal ID discs, proper dog tags, which I hadn't seen since Cyprus. We were also handed a pocket-sized copy of the Law of Armed Conflict, the kind of document you'd ideally never need but were expected to carry religiously, right alongside the laminated blood group card and a sense of humour.

As part of our final training in Germany, we sat through a series of lectures on Rules of Engagement (ROE). As expected, they hadn't been properly thought through. According to the lawyer, as aircraft commanders we were to issue three warnings before opening fire, something that might work on paper, but was utterly impractical in the air.

We confirmed our next-of-kin details. I wondered whether mine would even care if something happened.

Our personal weapons were bundled and loaded for transport.

On the morning of departure, the kids were allowed a late start to school so we could say our goodbyes. I gave each of them, and Emily, a hug and a kiss before heading out, I told I loved them very much.

At Hannover Airport, we checked in only to discover the RAF aircraft was unserviceable. The flight was postponed 24 hours. We were bussed back home.

Emily was less than thrilled to see me walk through the door.

She barely masked her irritation. Whatever excuse she gave, I felt I knew where she'd gone the moment I'd left that morning. What cut deepest wasn't the lie, it was the complete lack of surprise, the absence of even a flicker of warmth or affection, she appeared to be 100% checked-out. She had waited until I deployed, to finally make her intentions clear through actions, not words. I was no longer even part of the decision.

The pain of that realisation hit hard. But, as always, I didn't show it.

I slept on the sofa. The next morning, I was up early and got a lift into camp with Paul, too embarrassed and ashamed to tell anyone what had just unfolded at home.

We arrived at Split Airport and were met by members of the outgoing Squadron, then taken across the road to Divulje Barracks, an old, rundown complex functioning mainly as a transit camp. A few rear-based units were permanently stationed there, including Naval and RAF aviation detachments. It was also where Lt Col Coward would establish his headquarters, operations centre, and workshops.

Gary was no stranger to the conflict. In 1995, he'd made a name for himself as the spokesperson for the United Nations Protection Force (UNPROFOR). He wasn't a tall man, and his name often drew quiet sniggers, especially odd for an Army officer. But those who knew him, knew better. In my eyes, he stands shoulder to shoulder with George Hunt as one of the best soldiers I ever served with. He inspired action, had zero tolerance for fools, and led from the front.

The theatre induction began with a video. No welcome brief. No warning. Just a cold plunge into hell. Not a training video. A snuff reel. The greatest hits of depravity. Executions, mass graves, shredded villages, cruelty captured on grainy footage, then cut to the Annie Lenox song *"Why"*.

That song. That voice. That mournful wail stitched over footage of people doing the worst things imaginable to each other.

There were no explanations. No context. Just that haunting chorus: *"Why…"* as the images rolled on, men shot in the back, women wailing, children staring dead-eyed into the lens. It was manipulative, deliberate, and brutally effective.

By the time the screen faded to black, the silence in the room was absolute. No one spoke. Everyone stared. And I swear the temperature dropped. It must have got dusty in that room. And to this day, I can't hear that song without feeling sick.

We were there to hold the line between people who still wanted to kill each other; and we'd just been shown why that line mattered.

Why indeed.

Each aircraft commander had to complete two theatre Combat Readiness checks. After collecting escape and evasion kits, additional body armour, and a second 9mm sidearm, we sorted our charts, HLS guides, and maps before launching into the first check, a daylight sortie around the full operational sector. The second would be conducted on NVG.

We were shown where the most volatile flashpoints were; areas likely to reignite tensions between the warring factions. We also learned how to communicate with NATO and/or RAF AWACS aircraft operating overhead, despite never having had the right gear to practice the calls in training.

"Magic, Magic, this is Outlaw 22. Lynx departing Split for Prozor. Request Parrot and India check."

"Outlaw 22, this is Magic. You are identified. Parrot and India sweet. Squawk 5267."

"Magic" was the callsign for the AWACS. "*Parrot*" referred to our IFF, Identification Friend or Foe, which confirmed we weren't hostile and ensured the guys above didn't mistake us for something that needed a fast, noisy ending. "*India*" was the codeword for our secure crypto communications.

A clean "*Parrot and India*" meant we were positively identified, transmitting the correct crypto and safely inside the NATO surveillance net.

From Split, we flew northeast into Bosnia, using Prozor Lake as the visual gateway into central Bosnia, heading to our Forward Operating Base, callsign *'Eagle Ops'* in Gornji Vakuf (GV). 654 Squadron from 9 Regiment was there, waiting for us impatiently so they could finally rotate out. One final part of the handover remained: each of us paired with a 654 commander to fly a typical daily mission. These sorties introduced us to key fuel stops, danger zones, and the locations of known minefields.

I was paired with Mac; he'd left 651 for 654 about a year earlier. It was good to see him again.

In GV, we lived in a complex of 'Corrimecs', essentially insulated ISO-container-sized huts, each with a window at both ends, but critically, no heating or AC. Steve Priddle took the bunk at the far end, Paul the middle, and I took the one nearest the door. As we threw our kit on the beds, Mac and 'Chas' Challis were packing their final things.

Chas, like Mac, had transferred from the Royal Marines to the AAC. Built like a brick shithouse, with bratwurst fingers and a moustache that could filter krill, Chas was a well-known figure in the Corps. He didn't like many people and was famously grumpy. He was also very particular about his real name, something I still won't say, even now. On top of that, he had a noticeable tic. Paul, who'd done a six-month tour in Northern Ireland with him, warned me: *'whatever you do, don't mention the tic'.*

So, there I was, unpacking my kit, and Chas, who I'd never met before, was standing with his back to me, ticking away like Klunk from Stop the Pigeon. I tried; really tried to hold it in, but the giggles came bubbling up. Paul caught my eye, saw what was coming, and bailed from the Corrimec without a word. I followed seconds later, about to explode.

Our own crews, pilots and doorgunners, started arriving not long after. Each pilot went through CR checks with a QHI, followed by theatre familiarisation flights with us commanders. Then it was straight into daily tasking across the country. The 652 Ops team handed out the following day's jobs: one crew would forward-base to Banja Luka for the Divisional Commanders (GOC's) tasking. Another would be rear based in Split to support air testing or be on standby for medevac or compassionate evacuations, both 24-hour duties.

The rest of the crews picked up tasking lines from Divisional HQ, anything from bridge and powerline inspections with engineers, to flying war crime investigation teams to sites of atrocities or inserting patrols.

After about a week in GV, Paddy Roche, the Squadron 2 i/c, pulled me aside.

"You're the HWI," he said. *"You're being moved down to Split to manage all HWI requirements in theatre. You'll be with Regimental HQ as RHWI, oh and you're qualified to handle all the air testing."*

I didn't hear anything after *moved down to Split.* It was like someone had switched off the sound. Five months of graft, every ounce of sweat getting ready for this deployment, the stress at home, the marriage hanging by a thread, the kids, the mess of it all… and now I was being ripped away from the only stable part of my life: the Squadron.

No Paul. No Keith. No shared danger. No flying. Just air testing.

It hit with real weight. Not just disappointment, but that deep, hollow punch that drops you straight through the floor.

I walked away before I said something I'd regret. Past the aircraft, out to the far end of the HLS where the fuel pillow tanks sat in the darkness like giant forgotten slugs. No lights, no noise, just me and the sweet aroma of aviation fuel.

I sat on a pillow tank and the whole world caught up with me. Everything I'd been holding off. Everything I'd refused to feel. It just broke through. I cried. First time since I was seven, on that BOAC 747 leaving the States.

Someone must've clocked I'd gone missing. After ten, maybe fifteen minutes, a shape appeared out of the dark.

Paul. Grinning.

"You daft twat," he said. *"You didn't think that was real, did you? It's a wind-up. We put Paddy up to it!"*

For a moment I didn't know whether to laugh, keep crying, or chin him. Relief hit like someone had opened a valve.

I got up, wiped my face, and walked back towards the lines. As I passed Paddy, I mouthed something about him being a total cunt. He didn't say a word. Just grinned. He knew exactly how hard it had landed.

One of my first tasks in Outlaw 1, the GOC aircraft at Banja Luka was with Keith. He was the commander, I was flying. The HQ at the

Steel Mill had its own VIP helipad where we would pick up the General and his CP team, or whoever was on the tasking sheet. As it was a VIP pad, we had slightly different SOPs.

We would normally either be fully shut down for embarkation, or at the most, burning only; this was because there were often important conversations going on and we didn't want the VIP party hanging around a fully live aircraft finishing a chat.

The doorgunners always had a Mk15 passenger helmet ready and would ceremoniously strap it on the VIP. Watching a man who controlled divisions being fitted with a Buzz Lightyear helmet never stopped being funny.

Buzz Lightyear would then be connected to the aircraft intercom system; the Generals seemed to like having direct access to the cockpit and to listen in on our radios. The Lynx however had a switch that would isolate the cabin from the cockpit; we called it the 'safety switch' and would always check its position when we had passengers on the I/C – especially Generals.

On this duty, Keith and I landed at the VIP pad, the doorgunner did his thing and as I watched the party approach the aircraft, a dawning of recognition smacked me in the face. It was only Duckworth, the beast of Gutersloh! Although now a Major General, he was no leaner or smarter and he still had those antique glasses with the tape. If anything, promotion had only magnified the shabbiness.

I looked at Keith, *"Fuck me! It's Duckworth!"* I exclaimed.

As Duckworth strapped in, now with the ridiculous Buzz helmet on, guns connected him. My left hand slid slowly down to the safety switch and flicked it to '*cockpit only*'. I looked over my shoulder and said with all the subtlety of a brick through a window;

"Ayup Jack son! Canny to see you again, ya big scruffy bugger, how's our Vera getting on?"

Keith's eyes nearly detached themselves. For ten glorious seconds he thought I'd just ended both our careers. Then I cracked a grin and flipped the switch back. The look of relief on his face was priceless.

Duckworth, blissfully unaware, sat there like a badly-dressed owl in a flying greenhouse. A man whose entire bearing said, "*General*

Officer," yet everything about him screamed "*council estate pigeon fancier.*"

I loved being in theatre. The flying was going to be outstanding, no serviceability issues, no shortage of kit, apart from the chaff and flares. If we needed something, we got it. Every day was different. Only the duty crew days dragged.

Chapter 57. The Dawn of Tombstone and its Dogs

We shared the GV base with a battalion of infantry; the Queen's Lancashire Regiment (QLR). Their Battalion HQ was set up in a commandeered factory that the UN had repurposed. The main kitchen, mess and other facilities were in there too. In contrast, the AAC side looked like something cobbled together out of stubborn optimism: a sprawl of Corrimecs and some Harrier landing matting stretched out for the aircraft.

QLR had platoon outposts scattered around the area, patrolling in their Saxon 'chip vans' or Warrior Armoured Fighting Vehicles (AFV). The Saxon was a disgrace, a relic of Northern Ireland days. Unstable, unpopular, and unfit for anything other than delivering fish and chips on flat roads.

Just north of GV, on the main supply route to Banja Luka, sat Bugojno, home to one of QLR's platoon bases. Not long before we arrived, that platoon got stuck into the booze and rolled a Saxon. Their CO wasn't impressed. He slapped an instant alcohol ban across the board. The officers' mess, suddenly dry, found themselves with a surplus container of decent wines they couldn't touch.

In the finest traditions of British squaddie enterprise, we stepped in. We offered to take the wine off their hands, at a fair price, naturally, and suddenly found ourselves with an unexpected stockpile of Rawsons Retreat Cabernet Sauvignon. It quickly became a Squadron favourite and, for some of us, a lifelong taste.

But a wine like that demands more than just a plastic corrugated chair in the dust. If we were going to sip something with that much gravity, we needed somewhere worthy of it. We needed a bar.

Paul and I scouted a location. Chris gave the nod. Our callsigns in theatre had come down from the CAOCC in Vicenza; Lynx flight was the '*Outlaws*' and Gazelle flight '*Vikings*'.

Better than '*Gunfucks*', I suppose.

And so, we designed a wild west-styled bar: Tombstone. We built it between the Corrimecs and a row of ISO containers. Just over the fence, dangerously close to a marked minefield, was a random mound of earth. That became Boot Hill Cemetery. If someone annoyed us, we'd make a wooden cross, write their name and crime on it, and plant it firmly in the dirt. Morale through menace.

Across from the bar, the ISO containers holding our war stock missiles made the perfect backdrop for a dance floor. The BBQ, logically, went next to the cemetery, just in front of our missiles.

Paul and I made our way to the engineer camp across town, set up in an old factory. It was run by Gurkhas. We promised them a helicopter ride; no one said when, and in return, they gave us access to their building materials. Not just access. They loaded up a truck for us, sent a driver, and waved us off like VIPs. Corrugated roofing, two-by-fours, plywood, nails; everything we could possibly want.

Unfortunately, Johnny Ghurka; bless him, rammed his truck into our Hesco Bastion blast wall. The impact spilled all the hardcore from the protective bins, neatly creating the very dancefloor base we'd been planning anyway. And, even better, the poor driver took the blame for the whole incident.

Tombstone was operational soon after. It was a hit with everyone.

One of the many things we inherited from 654 was a pair of dogs. Neither was particularly likeable. Like people, some dogs are just arseholes; and we'd been gifted two of them.

The older one was GWA, a patchy, reddish-grey mutt, no matter how hard anyone tried with him, he was just a prick. GWA stood for Ginger With Attitude. The other was smaller, duller, and less sure of its breed. It was called FOG: Friend Of GWA.

They both made a sport of causing chaos. Their favourite trick was patrolling the narrow Corrimec corridors at night, quietly stealing trainers from outside doors and dragging them out to the HLS to chew them to pieces. Barking at nothing in the early hours was another treasured pastime.

The lads did their best to like them. Squaddies will try to adopt just about anything, given half a chance. But these two had the charm of a bad hangover and the manners to match.

Both came to untimely ends. FOG chased what was probably a rabbit and got tangled in the perimeter razor wire. GWA wandered out beyond the gate and was hit by a truck on the main road.

77 Homemade Corrimec furniture

78 Our Home. My bunk closest the door

79 Lt Col Gary Coward, UN Spokesman

80 Divulje Barracks, Split

Chapter 58. SA-6 and the Legacy of Scott O'Grady

Each morning began with a threat update, followed by the department heads' briefs. One early briefing mentioned an SA-6 threat near Teslic, east of Banja Luka, toward Doboj. A Serbian missile battery, although in barracks, had started lighting up NATO aircraft with its radar; they apparently had itchy fingers.

That day, James Gwynn and I were tasked to Banja Luka. Routine stuff. As we flew north, the Radar Warning Receiver (RWR) lit up, an SA-6 missile in surveillance mode had spotted us. I made the call:

"Magic, Magic, this is Outlaw 22. We have indications of Mud 6 bearing 035, 30 miles. Can you see it?"

"Outlaw 22, this is Magic. Standby... Negative active transmissions."

We were cruising at 3,000 AGL when the RWR tone escalated, the radar switched to Target Acquisition (TA). They were trying to lock us.

"Magic, Outlaw 22. Mud 6, bearing 035, 15 miles. Can you confirm?"

"Outlaw 22, copy Mud 6. Say duration and altitude. We're not seeing a firm picture."

This bloke, sat at FL300 over Hungary, brew in hand, wanted me to keep the lock, to give him time to figure out what we already knew. I told him I would.

A moment later, the radar shifted to Target Tracking (T6T). On the display: a stylised number 6 with wings. The enemy had a lock.

"Magic, Outlaw 22. Mud 6, Target Tracking, bearing 035, about 10 miles. Breaking the lock."

I took control from James, he was a solid pilot, but this was a commander's job and brought us into a steep descending turn. I brought the speed back to 40 knots, pushed the nose over, then let the speed rebuild to 120 knots, dumped the lift, spiralled down hard. As the rotors flirted with overspeed, I caught it with collective pitch.

I levelled at treetop height and threw in a few low-level steep turns for good measure.

SA-6s are old Soviet rockets, telegraph-pole size, and very capable. They'd downed an F-16, piloted by a US pilot, Scott O'Grady, nearby the previous year. In '94, Lt. Nick Richardson's Sea Harrier was also shot down not far away, but by a different type of missile.

The T6T alarm ceased. I climbed gently to 500 feet and radioed:

"Magic, Outlaw 22. Now at 500 ft, no further indications. Did you get a pinpoint?"

"Outlaw 22, MAGIC, negative. He's asleep now."

We pressed on to Banja Luka. I called the AAC liaison officer at Div HQ, callsign '*Martello*' and gave him a SITREP. On landing, he took me straight into the divisional Ops room in the steel factory; the Chief of Staff (CoS) was waiting.

Caught slightly off guard, and still with helmet hair, I gave him a full account. He nodded, muttered something like *"Good job, Sergeant,"* and moved on. Outside, the LO caught up with me, grinning. Apparently, the CoS was expecting something vague, what we gave him off my knee was solid intelligence.

James and I finished our tasking, then dropped into Squadron ops to hand the aircraft back. Chris Collett gave us the nod; word had already spread. Our morning's excitement hadn't gone unnoticed.

I was pleased. We'd handled it well. And we'd done the Squadron proud.

July 1996 was a full-throttle month for 652. Around the same time, 651 had been deployed to a forward operating base in Pale, near Sarajevo, where NATO HQ was located, they were mostly supporting VIP movements. We still saw them now and then, but the separation was beginning to show.

Chapter 59. The Mujahadeen v Monsieur SF

Then came a tasking that would stretch every bit of our training, and test Chris Collett's planning skills.

Back in '94, when Nick Richardson was shot down, he managed to eject and immediately went into full escape & evasion (E&E) mode. An SAS team was deployed to rescue him, and the eventual extraction was by French Puma. It worked, just. Thankfully, the outcome was a success. But not long after, the French withdrew their aviation assets from theatre, which meant that next time, it would be our problem.

And next time had arrived.

A mujahideen training camp had been identified in the French Area of Responsibility, and their Special Forces were tasked with making it go away. We were assigned to provide direct support.

Chris planned an operation that would push our NVG Cat 3 capability to its limits. Operation Parrot had three phases: phase 1: insertion by night to a location chosen by the French SF. Phase 2: standby for fire support, quick reaction or hot extraction if required, and finally, Phase 3: recovery of the team, again at night, and from wherever they decided.

We ran joint briefings and rehearsals with the French. The insertion and extraction were straightforward. The standby phase, however, raised some flags. We were prepared to go in, under fire if needed, but the bigger worry was being lured into an ambush. If their radios were captured or their comms compromised, we'd be flying straight into a trap. We suggested a duress code and a simple authentication system to verify that the people calling us in were, in fact, our people.

The French team leader looked puzzled.

"We may need to verify it's you before we come in, especially at night, coloured smoke won't help much."

He nodded sagely, and with complete sincerity replied, *""But monsieur, you will know eet ees us!"*

"How exactly?" someone asked.

He smiled. *"Frahhm owr oot-ray-joose Fronch ak-sawnce"*

Not exactly the elite tactical thinking we'd hoped for. But fair enough. It was said with such confidence, such pride, I genuinely wasn't sure if it was a joke or doctrine.

The insertion itself was exhilarating, a four-ship NVG Cat 3 run, clean and controlled. We were buzzing as we crewed out afterwards, once they were on the ground, we switched to QRF phase, ready to answer the call if things turned hot.

We didn't say we were playing the British cavalry, but we all knew what was going on.

As it turned out, our services weren't needed. Not a peep.

When extraction time came, we picked them up without incident. No contact, no engagement.

To this day, we never found out what happened at the camp. Maybe they redecorated the place and opened a bistro. Who knows?

What we do know is that Operation Parrot was a success. And that "*outrageous French accents*" are still not a recognised form of battlefield identification.

Chapter 60. NATO Firepower in all its Glory

And while all that was playing out, 652 was also tasked with supporting a NATO firepower demonstration. Our role was to provide two Lynx armed with FITOW missiles and door guns, along with a Gazelle for coordination.

These demonstrations were designed to be a very loud and dynamic reminder to the warring factions of exactly what NATO could do. The scenario was simple: short notice, overwhelming firepower, and the complete removal of a grid square.

The choreography was impressive.

First, a pair of US F-18s from Aviano dropped 500lb bombs and followed up with a strafing run. Then two RAF Chinooks underslung 105mm light guns into prepared fire positions.

The Gazelles coordinated the fire missions and the fast jets. Next, the Lynx pair came in for a live missile strike. The Gazelle then directed a second 155mm fire mission.

Finally, infantry stormed the position, shouting and stabbing things with conviction.

For the dress rehearsal, I was crewed with Cpl Jake 'Patty' La Belle. I was validating his FITOW qualification. Jake was a top bloke, went on to become an accomplished pilot and even featured on the TV show 'Flying Soldiers', though at the time, his map-reading still resembled performance art.

In the other aircraft were Keith and WO2 Paul Bennett, an aircrewman and our Squadron Sergeant Major, because of which he didn't get to fly nearly as much as he wanted.

After an eternity of sitting five miles from the range, we were cleared in. The Chinooks had done their bit. Now it was our turn.

"Target acquired come into constraints" Jake said as I flew a steady course, speed and attitude. The constraints indicator showing the cross hairs exactly in the small box. *"Fire on my command"*

"Outlaw 23, this is 22. Fire on three... two... one... firing now."

Two clouds of smoke. Two missiles launched in harmony. They gathered perfectly, no rogues and both hit the targets.

The actual live demo came soon after. This time I was flying with Cpl Eddy Smith. We'd known each other since Catterick when I was still badged RAVC on my Chieftain gunnery course, Eddy was 4 RTR.

Using the same plan, we were called in on the demo frequency that the spectators could hear.

Another pair of launches. Another pair of hits.

Four missiles. Four hits. No rogues.

We got a thumbs up from range control. No one made a fuss. No cheering. Just quiet satisfaction and the British military equivalent of applause, a polite nod and someone saying, *"Yeah, not bad, that."*

I was quietly amazed!

82 Me walking towards the camera at the FARP

83 Artillery Landing

84 Loitering, waiting to be called forward

Glamoc Firepower Demonstration

Chapter 61. Going Dutch

Tombstone was thriving. Word had spread that 652 had one of the best bars in theatre. We made a point of inviting key members of the QLR over for a glass of their, now our, wine. We knew we'd need favours from time to time, especially from their quartermaster, who soon became a closet regular.

Not far from us was another patrol base, manned by Joint Commission Observers (JCOs). These were essentially British SAS, or their equivalent. Their role was to gather intelligence, facilitate communication, and control aviation support when required. It was important that we built a solid working relationship with them.

The regular 22 SAS teams were supported by reservists, some of whom were on their first deployment. A few were overly enthusiastic. One, a mild-mannered solicitor in civilian day life and wannabe killer ninja by night, was being briefed on what support we could offer, and under what circumstances.

We explained the full range of support available. When we mentioned the option for armed action using missiles by day or night, mutually supported by doorgunners, his expression changed completely.

"What? You've got missiles?" he asked incredulously

"Yes mate. Several types. Top-attack, direct-fire. And if the weather's poor, we can use thermal imagery. If we can see the target, we can hit it, no lasers to get distracted."

"Can I see one?" came the wide-eyed response

"If there's a reason. But they're sealed in tubes."

"Wow."

It was hard to miss the expression on the patrol commander's face. A veteran of many operations, he looked like something inside him had just died.

We invited the JCO team over to Tombstone for a pivo. The solicitor wouldn't stop asking about helicopters. He was clearly overexcited, and not reading the room.

The commander, wearing a well-used fishtail parka and speaking in a strong Glaswegian accent, was visibly irritated. With a beer in one hand and the other in his pocket, he calmly began unloading and reloading a 9mm magazine using only his thumb, in about five seconds.

Then he snapped. We saw it coming. They left in a cloud of dust. We never saw the young lad again, although the JCOs would often stop by for a quiet beer, safe in the knowledge that they were safe from the SF groupies. They would bring fresh trout to BBQ to pay their way, which was gratefully received.

Not long after, Chris announced that Division was consolidating aviation operations. The only other Squadron flying at the time was a Dutch unit operating the Bölkow 105C. 299 Squadron of the RNAF would be moving in with us, though they'd remain self-contained.

The first to arrive were their engineers. Their job was to build the Dutch living quarters, operational buildings, and engineering workshops. Once their Corrimecs were in place, a digger began work on a massive hole. We assumed it was for a septic tank.

We were wrong.

It was a swimming pool.

The hole was lined with heavy plastic. They'd rigged it using a fuel filter from a B105, complete with pumps and fountains.

Because nothing says "ready for operations" like deck chairs and Dutchmen in speedos.

The Dutch Corrimec complex was enormous compared to ours. It had a fully stocked bar, a canteen area with actual furniture, a patio, proper laundry facilities, and private rooms for each person. Their technicians even had a full-size Rubb hangar, the same model 651 had in Gütersloh at the Western HAS site. Everything had heating and air conditioning.

They were clearly being well looked after.

The contrast was hard to miss. The UK had been there over a year, living in cramped Corrimecs, showering in cold water, and sharing rooms with at least two others. The Dutch had been in GV less than a week and were already living in a fully engineered resort.

Their helicopters didn't arrive until the site was completely ready to receive them. That alone told us everything we needed to know.

When they finally flew in, we walked over to welcome the crews. They climbed out, walked over, and then time seemed to slow. Helmets came off in unison. Long blonde hair spilled out and was casually flicked loose like something from a shampoo advert.

At least three of their pilots, and several of the ground crew were women.

Roughly fifteen jaws hit the ground at the same time. Someone dropped a spanner. No one spoke. It took a solid ten seconds before anyone remembered they were in uniform and then all eyes turned to the swimming pool; it wouldn't be just Dutchmen in speedos after all.

The Dutch brought more than a well-built Corrimec complex and a swimming pool. They brought a sense of energy, a breath of fresh air that cut through the daily grind. From day one, they were generous, welcoming, and easy to work with.

What had started as another routine tour suddenly felt like something different. Something better.

Chapter 62. Split

Paul had been handed the secondary task of managing the Lynx flight roster. Robbie thought he was being clever by offloading the job. In truth, he'd just made Paul the most powerful man in Tombstone.

Chris had approved a new policy; one or two glasses of wine allowed during the week. No more. Alcohol and helicopters don't mix, obviously, and it was hard to argue with the logic.

Fridays were the exception. Fewer restrictions, as long as no one kicked the arse out of it, guests were only allowed on Fridays. Routine still carried on, of course, crews still flew to Banja Luka for GOC tasking and to Split for duty crew.

That's where the roster politics kicked in.

Paddy Roche, Robbie and the other Officers started to notice a pattern. Every time something happened at GV on a Friday, usually involving music, laughter, and someone singing into a mop handle, they'd only hear about it afterwards. They were always in Split or Banja Luka.

Every time.

Paddy quietly had a word with Paul. Soon after, he started getting the occasional Friday back. Robbie didn't. He almost stopped pretending to be surprised.

Duty crew only had one real benefit, access to the civvy phone in the Regimental Ops room. That was it. On one of my first rotations, during the week obviously, and after the wind-up about moving to Regiment, I found myself with time to think. Proper time. And I realised, for the first time in years, I had space in my head.

It had taken a deployment and several months away from home for that to happen.

The only communication from home still came through Blueys and I wasn't getting any. Not from the kids. Not from Emily. Nothing.

So, I wrote. Two letters. One to Emily, one to Scott.

Emily's letter took effort. I treated it like a quiet introduction, not a plea or apology, just a human attempt at contact. I wrote with honesty, reached back to the good years, before it all cracked. I acknowledged my part in the slow collapse, the endless hours, the blind loyalty to a career I thought would carry us both. I didn't make excuses. I didn't ask for anything. Just laid it out plainly. We deserved that much.

Scott's letter was different.

That one came easily.

I kept it short. Tight. No pleasantries. I addressed him by name, which already said enough. I wanted him to know that I saw him, not the version Emily might paint, not as the grinning new chapter, but the man who stepped into someone else's hard earned life without so much as a blink.

I didn't threaten him. That's too easy and rarely necessary. But every line was written to land. I let him know I was still very much around, still watching the edges.

The replies came soon after. One from her. One from him. Cheerful. Breezy. A little too breezy. Clearly written side by side, both carefully worded to suggest normality, as if I'd imagined everything. As if they'd risen above it.

The message was clear: *We're fine. You're not part of this.*

But the timing, the tone, that much alignment doesn't happen by chance. I could picture them drafting them together. Her writing the first paragraph. Him reviewing it over her shoulder. Laughing. That, I imagined far too clearly.

Not rage. Just clarity.

And of course, they waited until I was stuck in theatre to say it. They knew I couldn't do a thing about it. That part stayed with me, it was deliberate.

Cowards.

It sank in slowly, like cold water. Whatever marriage we once had wasn't just strained, it was gone.

I was the only one who hadn't admitted it

Later, in the Ops room, I made a call home. Emily answered. She refused to let me speak to the kids.

The red mist came down. I remember shouting. I don't remember exactly what I said. I do remember smashing the handset against the wall.

The Operations Officer came over, calm and quiet. He gently took the phone from my hand and led me outside. Capt. Andy Lambert, a former Royal Marine Corporal who was now AAC, had been one course ahead of me at Middle Wallop. We'd always got on well.

I liked him. And in that moment, in Split, I needed someone like him.

A few days later, I was being driven from the HQ Squadron building in Divulje down to the flight line when I spotted a familiar face jogging towards me in sports kit. It was Lieutenant Alison Tetley of the RLC.

I'd first met her very briefly while sitting my educational exams, the ones required for promotion to Sergeant. She was invigilating, so just a short encounter.

The next time we had crossed paths was in Grafenwöhr, during Exercise Atlantic Resolve. We were both on the graveyard shift, working out of the same cabin. We drank far too much Folgers and chatted the hours away, she bore witness to the infamous 'Gunfucks' incident, something neither of us would forget.

We'd got on well. By chance, her uncle had been my SSM back in Junior Leaders and was now a Light Dragoon too.

I was heading up country after completing my stag as duty crew, but I waved as we passed and got the driver to stop. In a quick 20-second chat, she remembered me straight away and said she was also based up country, with 4th Armoured Brigade HQ in Šipovo.

It was brief, but it meant a lot to see a friendly face.

Back up at GV, trouble was brewing. Crews were returning from tasking in spectacular fashion, more airshow than operations, low, fast, and very noisy, to the point comms masts were being bent by downdraft. I'm surprised the Corrimecs weren't shifted.

Chris made it clear, in no uncertain terms, that it was to stop. Immediately and permanently.

Paul wasn't there for the brief, but I promised Chris and Robbie I would pass the message to him as clearly as they had to us.

Did I bollocks. I still owed him. I may have even encouraged more 'stylish' arrivals.

A few days later the RQHI was visiting GV when Paul came barrelling back with Trojan, low and fast enough to shake the chalk off the ops room tasking boards.

I didn't even have time to enjoy the view before the RQHI's expression changed to one of those quiet, murderous smiles.

The shit hit the fan.

The punishment? A week in split on duty crew.

I couldn't have been happier and to this day Paul still protests his innocence. Karma's a bitch.

Chapter 63. Skyhooks and Right Hooks

Once the Dutch were properly dug in at Gornji Vakuf, we found ourselves in a friendly rivalry with them over who could host the best Friday night. They were under similar restrictions to us, so the game was on.

Chris gave Paul and me permission to throw a party; strictly limited numbers, but we could invite a few guests.

Paul, being conveniently on punishment duty crew in Split, asked the Chinook crews, callsign 'Skyhook', if they fancied coming along. Free booze, food, music. All they had to do was run a scenic little shuttle around Bosnia to pick up a few people. They agreed instantly.

By the day of the party, we were ready. We'd made ice by freezing water bottles and smashing them with whatever blunt objects we had lying around. The beer was local Pivo, with Karlovacko the favourite, heavy glass bottles and all. Spirits were harder to come by, but someone pulled strings and came back with a crate of Jack Daniels. The QLR cookhouse providing plenty of catering.

The lads had set up a line of tents for people to crash in.

Then came the radio call:

"Eagle Ops, Eagle Ops, this is Skyhook 01, 2 minutes from the north with 57 POB."

I'll translate: the Chinook was arriving in two minutes, with 57 people. Now, a Chinook's big, but the official max is 55. We looked at each other, then at Chris, who just smiled with the weary pride of a father whose kids had stolen his car and brought it back washed and full of petrol.

We were stunned. On time and bang on cue, the RAF had delivered the goods. Literally. About 85 per cent of the passengers were women.

It was like watching a military version of Cinderella, except the ball gowns were cropped tops, short skirts, and jeans that left little to the imagination. One minute they were helmeted, rifle-carrying medics and logisticians; five minutes later they looked ready for a Friday night in Newcastle, all perfume and lipstick.

It was almost too much. Morale didn't just lift; it detonated. Blokes who hadn't said more than two words in a week were suddenly fluent in flirting. You could feel the atmosphere change.

The "bus route" Paul gave the Skyhook crew just happened to stop at every field hospital and logistics base where female personnel were stationed. An accident, we reassured Chris, who gave an all-knowing nod.

It was a resounding success. Everyone was well behaved.

Well… nearly everyone.

A female officer from the Royal Signals turned up wearing a revealing tee shirt over well-fitted combat trousers, the sort of combination that left little to the imagination and gave a lot of morale. Trojan may have taken this as some sort of unspoken challenge about who was better equipped, though that's purely speculation.

Either way, he was later quietly escorted back to his pit while trying to pull his trousers up with whatever dignity he had left trailing behind him like a deflating parachute.

Later that night, once most people were zipped up in their sleeping bags and snoring, a commotion came from Trojan's Corrimec. Paul 'Scraggs' Atherton went to check on him. Scraggs was our QHI having replaced Chris Keane, who sadly didn't deploy with us.

He opened the door, stepped inside, and immediately slipped in a giant puddle of warm piss that was still dripping from Trojan's bed. Trying to steady himself, Scraggs slipped again, this time landing squarely in it. He managed to soak up most of the contents of Trojan's bladder before he could crawl back out.

Skyhook departed the next morning with everyone accounted for; just. Then the clean-up began.

Conveniently, Paul and I were rostered together, so sadly we couldn't take part. Tragic.

Trojan, however, wasn't so lucky. I heard he got every shitty job going, including mopping up the five gallons of cookhouse coleslaw he'd tried to convince the Royal Signals officer to have a bath in with him.

She had of course firmly declined.

Chapter 64. Pranks, Skits and Bored Pilots

On our days off, we had to make our own fun. Tombstone was always good for a chat and a glass of wine, and whoever was duty barman got to hold court behind the bar without getting drunk. A battered CD player provided the soundtrack with everyone contributing discs from home.

Cholly treated CDs he didn't like as unexploded ordnance, dangerous, offensive and best lobbed into the minefield. His frisbees of musical shame included five copies of '*Jagged Little Pill*' and at least one Best of Celine Dion. No one dared complain; justice had been served. In a place where boredom was as lethal as enemy fire, pranks became our therapy. It kept the madness at bay.

In the Corrimec complex we had a communal room with a battered TV and a VHS player; no DVDs yet, this was still the dark ages. The only tapes we had were a '*The Fast Show*' box set, and within a week we could quote every sketch, word for word. It wasn't long before our radio calls took a surreal turn.

"Eagle Ops, Eagle Ops, this is Outlaw One-Five, two minutes from the south, we're cockneys, don't you know."

"Suits you, sir? Over."

The next day:

"Eagle Ops, this is Outlaw One-Five, today I will mostly be flying helicopters. Over."

One hot summer's day, Paul and I decided to build a "mess mortar", basically a bazooka made from a length of cardboard tubing, two baked bean cans, and duct tape. The cans formed a combustion chamber with a hole in the top. Squirt in some lighter fluid, apply flame, and the explosion would fire whatever you'd loaded downrange.

We fine-tuned and tweaked it until it could lob a Pivo bottle fifty metres. Then Cees, one of the Dutch engineers, suggested we try naphtha. He "just happened" to have some. Strange that.

We went full anti-tank mode. Paul took aim, I held the lighter. Tap on his helmet. Flame to the hole.

WHOOMPATA.

The tube caught fire. The bottle disappeared in a streak, last seen arcing towards the Bolkows a hundred metres away. Impact; inches from an aircraft canopy.

Our ears were still ringing when Chris burst out of the Ops Room, having heard the bang and seen the flash.

"ENOUGH."

Flames out, kit dumped, we found something else to do.

Cold, wet days had their own form of entertainment. The Sunday papers came with junky little magazines full of "SEND NO MONEY NOW" offers. An open invitation for bored aircrew. We'd spend hours clipping vouchers and sending gifts to people who didn't know they needed them. Which apparently turned out to be technically mail fraud.

The British Forces Post Office soon had its hands full. Chris Collett unexpectedly joined MENSA. Robbie Everett received so much post, apparently, he needed a separate mailbag. I don't know if that last part's true, but it's believable.

Among the highlights were gold-plated Star Trek phaser guns, a Dambusters memorial dinner plate, and eight summer dresses.

And one full-size greenhouse.

Robbie tried to convince us that the six-month supply of incontinence pants he received were *"actually quite useful."* We didn't buy it.

We had some total, full-on belly laughs, the kind that took your mind off where you were and what you were doing. For a few minutes you forgot about minefields, politics, and the fact you were a long way from home. It was just mates, mischief, and the ridiculous stories that, years later, still get wheeled out in bars with the same punchlines and the same laughs. Even now, all it takes is someone to say "SEND NO MONEY NOW" or "WHOOMPATA" and we're back in Bosnia.

Chapter 65. Low Routes and High Stakes

We were in and out of the town of Sipovo regularly. It was home to 4 Brigade HQ, a strategically important location tucked in a wooded valley, northeast of Gornji Vakuf. It wasn't much to look at, just a loose scatter of buildings between pine forests and dirt roads, but it mattered.

Geographically, Sipovo sat just west off Route Gull, the main north–south Main Supply Route (MSR) that threaded up through Bugojno, Donji Vakuf, and on to Banja Luka. It was a lifeline for NATO logistics. At Bugojno, the road carved northwards into a deep gorge, known as 'Cheddar Gorge', for its resemblance to the Somerset original, it offered a natural route through high terrain.

From the Gorge we found a series of re-entrants, narrow cuts in the ridgeline, that let us punch through to Sipovo even in poor weather. That re-entrant became known as "*the bobsleigh run.*" It was tight, fast, and thrilling. Flying it well gave me a shot of adrenaline.

We'd built these low-level routes for all-weather (well, 'most' weather) access. Getting in and out of Sipovo became routine. And for me, it became something else.

Sipovo was also where Capt. Pat McCardle, callsign '*Silverfox*', was posted. He was the AAC Liaison Officer, a fellow pilot from Northern Ireland, and one of the good guys. Pat had a feel for both the bureaucracy and the battlefield. He translated front-line requirements into language the HQ could digest.

He also happened to know Ali Tetley. They were both doing staff roles, both orbiting the same planning cells and working groups.

Ali had apparently asked Pat to let her know if I ever passed through.

And he had.

Early on, when we were still proving the *bobsleigh run* by day and night, Paul and I were crewed together, and I found myself conveniently routing via Sipovo for fuel more than once. I met up with Ali a few times over coffee in the crew room.

Nothing dramatic. Nothing formal. Just two people in the same time zone, swapping stories and trying to stay grounded.

I didn't know where it was heading. Everything at home was still cracked and crumbling. I was missing the kids constantly. I was angry, confused, grieving something I hadn't fully lost.

Was a friendship with Ali going to make things worse or better?

What I did know was that I had growing feelings for her. Whether they were real or just born of distance and distraction, I couldn't say. Whether they were appropriate was another question entirely.

I was on the verge of being a divorced Sergeant. She was an ambitious young officer, surrounded by peers and scrutiny.

But when I saw her smile, when we laughed about nonsense or just drank terrible coffee side by side, none of that seemed to matter. It felt simple. It felt like oxygen and I was lost in her eyes.

And for a while, that was enough to keep flying the bobsleigh run.

85 Lynx exiting the bobsleigh run near Sipovo

Chapter 66. The General and his Close Shave

The *Bobsleigh Run* earned its keep not long after. I'd been tasked with picking up General Mike Walker from Sarajevo and flying him to Sipovo, where Admiral Leighton was hosting a multinational gathering. A simple enough job on paper. It didn't stay that way.

After departing Sarajevo, I made the usual calls to AWACS. Normally, that meant 'MAGIC', the NATO/UK platform. They were superb. Rock solid. But today, MAGIC was off duty. In their place was the French contingent, callsign 'CYRANO'.

"Outlaw 22, 'ere is Cyrano, go-hed," came the reply in the sort of accent that would make a Bond villain blush.

"Cyrano, Outlaw 22, three miles northeast of Sarajevo for Sipovo, Parrot and India check."

"Outlaw 22: Parrot an' India… sweet, sweet."

That seemed to be that. We continued at about 300 feet AGL, entered the *Bobsleigh Run* and made for the designated aircraft parking area just outside Sipovo camp. There was already a line of aircraft on the ground; Blackhawks, BO105s, and another Lynx. We slotted in, landed and shut down. Every crew was waiting on someone senior.

I took my helmet off and wandered over to the nearest Blackhawk crew, mostly because they had better snacks. As I was chatting, an F-18 screamed overhead at about 2,000 feet. Inverted.

A few seconds later, it came back for another pass.

Trying not to sound too concerned, I nodded at the jet and said to the American I was scrounging from, *"What's with him? Getting brave, coming in that low."*

The American pilot, wearing a massive Stetson, just said, *"Woooo boy… reckon whatever you pulled done lit him up like a Fourth o' July firecracker. That boy come screamin' in after you like he was fixin' to tan your hide and nail it to the barn door."*

Or something like that. Probably.

Once the conference wrapped up, General Walker returned and strapped in. On departure I called AWACS again, this time thankfully MAGIC was back online.

"Magic, Outlaw 22. We had a CAP flyby about 60 minutes ago. Any idea what that was about?"

They came back quickly. Apparently, CYRANO failed to identify us on radar when we had departed Sarajevo. Given the sensitivity of the meeting in Sipovo, our presence had been flagged by the Combat Air Patrol (CAP). CYRANO, when asked, claimed to have no knowledge of our flight at all.

The CAP was cleared to descend for a visual ID.

They did.

They reported us as a Mil Mi-17, a Russian-built transport helicopter, inbound on a direct line to Admiral Leighton and the General Staff.

I relayed this back to General Walker over the intercom. He did not find it amusing and was not a happy General.

A few weeks later, Andy Lambert passed on that both F-18 crews had been grounded while an investigation took place. We'd come closer to a friendly fire incident than anyone was comfortable admitting at the time.

Chapter 67. Compassion and Minefields

By now we were approaching the mid-point of the tour. Paul and I were rostered as the duty crew in Split. I was the aircraft commander; he was the pilot. It didn't really matter.

We were sat in the Ops Room with a brew when Andy Lambert warned us to prepare for an imminent 'Comp A' flight. A Comp A, or "Compassionate A" is one of the few things the Army does well. If a soldier has a serious personal emergency, like a dying parent, the full weight of the military machine is thrown behind getting them home.

Andy gave us the bare bones. A Gunner from the Royal Artillery was stuck on a remote observation post high in the mountains. Our job was to get him off the mountain and fly him to Naples, where he'd transfer to a civilian flight back to the UK.

The weather was miserable. Low cloud, heavy rain, and little sign of improvement. We fuelled the aircraft to the gunnels and built a route: head north from Split, refuel at Tomislavgrad, push on past Sipovo, where we had a backup fuel option, and get a fresh weather check from Pat McCardle and the AAC det in Mrkonjić Grad.

If it looked flyable, we'd continue to the OP, collect the Gunner, refuel again at Sipovo and TSG on the way back, then into Split before heading across the Adriatic to Naples.

The plan was sound, and we were authorised for the full flight, minus the Naples leg, which had now been handed over to a Chinook.

Everything went well. At first. We refuelled at TSG and pressed north. Overhead Sipovo the weather wasn't brilliant, but it was good enough. We carried on, aware that we were already tight on crew duty hours. Normally we were capped at seven hours flying in a 24-hour period. The CO had granted us an extension to nine. We were already eating into it.

About fifteen minutes north of Sipovo, things began to tighten. The cloud was dropping, the fog was climbing, and the trees started vanishing into a white soup. The NVGs were good for contrast, but useless when you couldn't see more than fifty feet in front of you.

We weren't going to make it.

I made the call to turn back to Sipovo for fuel, then head back down-country. As we approached, we discovered that Sipovo, previously clear, had now vanished under a dense blanket of fog. Even the *Bobsleigh Run* was out. Visibility was so bad we couldn't see our own arses.

We couldn't land in a random field. Too many minefields; marked and unmarked.

Time for a rethink.

We knew there was a major power line running down from the ridgeline and passing just south of Brigade HQ. So, we came up with a plan. Paul would fly an approach to the pylon line, which we could still see poking through the fog, then we'd hover-taxi slowly down the right-hand side at six feet, following the line until we reached the road leading to the HQ front gate. From there, we'd hop over the fence, follow the internal road up to the HLS, land, fuel up, and reassess.

It started well. We shot the approach to the pylons, but as soon as we entered the fog, the rotor wash kicked up a full white-out.

Paul did exactly what we'd briefed. A hard climb-out to the right, the only direction we knew was clear. As we broke out of the murk and into clearer air, our fuel warning lights changed from steady to flashing. That meant we were down to fumes.

Minimum fuel for a safe two-engine landing was 100kg.

We were at 75kg and dropping.

We had one more approach left. After that, it would be one engine and even worse options.

Then the radio crackled:

"Lynx overhead Sipovo, this is Silverfox, do you copy?"

"Silverfox, Outlaw 22, we copy. We're low on fuel and attempting a second approach. Have rescue ready."

"Outlaw 22, Silverfox: I have you. Follow my beam."

Pat - lovely Pat - had managed to dig out a Dragonlight, a monster of a searchlight that looked like it belonged on a naval gun deck. He stood in the middle of the helipad and pointed it up through the fog, trying to angle the beam at roughly 3 degrees, our usual descent profile.

We saw it pierce the fog like a lightsabre, a glowing shaft pointing directly to us.

Paul came off NVGs and edged the nose into the tight beam. I stayed on NVG and called altitudes at 25 feet intervals. The trick was to keep the beam centred on the nose and ride it like a rail.

I set the radar altimeter to 15 feet. If we didn't see the ground by then, we'd have to go around again – possibly into a minefield, on minimum fuel, possibly on one engine.

Just as the Rad Alt started to warble, we broke visual. Pat stepped promptly out of the way, and Paul brought us into a perfect landing.

We shut down the engines before they shut themselves down.

Still shaking, we climbed out to find Pat waiting for us, laughing his head off.

"FUCK. THAT."

He reported back to Regiment that we were safely on the ground in Sipovo, the aircraft was secure and would be looked after.

This saga doesn't end here.

86 At FOB Red Dragon

87 Callsign 'Vortex'

88 Meeting the Locals

Around Bosnia

Chapter 68. Debrief, Bibles and the Clap

Pat arranged a couple of bedspaces for us, not hard to do in a field hospital. We bedded down and chatted about what had just happened: what we'd done wrong, what we'd done right, and why we were still breathing. The informal debrief mattered, not for blame, but for lessons. Truth was, given the information we had at the time, there wasn't much we'd do differently.

We could have turned back earlier but earlier might have been too soon.

The next day the weather hadn't improved, and in truth, we needed a bit of time to decompress. And where better to do that?

I went looking for Ali; Sod's law, of course she was down in Split for a few days.

Wandering around the site, we found the chapel set up by the Padre from the Royal Green Jackets (RGJ), who were the resident battalion in the sector. On a desk sat a tidy pile of RGJ-branded bibles, free to a good home. So, we helped ourselves. Generously. If anyone needed spiritual reinforcement, it was us.

From there, and I'm still not sure why, we ended up in the hospital's sexual health clinic. Just inside the entrance was one of those tourist-style information stands, stocked with glossy leaflets offering urgent advice on gonorrhoea, herpes, syphilis, and their many colourful, drippy friends. We gathered a mighty handful.

Eventually we ended up in Pat's office. We talked through the previous night's adventure and promised him quite literally anything he wanted as thanks for saving our skins. It was then he sheepishly admitted that he'd been halfway through a bottle of something with a mate when he heard us in the overhead and realised something was very wrong.

He sobered up very quickly. We didn't judge. The Dragonlight beam was straight as a die.

Pat was called away for a briefing. Left to our own devices, and being self-respecting Sergeants, we rifled through his office.

That's when we struck gold: a pack of A4 internal mail envelopes, the kind with the multiple address boxes. Inspiration struck.

We selected appropriate STD leaflets for various lucky recipients, tailored, of course, for maximum effect, and supplemented each one with a pocket-sized RGJ bible. The result was a series of highly personalised spiritual and medical care packages, thoughtfully addressed and quietly popped into the internal post system.

And we then forgot all about them.

We ended up staying another night in Sipovo as guests of Pat and the Field Hospital team. That evening, we joined them at the bar for a few beers. They were a great bunch, some NHS reserves, but mostly full-time RAMC combat medics.

A few were curious about how Paul and I had become pilots, coming from the infantry and cavalry respectively. The AAC had its fair share of RAMC pilots, Jon 'Centreline' Slade was one of the more notable ones. We offered them a tour of the Lynx, which was still sitting on the HLS, in case they were interested in joining the dark side.

We set off on foot, taking the shortest route, a straight line across a large, open area. There were six of us. Then suddenly… there were only five.

Paul walked straight into the hospital burns pit, the disposal site for medical waste: syringes, gall bladders, dressings, and on more than one occasion, severed limbs.

He stood up slowly.

"Ah man! I'm covered in shite!" His voice an octave higher than normal.

Then he looked at his arms. He'd landed on a batch of burnt and blackened hypodermic needles. Luckily, they'd only pierced his flying suit and cold weather jacket, not his skin, he looked like a scorched version of Sonic the Hedgehog. His normally immaculate crew cut was peppered with black sludge from God knows what.

The only time I'd ever laughed harder was at Robbie's house, when he took the fruit mélange on board.

We made it to the HLS. Someone produced wet wipes and bandages, and Paul began the grim clean-up. It was now late, and cold. The only way to keep warm, and importantly to impress, was to make use of the aircraft cabin heating system.

That meant lighting the No. 1 engine.

I climbed into the cockpit and cranked it. So, there we were: in a Lynx with an engine running at 02:00, a crowd of off-duty medics, Paul still covered in mystery gloop, and the faint hope that this might be a good recruiting moment.

The next morning, the weather cleared. We were told to head straight back to GV rather than return to Split.

Roughly a week later, the reactions to our kind and thoughtful mailing list started rolling in.

"Who's the phantom clap-leaflet dealer?"

"Why are bibles turning up all over the theatre?"

Most of RHQ 1 Regt AAC saw the funny side, especially RSM Terry Tregaskes, whose leaflet selection and bible had been meticulously personalised. Clearly, we were high on the suspect list. Terry played detective and put two and two together.

The RGJ bibles were only available in Sipovo. He rang the Padre to find out when he'd run out of stock, matched the date, and just needed to know which AAC crew had been there.

I still reckon Pat touted on us.

We never found out what Gary thought of his gifts.

Chapter 69. In a Mess

Everyone on a six-month tour gets two weeks Rest & Recuperation. My slot came just past the midpoint, and I wasn't sure I wanted it. I had no idea of what was going to happen. Going home wasn't an option, I wasn't going to sleep on a sofa for 2 weeks, so I booked myself into the WOs' & Sgts' Mess in the barracks.

The mess was mostly empty, so I took a large attic room in an annex. It gave me privacy and enough space to live in my own head.

I was itching to see the kids, so went around to the house in a taxi, I would collect my bike and use that for wheels for the remainder of my time off. As I recall, and it's hazy to be honest, Emily wasn't there, but the kids were expecting me.

She told me it was my job to tell them we were divorcing.

I do remember, and will never forget, telling them that "something awful had happened to our family", and that I couldn't be with them as much as we all wanted.

It's hard to talk when it feels like you've got a baked potato in your throat. Tears were streaming down my face. The eldest three understood what was happening. Fran didn't. She was busy adjusting her rabbit ears, more focused on the happy world she lived in. God, I loved her for that innocence.

It wasn't a restful two weeks. I didn't recuperate. This was far more stressful than anything operations could throw at me.

Sitting in my attic room in the mess, festering, I really wanted, no, needed, to hurt Scott. And I do mean real, physical damage. I hatched a plan over a one-man pizza.

I found a squeezy bottle and filled it with brake fluid, grabbed a ball-pein hammer, and went to his car showroom late one night. The plan was simple: spray every car with brake fluid and take a hammer to each one. Total write-off.

I pulled up outside, took the bottle and hammer from my tank bag, and, deliberately leaving my helmet on, walked up to the first in a long line of brand-new cars, geeing myself up.

I stopped.

Not because I'd calmed down. Because I hadn't.

What was I thinking? I'd just lost my family. Did I really want to lose my career as well?

I threw the hammer and bottle under a car and went back to the mess.

In a mess.

Just before I flew back to theatre, I went to see the kids. Matt and Alex each gave me an envelope; I still have them. Alex's was a drawing of a horse she loved, with a beautiful message inside a card, Matts was a collection of pictures of aircraft and a Lynx that he had cut out from books.

Hugs, kisses and tears. I knew that was it, there was nothing more I could do. I hadn't wanted my family to disintegrate, I had tried with Emily in every way I knew how, letters, phone calls, acceptance of my role and her actions.

I tried to forgive. I wanted my family back. But between Emily and Scott, that was never going to happen.

I flew back to Split and slipped straight into the rhythm of tasking. No one asked how R&R had gone, and I was grateful for that. What was done was done.

Bosnia didn't care if my life was falling apart. The focus now was the upcoming elections, a major effort involving logistics, security, and not enough aircraft to go around.

Chapter 70. Elections, BATs and PIFWCs

A key aspect of the Dayton Peace Agreement was the requirement to conduct free and fair elections.

When the elections took place, we were tasked with securing polling stations and ensuring the safe collection and transport of the ballot boxes. The most secure option was by air, but we didn't have the aircraft numbers to do the job properly without compromising routine operations.

The decision was made to reinforce us with another Squadron from the UK. 659 Sqn was selected. They flew the Lynx Mk 9.

The Mk 9 was wheeled, unarmed, and ideal for moving ballot boxes. Strictly utility, which made it a perfect fit for the task.

659 flew from the UK and arrived in Split toward the end of August, none the worse for wear. On arrival, their crews completed the same theatre readiness checks we'd all done, a sign-off sortie with a QHI, followed by a familiarisation flight with a line aircraft commander.

A few days later, the first Mk 9 landed at GV. It was crewed by Sergeants Dom Giles and Phil Rawnton. I knew them both, but Dom especially well. We'd first met on Lynx conversion and again later during the command course.

Dom was a tall Londoner who'd taken the scenic route to flying. Ex-Household Cavalry, then a London policeman, then back in green, this time with the AAC. He'd made it through selection and trained as a pilot. A solid operator. Calm, very capable and never any drama.

The plan was for me to fly with Phil the next day in his Mk 9, and for Dom to pair up with Paul in one of our Mk 7s.

But first, in true 652 Tombstone fashion, there was a welcome BBQ.

Morning came, the aircraft were ready, the flights were planned, but there was no sign of Dom or Phil.

After a brief hunt, Dom was found asleep under the BBQ, wrapped in empty charcoal bags in a failed attempt to stay warm. Phil had tucked himself into the back of an aircraft and called it a night there.

In all the usual theatre arrival chaos, and our eagerness to make them feel welcome, we'd somehow forgotten to assign them accommodation. They hadn't realised until everyone else had disappeared for the night, and by then it was too late to ask.

To their credit, they adapted as any good squaddie would, make do, find somewhere quiet, and get your head down. No harm done and we all flew as planned the next day, on time, fully switched on.

I was in the left-hand seat of the Mk 9 with Rawny flying from the right. Because the Mk 9 didn't have a missile sight, the roof above my seat was just a plain window. Normally not a big deal. That day, it became quite useful.

Part of our tasking was to get across to Sarajevo.

Sarajevo lay to the southeast of GV, but between us and the city was a diagonal ridgeline running northwest to southeast, thickly forested and steep in places. Tucked through it was a good low-level route we called 'Woodcutters'. In decent weather, the route was stunningly beautiful, sharp, clean, and green, like something from the Swiss Alps. That's where the name came from.

The only serious hazard was a set of high-voltage power lines running parallel to the ridge. We had plenty of clearance to pass beneath them, but only if you knew exactly where they were. I was showing Rawny the safest place to cross and how to pick them up visually.

The weather wasn't perfect, but was still within limits, enough to make it worth using Woodcutters for real. It was a good opportunity to talk Rawny through the route: entry points, navigation features, where to turn, what to expect.

We crept up the re-entrant. My eyes were scanning everywhere as we neared the wires, but I couldn't see them. I told Rawny to come to a hover. I didn't want us going any further without a positive ID.

A previous crew had had a wire strike during their tour. They got away with it, just, but it was a sharp reminder of how unforgiving this place could be.

I leaned back, looked up through the roof, and there they were. The wires.

Directly above us.

We were exactly on route, with no danger of a strike, but seeing them that close and that suddenly nearly stopped my heart. There's something about wires, they stay still, they don't move much, and they'll kill you just the same as ground fire.

I didn't realise it at the time, or for quite a while after, but that minor incident burrowed into my head like some kind of worm, and stayed there, festering.

Even now, I sometimes see them in dreams. And in real life, when I find myself beneath a web of overhead wires, I get an overwhelming sense of dread. It's not rational. But it's real.

Once we'd reset, we carried on. I pointed out the rest of the route, including one of its landmarks, a remote farmhouse in the open. Or rather, what used to be a farmhouse.

The building had been destroyed completely. Not damaged or abandoned. Flattened. Just rubble, dust and twisted metal. There were no other structures for miles. No road. No obvious vehicle track.

Whoever had done it had carried in enough explosive by foot to erase it, deliberately, methodically, and without hesitation. Not for military effect. Not to deny cover. Simply to remove it from existence.

It reminded me of Northern Ireland, not in scale, but in intent and pure hatred. In Northern Ireland, violence was often calculated. This wasn't. This was erasure and it was disturbing.

We arrived back in GV quite late that night, Dom and Paul had been back for a while. The four of us had a chat over a dobra Pepsi and picked apart the day's events. Brutal honesty was the rule; what had gone wrong, what we'd do differently, and how to get better.

We called a Jack and Coke a "*dobra Pepsi,*" good Pepsi in Serbo-Croat. It wasn't clever, but it kept the pretence alive that we weren't bending the rules. In truth, it was just an excuse not to drown in another warm Pivo, the only other option Bosnia had on tap.

About a week later, I was back in Split on duty crew again. Being duty commander meant you were a jack of all trades. Not just air testing and ad hoc tasking, you also helped the Ops Officer pull together the daily CO's briefing. The agenda was always the same: weather, intelligence, ops, then the REME would give the aircraft state.

Gary always expected his officers to deliver these briefs properly. If they slipped up, they were dismantled, no shouting, no swearing, just a verbal ninja attack that left you needing to sit down.

That morning, the REME technical officer forgot to include the Mk 9 aircraft serviceability state. Gary's eyes narrowed. I was 'dispatched' to get my arse to the hangar and get it. You didn't mess around when Gary was annoyed.

I sprinted, not ran, sprinted, arms pumping; Jockey the PTI would have been proud, to the -700 office and burst in. The 659 REME Staff Sergeant was leaning back in his chair like he owned the place, boots up on a battered filing cabinet, tea mug resting on his chest. He didn't even look at me at first.

"Mate! I need the Mk 9 state immediately," I panted out of breath.

He raised his head just enough to make eye contact, then gave me the slowest possible once-over.

"It'll take me a while to pull that together," he said, sucking his teeth.

I told him he needed to pull his finger out of his arse and give me the details for the CO: right fucking now.

That's when he grinned.

"Look Pal, there's not an AAC officer I'm worried about. Tell him, from me, he'll get the aircraft state when I'm ready, and not before. I shit AAC officers for breakfast."

As he spoke, I caught movement out of the corner of my eye. Through the open door I could see Gary jogging across the dispersal. Not sprinting, just a steady, controlled jog, like a bomber on its final attack approach with his laser locked. The Staff Sergeant kept talking, unaware.

Gary was closing the gap, no noise until the very last moment, and then all hell broke loose.

"That's okay, mate," I said, stepping neatly to one side. *"You can tell him yourself."*

The Staff Sergeant turned, still half-smirking and found himself face-to-face with an irate CO who radiated absolute authority.

What followed was the single most awe-inspiring bollocking I have ever witnessed. Gary didn't raise his voice once. It was like watching a precision strike: deliberate, devastating, and impossible to walk away from unscathed.

The Staff Sergeant collapsed in on himself like a deflating blow-up sex doll until there was nothing left but a nodding head and a muttered *"Yes, sir."*

My personal relationship with the REME was hit and miss. Some Air Techs were superb; Sgt Rab Steel was an example of an Air Tech (AT) I respected and liked very much. There were plenty of others though, like this SSgt, that had a 'unionised' mentality. And, as they controlled the serviceability state, a flying aircraft meant more work for them.

Those Air Techs were known as BATs, Bastard Air Techs, and every Squadron had a few.

659 cracked on with their ballot box taxi service, we cracked on with Divisional tasking. When all the electioneering was done and dusted, 659 self-deployed back to the UK.

Steve Priddle and I were duty crew in Split when they departed. A few hours later we received a tasking. Fly an aircraft technician up to Pula on the Croatian/Italian border, a 659 aircraft had gone unserviceable at the airport on a refuelling stop.

Steve and I hooked up with the technician, helped him load his kit into the back and set off for Pula. We arrived and were met by the 659 crew who took us to the aircraft requiring attention. It was an easy fix apparently.

We stayed with the tech until he finished, just in case an air test was needed; it wasn't, so when the repair had been signed off the 5 of us went to a local hotel. We checked in and agreed to meet in the lobby in 15 minutes so that we could get something to eat. We were all looking forward to a decent hotel meal!

We moved to the bar for a quiet glass of red. Also in the bar was a large, loud, drunk local. We were still in flying suits; we didn't have any civvies with us. This man took a great interest in us and shouted us over. He asked us in very broken English what we were doing, and could he buy us a drink?

So as not to offend, we agreed to have a quick drink with him, toasted IFOR and then NATO, and pretended to laugh with him.

Early the next morning the 659 crew set off, hoping to catch up with the rest of their Squadron somewhere in France. Steve and I cruised back down the Adriatic coast to Split and reported back to the Ops Room.

On the wall of the Ops Room was a large poster filled with mugshots of Persons Indicted For War Crimes (PIFWCs), or PIFWICS.

My stomach dropped when I saw one of the faces near the top. I knew that face.

I called Steve over.

"See anyone you recognise?"

He didn't answer straight away. Then:

"Oh shit."

We'd shared drinks the night before with a man wanted for crimes that had wiped entire villages off the map. We filed a report with the Regimental Intelligence Officer and left it at that.

It's a strange feeling, realising you've laughed with someone whose name sits beneath the words mass murder. War criminals don't always look like monsters. Sometimes they're just the loud bloke at the bar.

Chapter 71. Wine, Squid and Alanis

After the duty crew stint, I had a couple of days off. By then, I'd been talking to Ali quite a bit. She didn't want us meeting too often, or openly, in Sipovo. The only person she trusted to know we were in contact was Pat McCardle. We spoke about my R&R and how it had gone. She listened.

The more we talked, the more I realised how different she was. Not just from Emily, but from most people, full stop. She was funny, sharp, ambitious and grounded. Easy company. And as a bonus, she was beautiful.

Music has a way of pinning a moment in time. For me, Alanis Morissette's '*Hand in My Pocket*' does it. One line of the chorus takes me straight back to Trogir, a fishing village just north of Split.

Ali had arranged to be in Split while I was off. To keep things discreet, she hired a car and we drove out to Trogir. Somewhere along the way, the radio crackled to life with an Alanis Morissette song that could have been written for her. It was all contradictions and confessions. Broke but happy, grounded but overwhelmed, lost yet hopeful.

Every line seemed to sketch her character better than I ever could. I looked across at her, knowing the song had summed her up in ways my words never would.

We spent the day on a rocky beach overlooking the still Adriatic. We drank wine, shared a picnic and talked about everything, barely pausing for breath. As night fell, the fishing boats came out, stringing lights across the water and dropping lines for squid.

Later, we climbed up to a restaurant on the hill, just as a fisherman was hauling in a fresh catch. It was cooked and on our plates within the hour, the best meal of my tour for more than one reason. Then we drove back to Split in silence, lost in our own thoughts, and headed our separate ways.

I wanted to feel her touch. Properly. Not just out of desire, but because I needed something real and warm to remind me I wasn't beyond saving. It went deeper than the physical.

I craved comfort, connection, the sense that something good still existed in my life. I was hurting, raw and alone, and somehow, she made me feel less broken.

Ali had her world, her rules and I wasn't about to mess that up. So, we didn't. We said goodnight quietly, politely, like nothing had passed between us. But something had. We both knew it.

In the Squadron, only Keith and Paul knew about Ali. They were both solid, supportive without fuss, as you'd expect, and without judgement. They were also the only ones who knew how R&R had really gone, and the state I'd come back in.

With their wives still in Gütersloh, they were hearing things. Stories about Emily being out most nights, living like a teenager in bars and clubs with Scott.

They never said much.

They didn't need to. We knew each other well enough for that.

But they knew enough.

Chapter 72. Tail Rotors, Fences and Ranges

About a week after returning from Pula, I was back on routine tasking. I was also planning a range package for the Squadron. We still had a healthy allocation of missiles and 155mm artillery that were begging to be fired. The war stock missiles had a shelf life based on flight hours. Some were near their limit, so we'd be helping them retire with a bang.

Chris Collett and I did the range recces. I needed to make sure my plan fitted the safety template at Glamoc; the same place we'd run the firepower demo. Live missiles meant a different safety template.

My plan had Lynx crews firing by day and night, which meant we needed heated targets for the sighting system. The Gazelles would direct their full allocation of artillery. Finally, the doorgunners would engage infantry targets on battle runs, controlled by their Lynx commander.

On the first recce, Chris and I bumped into a US Kiowa Warrior crew planning a similar package for themselves. They asked why we were using the callsign 'Outlaw'. Their unit were the 'Comanches'.

We explained we had been assigned it and Chris suggested that they should have a Squadron fly-in to GV for a joint 'team building' experience. Their OC explained that even on a 12-month tour, they were forbidden to drink unless refusing would offend the host.

Chris removed his glasses with a theatrical flourish:

"As a Major in the British Army, I would be personally offended if this rare opportunity to share experience was rejected."

They immediately agreed and we exchanged numbers.

Before the range package could go ahead, I was conducting routine tasking. For one such tasking, Eddie Smith and I were crewed together, with L/Cpl Brian Luke as our doorgunner. The weather was poor and getting worse. It was a mundane run to and from Banja Luka. With the Cheddar Gorge route off Route Gull, it wasn't too much of an issue.

Going northbound, Cheddar Gorge was busy because the weather was on the deck over the high ground. There were a couple of sharp turns in the gorge. SOP was to call before each one, advising other aircraft of our position and routing. We heard a lumbering German

CH-53 making calls.

We dropped our passengers in Banja Luka and lifted for the return leg to GV, via Cheddar Gorge southbound. As we pushed south, the weather deteriorated. Cloud was now hanging down into the gorge. With the amount of two-way traffic, it was no longer safe. As we passed a platoon base at Krupa, I made the call and Eddie landed on their nominated HLS.

The platoon base was in a commandeered shoe factory. Their HLS was outside the base, across Route Gull, and it had little protection.

As we landed, a young officer came over to see what was going on. An unplanned Lynx arrival wasn't normal. I explained we were using their location as a diversion. We'd need somewhere to sleep and a security detail for the aircraft. We were carrying NATO secret crypto fills, defensive systems and the aircraft itself was worth stealing.

He only had a platoon. Half were asleep and the rest were out on patrol or busy elsewhere. We could though, move the helicopter inside his small compound.

Eddie, Luke and I walked over to survey the suggested landing spot, he was right, it was small. There was a small tree that would fit under the rotors on the left side, so I told Eddie that I would fly it as I would be able to monitor it. Luke would act as a marshaller and guide us in.

While Eddie and I went back to the aircraft, Luke swept the area for loose items.

It was tight, but it was workable. A clean hover and a steady set-down would do it. The tree to my left was manageable. The fence behind? Luke had my back.

I came into the hover and hopped over the MSR to the new spot. Eddie gave me a running commentary from his side. Keeping an eye on the tree and Luke's signals, I shot the approach, landed and started the shutdown. As I did, Luke motioned to me to look behind the aircraft.

I hadn't applied the rotor brake yet. The tail rotor was sparking like hell, buried six to eight inches into the rusty chain-link fence!

I applied the rotor brake to stop it as soon as possible.

As I jumped out to inspect the damage, I ripped off my helmet and launched it at Luke. I called him every name under the sun, most of them unfair.

All four tail rotor blades were scratched across their width, paint stripped but no obvious damage to the skin.

We then heard another Lynx overhead. I got on the radio.

"Outlaw callsign in Cheddar Gorge at Krupa, this is Outlaw 22, do you copy?"

"Outlaw 22, this is Outlaw 01, go ahead"

"Outlaw 01, thanks. We have a tech issue at Krupa. Can you assist?

The Lynx landed where we'd been. I walked over to explain that we'd need a tech to assess the tail rotor and any possible shock loading through the transmission.

Climbing out were the CO and the Regimental QHI, Andy Fellows.

Not ideal.

I explained what had happened, calmly. The mantra was simple: if something went wrong, if you'd done everything right and could account for your actions, you'd be OK. I had. And I could.

Within a few hours, a vehicle arrived with technicians. I was relieved to see Sgt Rab Steel, a tech who wouldn't exaggerate the problem. Rab took a look and decided to replace the blades.

He also confirmed there was no shock loading. The fence had been so fragile that nothing substantial had been struck. Had it been stronger, or we'd hit a post, it would have been a very different story.

The next day, Eddie, Luke and I completed the tasking. I had a long conversation with both. I apologised to Luke for bouncing my helmet off him. I filed a report, as was required, and thought nothing more of it.

There was no time to lick wounds. The range package I'd poured myself into was finally happening. Paul and I deployed up to Glamoc early to set the targets up. We'd arranged for a local artillery unit to collect abandoned cars, remove the engines and fill them with anything flammable, including big plastic cans of petrol.

A long rope was attached to the car and the other end had a loop that would fit over the aircraft hook. Paul flew a long, slow transition, took it to a predetermined location and jettisoned it. Eventually we had four stacks of four cars, all stuffed with highly flammable rubbish.

The next job was mine. In the back on the gun, I loaded 7.62 in a one-and-one mix: one ball, one tracer. Tracer ignites at about 100 metres from the muzzle. Paul held us just over 200 metres from the target and I opened up. The tracers took longer than expected to ignite the contents, but each stack went up impressively. The targets burnt ferociously for about an hour. Very satisfying.

The day shoot went well. The Gazelles directed plenty of 155mm from the AS-90s guns. The Lynx crews fired at their car stacks, then ran battle drills against rock fields marked as infantry. But the highlight was an NVG Cat 3 missile attack. The cars held heat well into the night, so when the NVG shoot came, every commander had a proper thermal target.

A few days later, I was in Split and bumped into Terry, the RSM. *"You're in deep shit,"* he said, and clearly wasn't joking. I was stunned. No one had said a word to me.

Terry told me he'd overheard Andy Fellows talking. The RQHI wanted to formally ground me. Andy Lambert and a few others were pushing back.

Eventually, I was brought into the conversation. Andy Fellows made it clear he thought a grounding was warranted. His phrase was *'culpable negligence'*. I was furious. I told him straight. If that phrase was applied to me or anyone on my crew, I'd take it to courts martial.

I meant it.

Gary then spoke to me, and I told him the same. I'd taken enough shit recently, but I wasn't taking that.

Later that week, Gary confirmed I'd lose my commander status for a short period, a couple of days. I'd be limited to flying with Chris Collett or a QHI, and I'd need a check ride with the RQHI to be reinstated.

Then he asked me for my thoughts.

"Sir, I think this is wrong," I said. *"It does nothing to promote a just culture, and it ignores the circumstances of what was a minor incident."*

He nodded. *"Let's have this chat again in six months. Let's see how you feel then."*

I always respected Gary. He was a solid CO, calm, fair, and trustworthy. I felt aggrieved, but I respected him more after that conversation. Not everyone came out of it the same way.

The first person on the phone to check I was OK was Ali. Pat had told her what had happened. A friendly voice was exactly what I needed. She didn't ask stupid questions, didn't pry or lecture. She just listened.

Then she said, *"I heard. That's bloody harsh. Are you alright?"*

It was a pathetic process, but that didn't stop Keith, Paul, and the rest taking the piss at every opportunity. I was quickly christened *"Unlucky Jim"*, a nod to *The Fast Show's* Unlucky Alf and his ever-growing list of disasters. From that point on, I was Unlucky Jim.

It didn't matter if I was on time, in the right, or stone-cold sober. *"Ah bugger"* followed me like a second callsign.

And now you know. That's where it came from.

89 Rab Steele replacing the TRBs

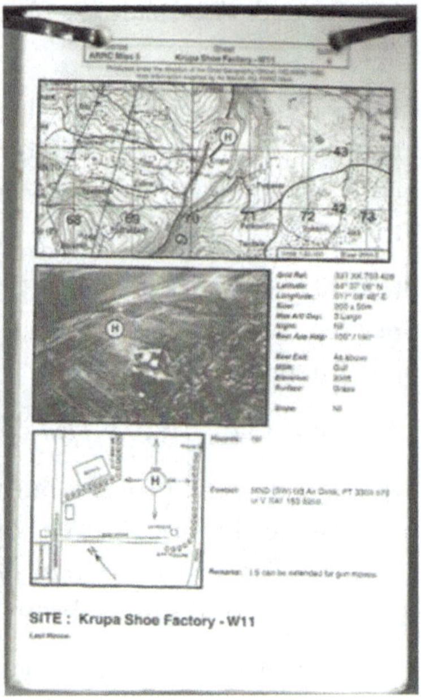

90 Landing Chart, Krupa

91 Taking a bite

Tail Rotors & Fences

Chapter 73. Outlaws, Comanches and Apache's

The Friday arrived for our gathering with 3rd Battalion, 4th Aviation Brigade (3/4). We'd also tried to get Vince and the rest of 2/6 involved. They were just up the road in Tuzla, but their Apache's weren't allowed outside their area of responsibility. The invitation was declined; offence or no offence being taken.

This wasn't going to be anything like the welcome party. No nurses. No Chinooks. Just a quiet drink between professionals.

As they landed one by one, the HLS filled at an alarming rate. Lynx, Gazelles, BO105s and now Kiowa Warriors. We had to push out into the Chinook strip just to fit everyone in. Chris and I met up with the guys we'd originally spoken to and introduced them to the rest of the Squadron.

Seeing all those Kiowas armed to the teeth, bristling with impressive kit, definitely gave Robbie a twinge in his pants.

They went quiet when they saw how we lived. The contrast was stark. They couldn't believe how badly we were accommodated. They may have had a deployable pizza joint and McDonald's.

But when they saw Tombstone…

priorities shifted.

Their faces lit up, massive, cheese-eating grins all round. They'd been dry for almost a year.

Not a single drop.

They were shown to their sleeping spots, usually a patch of floor in someone's Corrimec. No one cared. There was beer, and they were allowed to drink it.

Some changed into the standard grey US Army sweats, others stayed in flying suits. Within a couple of hours, most were completely paralytic. The pivo was strong, and this was their first proper drink in months.

By 20:00, Tombstone was packed. The Cloggies arrived en masse, including the women. When the Americans saw them, more jaws dropped. These lads hadn't seen a woman in lipstick in almost a year.

It didn't go unnoticed.

Ronald, the Dutch engineer and washboard virtuoso, led the inevitable rendition of *Kedeng Kedeng,* an absurd Dutch train song that, along with Alanis, became one of the defining tracks of the tour. Every time it came on, the whole room bounced like a rave in a railway station.

The Pivo was flowing like it had been uncorked from a fire hydrant. Stories flew across the bar. The battered CD player pumped out Oasis, Alanis and various crimes against techno. A few of the Americans just stood there with beers in hand, eyes wide, beaming with that perfect blend of disbelief, gratitude and cultural confusion. It was a frat house party meets NATO mess night.

By 01:00, the floor was a battlefield of boots, empty cans, ration wrappers and half-dressed aviators using body armour as pillows. An exorcism of stress and deployment fatigue. The kind of evening that wouldn't show up in any report but would be remembered far longer than any patrol or tasking.

At least one of them, unfortunately the one bunking on our floor, failed to hold it together. Fully unconscious and very much the worse for wear, he managed to shit himself. In true military tradition, we did the honourable thing, took photos, invited his mates in to admire the scene and ensured it would live on as Squadron legend.

The Comanches departed the next day in dribs and drabs; basically, as they became fit to fly. That evening is still talked about in depth and always will be!

To make up for not being able to get 2/6 Cav down, Keith and I took a training sortie up to Tuzla. I booked us in through Eagle Ops so they were expecting us. I had the correct field joining instructions and we joined the Tuzla arrivals procedure correctly.

"Tuzla Eagle Base, this is Outlaw 22, good morning" no response, I double checked the frequency, it was correct.

"Tuzla Eagle Base, this is Outlaw 22, good morning", again no answer.

The third attempt finally raised a response.

"Ghostrider, state your intentions" ('Ghostrider' is the callsign the US uses for an unidentified aircraft.)

"Tuzla Eagle Base, callsign is <u>Outlaw 22</u> on PPR 0428, request parking instructions"

Eventually we got back *"Ghostrider, park spot 6"*

We parked between two Apache's. A HUMVEE screeched to a halt next to us and out jumped Vince.

"What the fuck was that all about?" I asked him.

The woman in the tower had ID'd us as an Mi-17 Hip, a big Russian troop carrier, and thought our callsign was a spoof. It wasn't until Vince stormed in that she realised we were a British Lynx, not Soviet leftovers. We had some lunch and took a few pictures.

A theme was developing. Americans simply could not recognise a Lynx from a Hip.

92 Me, Vince Stella and Keith at Tuzla Eagle Base

93 Me and Paul at FOB Red Dragon, Klujc

94 Mission Planning Session, Gornji Vakuf

95 Me and Paul with The Light Dragoons at Banja Luka Wood Factory

Around the Country

96 Me and Ronald the Washboard King

97 Paul and Keith

98 Me, Guest, Trojan

99 Tombstone in full flow

100 Eddie Smith in the Middle

101 Kirsty and Chris Collett

102 Me and Keith with an American

103 Kirsten and Cholly

The Comanches meet the Outlaws

Chapter 74. Running Down

As the year drew to a close, so did our tour. The weather was changing; snowfall increased and the days shortened, meaning we spent more time on NVGs. The trickiest part of flying at that time of year was the transition between day and night. That in-between period where you could still fly unassisted, but NVGs were starting to help.

The snow changed everything. It threw light back at us, making the world glow but killing depth perception. What looked flat could be a slope; what looked high could be a ditch. Our radar altimeters became vital. Ironically, even they were temperamental in snow, the return signal often distorted or lost altogether.

Late in the tour, I was flying with Trojan. We'd been on a routine sortie around the country and were returning to GV from Banja Luka. It was one of those nights when the sky was heavy with snow clouds. Not snowing, but you knew it was coming in hard. It was incredibly dark, with little ambient or man-made light for the NVGs to amplify.

Approaching the exit of Cheddar Gorge at Bugojno there were several sets of wires. We'd been under them hundreds of times and knew exactly where they were. Despite my experience with Rawny, wires weren't the problem tonight.

Trojan positioned us exactly where we'd cross under the wires and flew slowly and deliberately. My eyes were on the wires, which stood out clearly with the crust of snow. I was giving Trojan a running commentary when the doorgunner leaned forward between the crew seats and said:

"What was that? There was a flash on the warning panel"

Neither of us had seen anything. But we had heard a quick 'bleep', which we assumed was the Rad Alt. Entirely normal. We continued to cross under the wires. When we were clear, Trojan applied power and climbed to about 300 ft, increasing speed as he did so.

I matched the torques. The Lynx's twin engines feed a single gearbox, so the inputs must stay within a set range. I twisted the lever grip above my head, watching the gauge pointers come together, and saw it. A flicker of red, almost purple under the blue

ELP lighting, on the CWP where it didn't belong. Just for an instant.

At the same time, the No. 1 engine oil pressure needle dipped into the red, then sprang back to normal.

It did it again. It was hard to notice due to the blue lighting shrouding the instrument panel, but it was enough to get my full attention.

I blipped the 'lip light' attached to my boom mic. It was white light, but we needed it to read anything properly. I reached into my door pocket and brought out my Flight Reference Cards (FRCs):

Consulting them gave:

Immediate Actions:	*Establish safe single engine flight condition.*
	Retard Faulty Engine ECL to GRD IDLE.
Subsequent Actions:	*Shut Down Affected Engine*
	Land as soon as practicable

The problem was our indications were sporadic. The FRC actions were for a confirmed **ENG OIL PRESSURE** in the red.

Ours wasn't.

Assessing our situation was the critical factor. I had memories of Mac and Sean's failure in Northern Ireland at the back of my mind. Different circumstances, but the same lesson: get the diagnosis right. Get it wrong, and it could kill you.

The FRCs were written for specific, foreseeable malfunctions, not a chain of inter-related failures. This was where that air testing qualification, and my time with the RQHI back in Gütersloh, stopped being dull and started being useful. It taught you to think, not just react.

Was it a faulty gauge or sensor? If I shut down the engine, we'd be committed to a single-engine run-on landing, at night. Normally manageable. Steve Priddle had one not long ago in TSG. GV even had a nominated area for it, but it was far from ideal given the proximity of the minefields.

Depending on the wind direction over the designated area, there was either a line of trees at one end, or a set of wires at the other. Using it would probably mean cooking the good engine to ensure a safe landing.

Pilots use a decision-making model called DODAR: Diagnosis, Options, Decide, Action, Review.

Diagnosis: a possible imminent single-engine failure.

Options: shut it down, with all that entailed, or keep it running and confirm the failure.

I chose the second.

Action: ease into a gentle climb using speed, not power, and gain some height. Turn for GV. Warn Eagle Ops of a possible emergency and have assistance ready. Plan an expedited twin-engine landing as close to the Ops room as possible. If the engine failed en route, shut it down and execute a single-engine run-on in the designated area.

Trojan flew a flawless approach into GV. I followed through on the controls, ready to take over if the engine gave out at the worst possible moment.

As I called *"committed",* I remember thinking, if it's going to go, it'll be now. It didn't.

The Air Techs later diagnosed a faulty oil pressure sensor on the No. 1 engine. We'd made the right call.

Did anyone in the chain of command say well done? Not a chance. It was just us doing our jobs in circumstances you can't train for. Experience got us through it.

I didn't expect anything, not even a *"good show, young man",* but after Krupa, it would have been a nice touch.

My confidence was back.

Chapter 75. Homeward Bound

Our handover date was set for mid-December. I'd volunteered to be one of the last out, nervous about what Christmas "at home" would mean for me. Home wasn't home anymore. Not even a house.

Terry, the RSM, arranged a farewell mess dinner for as many of the Regiment's officers and SNCOs as he could herd. He'd found an old Croatian farmhouse-restaurant on a remote hillside south of Split.

The food was incredible, the pivo cold and the slivovitz potent, a plum brandy that could strip paint and sometimes clocked in at over 70% ABV. Conversations flowed from perceived slights to actual slights and any other grievances worth airing. It was a great night for everyone.

Keith was one of the first to leave theatre. He had good reasons, but that didn't stop the disappointment of seeing him go. We never told him that, obviously. We just took the piss and called him a part-timer.

On the day Keith was due to board a C-130 out of Split, I was on task with Trojan, pulling a couple of soldiers out of the mountains and down for their own flight out.

As we shot the approach to runway 23 at Split, we could see the Hercules with the ramp down and a long line of passengers waiting to board. Then I spotted him. Keith was in the queue, bergan over his shoulder, waiting to board.

"FUCK me! It's Keith!" I yelled over the intercom.

We hover-taxied towards the Fat Albert to drop our own passengers. Keith saw us. I opened my window and waved. I also gave him a few wanker gestures and a dismissive thumb over the shoulder, as if to say, *"Go on then. Fuck off, wanker."*

I was delighted to see him. He waved back making his own gesticulations, equally immature.

Then I saw the others.

That's when it got awkward.

In front of Keith was the CO, the Ops Officer and the RSM. I can only imagine what they thought, watching one of their aircraft taxi in while the aircraft commander hung out the window making rude gestures at the queue they were standing in!

Not my finest moment.

I left GV about a week before Christmas. I'd handed over my HWI duties, the war stock missiles and my Corrimec. I was the last one out of it. Paul and Steve had already gone.

What did surprise me was the amount of spare ammunition Paul had collected "just in case". He'd thoughtfully hidden it in our Corrimec for me to dispose of. There was a 100-round belt of 7.62, a healthy mix of 9mm and 5.56mm and, for good measure, a few smoke grenades. Thanks, wanker.

The only way I could get rid of it without dropping anyone in the shit was simple. Ditch it into the minefield.

My last duty in Bosnia was the duty crew stint at Split. Ali was also there, preparing for her own departure. She'd been posted to Tidworth in Hampshire. We said our goodbyes. We kissed.

I landed back in Hannover and got bussed to Gutersloh.

As we pulled up in front of RHQ, there was a crowd of families waiting to welcome their husbands and dad's home. Kids in elf costumes, Glühwein flowing, carols playing. A picture-postcard welcome home. The reunions were heartwarming.

I got my sleeping bag, helmet and bergan and made the lonely walk over to my room in the mess attic. Alone.

She couldn't even bring the kids.

I was devastated.

That was my Christmas. Alone.

I couldn't eat. For three or four days, nothing. No hunger, no appetite, just… nothing. By day six, I knew I'd be in trouble if I didn't get something down.

I bought a plain pizza from the German vendor on camp. I wasn't ready for the mess, and most people were back in the UK anyway.

I forced a few bites down.

It felt like eating cardboard.

After that, I managed a bit more each day. Enough to function.

I must have blocked most of that period out, because I don't remember Christmas Day at all. Not a single detail. I knew Emily had moved out of the married quarter and taken the kids in with Scott.

The days around Christmas are just blank.

An empty space in my memory.

Probably just as well.

There wasn't much mail, but one letter stood out. It was from my cousin Jeff in the US. He worked for American Airlines and had sent me a 'Buddy Pass', a blank ticket. I planned to use it to go to New York for New Year, do the Times Square thing, then head up to Pittsfield to see the rest of the family.

Mobile phones were only just becoming popular, so I bought a Nokia 2110 and called Ali in Tidworth. I told her my plans for New Year and asked her to come with me.

She couldn't. She already had plans.

I rode to Oxford to stay with friends before heading to Heathrow for my flight to NYC. Ali called me back and said she'd try to change her plans, but no promises. Hoping my luck would change, I went to a travel agent and booked her on my flight anyway. Just in case.

Chapter 76. Reflections

Operation Resolute 2 had been the best and worst six months of my life.

I'd met up with old friends. Mig Gomez from JLR, now a Sergeant in the Royal Engineers, was escorting Martin Bell, the BBC's man in the white suit, around the minefields. I'd been tasked to fly them. The Light Dragoons had squadrons in Glamoc and Banja Luka, and I supported them too. It was genuinely uplifting to work alongside them. They'd done the Regiment proud and maintained a reputation for excellence matched by few.

Keith and I linked up with Vince and the men of 2/6. Paul and I had become almost inseparable, our reputation now preceding us wherever we went.

I logged nearly 300 flying hours, about 50 of them on NVG. I passed 1,000 hours total time, then 1,100. I only fired two missiles, but oversaw the firing of about twenty.

We hadn't lost anyone, despite some close calls.

The pilots of 652 had become an impressive team. We knew each other's strengths and weaknesses and used that to our advantage. Chris Collett had been an exceptional boss, I would have followed him anywhere.

Lt Col Gary Coward was a beacon, an inspirational leader.

We learned more about human nature than any of us wanted to. Some of us flew war crimes investigators to massacre sites. We saw towns emptied and homes smashed to rubble because the roofline gave away the owner's faith.

We saw kids, half-starved and wide-eyed, usually in a minging Man Utd or Liverpool shirt, racing towards us to beg for Penguin biscuits from our lunchboxes. We gave them gladly. It only made them run faster next time, dodging skids and tail rotors for a biscuit. They did at least have the good sense to leave the dried-out, grease-soaked ham sandwiches untouched.

It became dangerous, and we felt terrible having to tell them to stay away.

They couldn't understand.

I think we blocked out more than we admitted. It was easier for us than for the poor bloody infantry, who lived even closer to the constant exposure to death, destruction, and sadism.

Bosnia has since been overshadowed by Iraq and Afghanistan, but it was the UK's largest overseas deployment since the Second World War.

The opening act in nearly two decades of continuous operations.

Throughout it all, I lost my family. That devastated me. From the age of eighteen, everything I did had them at the forefront of my mind. I wanted my children to be proud of me, but what I sensed coming from Emily didn't bode well.

652 went to Bosnia thinking we were ready.

We left knowing we were operating at a different level.

What came next?

Many of us were due postings. The IRA ceasefire announced in 1994 had broken down earlier in the year.

It was getting interesting again.

Figure 104 652 Sqn at T55 Ridge

Germany, Gutersloh: 1 Regt AAC, 652 Sqn, Lynx Flight, Jan – Apr 97

Chapter 77. New Year, New York, New Life

By the time Ali called me back, Bosnia felt behind me, but it was still in my head. The winter, the constant NVG flights, the wires, the near misses. The distance from home. The marriage that was over, bar the paperwork. I'd been running on duty, adrenaline, and stubbornness for months.

She'd changed her plans. She could come to New York. I was delighted. And nervous. Up to this point we'd genuinely only been friends. Now we were going away together. What if it was a huge mistake and I was just on the rebound?

We met at Heathrow and boarded the American Airlines flight to NYC. We talked the whole way about our different Bosnia tours, inter-rank relationships, and what would happen if ours went anywhere. Any fallout would land on Ali. She was the officer. I was just a bloke. We were realistic about the risks.

We didn't talk about feelings, as if they were already decided. But there were moments on that long flight when our knees brushed and neither of us moved away.

A yellow cab took us to our hotel, a few blocks from Times Square. It was nothing to write home about. In fact, it was a bit grim. Neither of us noticed, but that was the moment it all became real.

The next day we wandered the city, did the tourist bit, mostly shopping, then headed to Times Square around 18:00. The hotel receptionist had advised us to get there early if we wanted a decent spot. It was cold. The kind of cold that hurts. Almost Army-cold, but not quite.

By 20:00 the place was heaving. Ali stood in front of me, my arms wrapped around her, swaying slightly. At first it was to keep warm, but before long the warmth wasn't the only reason I didn't let go. Her hair brushed my cheek; I could smell her perfume even through the freezing air.

The temperature steadily dropped, smashed through Army-cold, into some kind of Arctic death-march cold. By 23:55 we just wanted that bloody ball to drop. Not for the celebration, just so it would finally be over, and we could get out of there.

Midnight. Cheers. Confetti. We tried to move. Cramp and cold had locked our legs solid. We hobbled away from Times Square like a pair of geriatric cripples, propping each other up all the way back to the hotel, where, despite the state of our legs, we finally got round to what we'd both wanted all night.

On 2 January, we drove up to Pittsfield. It wasn't the first time I'd been back, so I knew my way. We stayed with my Aunt Fran, the very image of an Italian woman, short, with dark hair in a 60s beehive.

We spent the week catching up with family.

Aunt Fran showed us to our room. She saw us glance at the double bed and said, *"Oh, sorry, I thought you were together."*

Ali, quick as ever, saw me hesitate and stepped in. *"Oh, it's OK. We absolutely are."*

Hearing that stopped me for a moment.

It was such a small thing, but it meant more than I expected. For the first time in a long while, I felt... settled.

Wanted.

While we were there, Ali and I bought my cousin John's kids some belated Christmas presents. A drum kit, a trumpet and a range of other loud stuff I knew wouldn't last long. I'm a generous uncle. What can I say.

In a childish display of revenge, a couple of days later, at about 05:00, he burst into our room with a Super Soaker and drenched us both. Fair enough, but that meant war.

Before leaving Pittsfield, I saw a Sunday magazine with more of those "send no money now" adverts. Remembering the hours of fun we'd had in GV, I found the perfect one for John.

It was for a bulletproof door. A very expensive bulletproof door. The difference was that these adverts in the US put the recipient on the hook for the full price, unlike the UK versions. He was not happy with Jimmy at all.

We drove back to New York having survived Times Square, Aunt

Fran's hospitality, and John's pre-dawn super-soaker assault. Then it was back to Heathrow. Ali headed for Tidworth; I picked up my bike from friends in Oxford and rode back to Germany.

The long flights and long drives hadn't dulled what was there between us. If anything, they'd sharpened it. I could still feel the press of her against me in Times Square, still smell her hair in the cold air, still hear the way she'd said, "we absolutely are," without hesitation.

Neither of us knew what the future would hold.

But I knew this much. I wanted her.

And for the first time in a long while, I felt wanted too.

Chapter 78. In a Mess. Again

The Regiment started to drift back to Gütersloh after an extended Christmas leave. We had post-operational leave on top of the standard block leave. I hadn't flown for nearly six weeks, along with others who'd been on the ground even longer. We were all itching to get back in the air.

We went straight into air testing. Anyone needing a currency check was booked in with Regimental Standards.

I needed to be flying. Any reason would be good enough. Anything to stay in the air and out of that attic room.

Luckily, we were heavily tasked almost immediately. The Division was holding a huge multinational exercise focused on the Bergen-Hohne ranges. Exercise Shelldrake Spear would see the Divisional Artillery Group (DAG), supported mostly by UK heavy artillery, coordinating with the Germans and the Danes. The DAG also controlled any fast jet action, so the 'crabs' were involved as well.

The Gazelle crews would be doing their thing providing recce, directing artillery fire and controlling fast jet attacks. Our Gazelle crews really were excellent, although I'd never say that to a floppy puke.

I often listened in on their tactical freqs, the back-and-forth between them and, usually, a Tornado crew. Very professional.

The Lynx crews provided direct anti-tank fire within a JAAT, a Joint Air Attack Team. We'd toyed with the concept before Bosnia. I was never a fan. My command troop experience shouted that JAAT was too complicated, too fragile. One small slip and the whole thing fell over.

A typical JAAT might call for jets for two minutes, helicopters for five, artillery for six. That meant we'd loiter nearby, wait for our slot, dart in, acquire targets, and shoot in our tiny window before the next element rolled in. We always tried to get the artillery to go last. Their smoke and obscuration could blind both the jets and us.

9 times out of 10, the RAF would be late, with a knock-on effect that ran a coach and horses through the concept.

On this very rare occasion, a couple of Tornado jockeys did the unthinkable. They left their five-star hotel and ventured into the field to see how we operated. It was supposed to aid their understanding of the JAAT.

Keith and I were crewed together and were tasked to pick up one of these pencil-tashed RAF Romeos from Div HQ and take him to RHQ.

"J" for Johnny climbed into the back and, instead of using a Mk 15 Buzz helmet, plugged his own electric hat into the intercom. We had a good chat with him on the 30-minute journey at about 60 ft, which was quite high for us. He chatted happily about the wonders of precision-guided munitions and how the RAF version of "*the field*" involved a hotel lobby with a bar.

Then we reached the CATO entry point to RHQ: Concealed Approach and Take-Off. The idea was to mask a FOB or FARP by entering from miles away, skimming in low, fast, and on radio silence, buried in radar ground clutter.

The CATO is designed to be low and fast. 120 knots is slow to a jet driver, but below treetop height, hopping fences and ducking under wires, it's damned fast. Everything is relative.

Landing at RHQ, our budding Chuck Yeager had developed a noticeable stammer.

He climbed out looking like he'd just seen the afterlife. Eyes on stalks, pupils so dilated we could practically see his brain, sweat already beading under the rim of his helmet. *"You're dangerous. Do you know how close your rotors were to those trees?"*

His voice had climbed an octave, moustache twitching like it was trying to make a break for it.

Keith replied, tactfully as ever. "*Yes. Yes, I do. And they weren't close at all. We took it easy.*"

We hadn't.

Johnny walked off to the command post muttering something getting a Land Rover back to Division.

We did several JAAT runs. I doubt any met the textbook definition of success, but Brigadier Keeling, the CRA, was happy, so the boxes got ticked. I enjoyed working alongside a recce troop from The Light Dragoons. Hard to miss a Geordie accent over the net.

Life in the mess attic was depressing. Only the flying kept me up. I couldn't see the kids, despite them being in the same town.

I went to their local stables as often as I could, to watch gymkhanas or riding lessons. It took me back to Bovington. Emily walked around as if she owned the place, chatting to the locals in her broken German. I could understand just enough to know she was telling them I was her ex-husband, adding an eye roll as if to say, *"Just look at him."*

Scott kept well out of my way. He may have found the hammer and bottle of brake fluid and joined the dots.

Matt and Alex were sent to separate boarding schools in the UK. Matt's was a civilian one in Norwich. Alex went to the Duke of York's Royal Military School near Dover. I had no say. No consultation.

I found out after the fact. That hurt more than I expected.

It worked out very well for Alex. She excelled, and that school helped shape her into the kind, caring, responsible adult she is now, and a fantastic mum.

Matt wasn't particularly academic, and I don't think being away from home helped him.

Jessica and Fran stayed in German schools. Fran did well, she was young enough to adapt quickly.

It proved disastrous for Jessica. She started rebelling early.

And she knew exactly how to do it.

Flying was still the only place everything made sense. On the ground, life was messier than any JAAT.

Chapter 79. Auf Wiedersehen Pet

Since the IRA ceasefire collapsed, Northern Ireland was back on a knife edge. I was due a posting and requested 655 Sqn, based at RAF Aldergrove just outside Antrim. Confirmation came through in the form of a posting order, my first since joining the AAC. I was to report to 655 in mid-May.

But there was still flying to be had out of Gutersloh. Following Shelldrake Spear, Exercise Flying Rhino '97 was another divisional artillery exercise, this time involving newer NATO members, Lithuania and the Czech Republic.

Knowing I was posted and needed to arrive in Northern Ireland fully current, I squeezed in flying time with Chris Keane to recertify my NVG Cat 3 and instrument ratings.

My last flight in Germany was with Paul. We were back at the ILRPS school in Weingarten, running a limited hunter-force role for the red side, just a single aircraft this time.

Professionally, I left Germany an experienced, highly qualified aircraft commander and HWI, with an operational tour behind me. I had nearly 1,200 flying hours.

During my exit interview, the CO, true to his word, brought up the Krupa incident. He asked me to reflect on the aftermath. I answered honestly.

"Sir, I think it was handled poorly."

I didn't sugar-coat it. I explained how it unfolded, how it affected me, and what it meant for trust and morale. To his immense credit, Gary didn't deflect or defend the chain of command. He listened. And then, with the sincerity that marked him out as a superb officer, he quietly agreed: *"It could have been handled better."* He acknowledged I'd been poorly treated.

In that moment, my respect for him was cemented for life. It wasn't just that he admitted the failing. He cared enough to ask the question and had the humility to accept my answer, right or wrong. That takes courage, and real leadership.

As a father, husband and man, I was leaving behind my beautiful, funny, loving children, whom I adored. I was also leaving behind a soon-to-be ex-wife who, I thought, was rapidly turning into her mother: selfish, self-centred, with the same nasty streak.

Although I'd thought about it, and discussed it with Ali, I just couldn't stay. Every day was torture. A pilot in the wrong frame of mind is a liability.

I had embarkation leave before I needed to be in Aldergrove. Ali's sister worked for a travel agent and could get good rates. Not Cousin-Jeff free, but good enough. We booked a holiday in the Dominican Republic.

I packed up what little I owned, mostly flying kit, and boxed it for shipment to Aldergrove, keeping just enough to get by for a few weeks. I left Germany with little more than what I had on my back, in my tank bags and panniers. That Honda VTR was severely overloaded, but I made it to Oxford where I stayed overnight with friends.

The next day, I met Ali at Bicester shopping park, bought some holiday clothes, and was just happy to see her. That night, in a hotel near Gatwick, it finally felt like we had a bit of space to just be together.

The next day we flew out to an all-inclusive resort in Puerto Plata.

The resort was nice. Hot, sunny, and exactly what we needed, away from the Army. Ali had been getting sideways remarks about her relationship with a Sergeant. Despite our best efforts, word was out.

We spent our time in the resort, we had been warned it wasn't safe outside. The casino became popular with us, largely because the croupiers couldn't count to 10 using both hands. We could. And soon realised that if we disputed a losing hand, rather than argue, they would simply pay out.

No CCTV made this a sport.

Getting cabin fever, we rented a motorcycle and headed out. We'd heard we might catch the humpback whale migration at the tip of the island. Ali had her hair in braids, which looked great, but was a mistake in that climate. Perched on the pillion, she started to feel her scalp burning.

I pulled over to help her put some sunscreen on. Within seconds, another motorcycle with three scumbags pulled up beside us, two of them drawing machetes. Really fucking big machetes with an edge that glinted in the sun, and you could clearly see your immediate future.

""Bag, bag. Give me your bag."

Ali had a small day sack on her back that fortunately only had sunscreen, a couple of bruised bananas from breakfast, a bottle of water and a cheap digital camera.

This is the bit where I'd love to say, *"the red mist descended and I dismantled them all with ninja precision for daring to menace my girlfriend"*. In reality, we handed over the bag while politely suggesting they take it and leave us alone.

They rifled through the bag, saw the bananas and the camera, and looked at us with the pure disappointment of a child unwrapping socks at Christmas. Then, in a fit of pique, they slashed both tyres before wobbling off on their own bike, machetes waving.

We couldn't ride two-up on flat tyres, so I pushed the bike about a mile back to the last village. A garage owner, built like a wardrobe, took pity on us. Using bare feet and no tools, he removed the tyre from the rim with his seemingly hydraulically powered hammer toes.

His ability to control them individually was a feat of evolution-defying anatomy.

He told us to pay him later, which in the Dominican Republic is either incredible generosity or a long-term investment in seeing how ridiculous British people can look on an overloaded motorbike.

Back at the resort, we showered, headed for the casino, and 'won' enough for a replacement camera. We laughed about what the thieves would find on the one they'd taken, mostly daft pictures of us pulling faces and my white, spotty arse. I hope they enjoyed the view.

Another flight back to the UK with Ali, knowing and not knowing where we were going at the same time.

A last night together near Gatwick and we said goodbye. Again.

I went back to Oxford, collected everything I owned that wasn't in a shipping box, and rode up to Stranraer in Scotland to get the ferry over to Larne in Northern Ireland.

Northern Ireland, Aldergrove: 5 Regt AAC, 655 Sqn, Apr 97 – Jul 99

Operation Banner

Chapter 80. The Jock Nag

I pulled up to the gate at RAF Aldergrove, manned by White Caps, the RAF Police, and presented my ID card after a tortuous ride north. I hadn't realised just how long and boring the A75 was. It ran west from Dumfries to the port of Stranraer, a popular route for trucks. Get stuck behind one or two and it was all over, even on a V-twin torque monster of a bike.

I was fully loaded. I had a military bergan packed to bursting, stuffed with all the holiday kit I'd tried to compress into it. I had a tank bag and a set of panniers; I'm surprised the thing didn't bend in the middle. I'd pulled a large white bin liner over the bergan. The forecast was wet, I didn't want everything soaked, and I needed to disguise the fact I was military.

The ferry was a respite, but it was obvious, despite my best camouflage efforts, that I was military. The short hair and English accent were more than enough evidence.

The RAF cop on the gate said I looked like I had R2D2 on my back, then gave me directions to the Sergeants' Mess in Alexander Barracks, a smaller compound within Aldergrove that housed 5 Regiment AAC.

May, the mess manager and unofficial "mother" of the mess, gave me a choice of rooms. She took all of us single lads under her wing, a wonderful human being.

We spent many a night talking shite with May on the other side of the bar, dispensing wisdom and relationship advice like she'd seen it all before.

We were lucky to have her.

She gave me a choice of rooms. I picked a large one, technically a two-person room, currently unoccupied. Possession being nine-tenths of the law, I moved in and set about turning it into a flat.

5 Regiment was unusual in its composition. It had only two flying Squadrons: 665 Squadron with Gazelles, and 655 (The Scottish

Horse) Squadron with modified Lynx. There was also 1 Flight AAC, a specialist unit operating the fixed-wing Britten-Norman BN-2 Islander.

1 Flight's primary task was photo reconnaissance under the direction of the Reconnaissance Intelligence Centre (Northern Ireland), an exceptionally important counter-insurgency role.

The Scottish Horse link began in the Second World War, when 655 supported operations at Anzio. As a result, 655 aircrew wore the Duke of Atholl's tartan, Murray of Atholl, behind the Squadron crest and marked the Battle of Anzio each year with a combined Officers' and Sergeants' Mess dinner. A spectacular orgy of Scottish whisky, Islay lobster and tomfoolery.

The Squadron also had its own bar, the infamous 'Jock Nag', wherever it was posted.

The change in tempo from anti-tank operations on the north German plains to operational counter-insurgency flying was all-consuming and exciting. I'd loved my time in Bosnia. Proper flying. Every sortie was real, not an exercise. With 655 I could expect this for as long as I was posted there. This was a permanent posting, not a six-month roulement.

Unlike Bosnia, where we knew the end date, the intensity here was constant. This was home now.

I reported to the sqn lines and was informed I was in 'B' flight.

Bessbrook Mill, South Armagh, was an infamous posting for all Lynx pilots in the British Army. Allegedly, and quite believably, Bessbrook Mill was the busiest heliport in Europe throughout the 1980s and 1990s, handling all three services around the clock.

The main bases in South Armagh were assigned codes: Bessbrook Mill was Red 850, Crossmaglen Red 650, Forkhill Red 552, and Newtownhamilton Red 350.

The Lynx tasking lines were fixed by role and callsign:

Lynx 1 – 4 Dedicated to 'C' flight. A specialist, hand-picked flight that were committed to Special Forces (SF) and other intelligence operations. C Flight had its own flight commander, Grant Better, a Warrant Officer who was unflappable, steady, and utterly reliable. A gentleman to boot.

Lynx 5 Forward-based at Bessbrook Mill, with overall command of tri-service aviation in the South Armagh AOR. Lynx 5 commanders were the most experienced commanders.

Lynx 6 Air Test/Maintenance flights.

Lynx 7 Forward based at Bessbrook Mill alongside Lynx 5, acting as its deputy.

Lynx 8-9 Province wide tasking as a pair, usually in support of specific operations.

Lynx 10 Dedicated for training.

Lynx 11 A specialist surveillance aircraft fitted with Project Chancellor, a Wescam camera system vastly superior to the older Heli-tele.

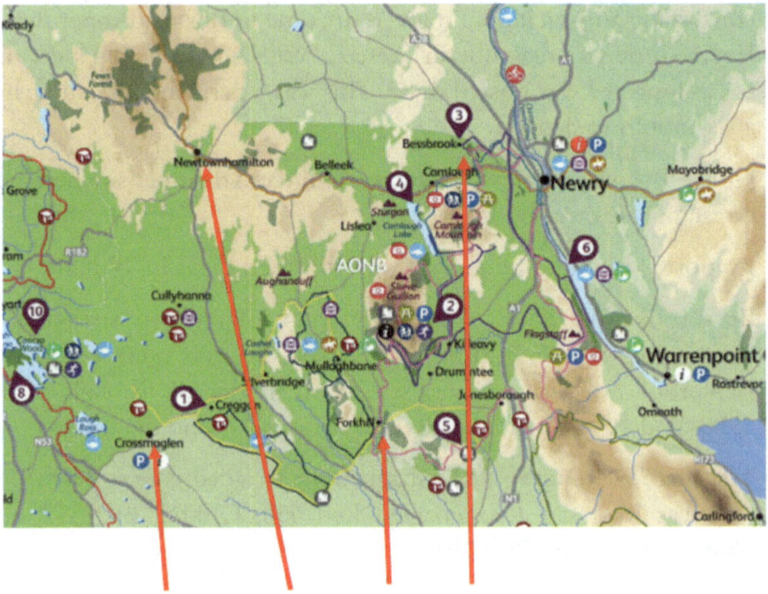

105 South Armagh Main Locations

Chapter 81. Flights, Friends and Fuckwits

The Squadron dynamics deserve a short explanation at this point.

Historically, a pilot wouldn't be posted straight to Northern Ireland without solid experience first, certainly not straight from the Army Pilots' Course. Some were, though. 'Hagar' Hatchard arrived straight from APC 342.

Following the first ceasefire, that unwritten rule began to erode. Inexperience on its own was usually not the problem. Everyone must learn somewhere, and most did so with humility and care. The issue was structural. The Squadron was split into 'A' Flight, the aircraft commanders, and 'B' Flight, the pilots, instead of the usual mix of experience within each flight.

I never forgot being handed a red grading on the pilots' course after a sortie I thought was *"easily a blue"*. Paul Connor brought me back to reality, brutally but without drama. Confidence was fine. Overconfidence would be punished.

The lesson stuck. South Armagh was not a place for learning that lesson for the first time. The real problem here was inexperience paired with confidence far in excess of ability. That combination didn't create learners. It created liabilities. And that was where the fuckwits resided.

South Armagh was the examiner now. There was no review time if you got it wrong.

As mentioned, C Flight was a part-time flight of crews hand-picked by Grant and the boss, at that time a Major from the Grenadier Guards. When on roster, they carried a pager on 60 minutes' notice to move.

If called, they'd muster in the Squadron lines, brief, then hop the aircraft over the boundary and land in "the Hangar", a sealed area used by the SAS, 14 Int and other spooks, some of them more civilian than soldier.

When I arrived, I had no clue about this bizarre structure. To compound it, the rostering was done by an inexperienced Sergeant fresh out of pilot training. A good man, but out of his depth.

655 did have serious pedigree. Unusually, it had a Captain as SQHI: Marty Gilbraith, a former Para and late-entry officer. His deputy,

Dai Netherbridge, was a schoolmaster in the cockpit, precise to the inch. High-tempo ops didn't suit him in the same way.

With so many inexperienced pilots and the tempo rising, 5 Regiment asked the other Regiments to send experience. So, I rapidly found myself joined by old friends. 9 Regiment sent Mac and Chas. 4 Regiment sent Dom and Dan Brodie. 3 Regiment sent Gary (Fucking) Smythe (GFS) and Jon Hadlow. 1 Regiment had sent me, and the Badger was on his way. There were others, but this cohort meant business. It was a shock to 655.

Our personal weapons were different here too: the Heckler & Koch HK53, a compact 5.56mm carbine we had to qualify on before duty, and the trusted 9mm pistol carried in our LCJ.

My first flight was fully operational, with Marty Gilbraith as the Lynx 7 crew. And my first time in South Armagh turned out to be memorable.

Chapter 82. The Hitman's Wife

The crews in Northern Ireland were far more accustomed to operating with a doorgunner. In NI the doorgunner was more than just a gunner: medic, loadmaster, and supplier of essential Sharps Extra Strong 'mission mints'. Their day started before ours, collecting GPMGs, spare barrels, ammunition, medic bags and everything else we needed. They were a fully-fledged, critical crew member.

With this Lynx 7 duty being as much training as operational flying, the doorgunner would be relied upon heavily.

On the way down to Bessbrook we varied our turn-around times to avoid pattern setting. On our allocated VHF frequency, we picked up two general aviation pilots chatting somewhere in the Republic:

"Seamus, where the fuck are ye'?"

"I'm right here, near the town" came the reply

"Which fookin' town would that be?"

"If I knew that, I would be fookin' lost would I now!"

I roared with laughter. Sgt Aidan 'Logic' Melly, the Lynx 5 commander, came up on frequency in his trademark dry tone:

"Unknown callsign on 123.4, this is a military frequency, please change to another common frequency now"

"Ah go fook yersel"

That did it for me. I had literal tears rolling down my cheeks. Marty didn't find it funny at all and thought I should be more controlled. He was right.

On arrival at the Mill, Marty walked me through the landing points, or 'spots'. Spot 1, closest to the Ops room, was reserved for Lynx 5. Spot 3 was for Lynx 7. Further down the tiered slope the RAF Puma and Royal Navy Sea King had their own areas, while the Gazelle used Spot 9, better known as 'The Well'. Knowing these procedures mattered enormously.

In 1992 there had been a mid-air collision between a Gazelle and a Puma during final approach. The Gazelle's main rotors struck the Puma's cockpit. Both aircraft fell to the ground, killing all four aboard the Puma and seriously injuring the Gazelle crew.

The accident was deemed to have been caused by inadequate use of the correct radio procedures and a breakdown in the passage of information between the RAF and the AAC.

Marty walked me through every procedure. R850 had two standard approach and departure lanes for deconfliction of arriving and departing aircraft. They were wind-dependent and known as the *Eastern Fence* and the notorious westerly *Tube.*

The Tube directed aircraft into the heliport over a housing estate that famously had the *"Snippers at Work"* warning sign.

We were never sure what the 'snippers' were working at! The Tube was flown extremely low and as fast as was safe to reduce exposure to machine-gun and sniper fire.

Multi-weapon shoots against aircraft arriving at a base were a real threat. In 1991, by sheer coincidence or impeccable timing, a French TV crew wandered into Crossmaglen square, cameras rolling, just as a Lynx came into land at the base. The IRA opened fire with heavy machine guns from their hidden positions, and the Lynx went down in a field nearby.

Miraculously, everyone survived. The footage later made the journalists' careers. Nothing like frontline action to spice up a travelogue.

Additionally, there was a third option to arrive at Red 850, and that was *'from the overhead'*. The procedure was simple in theory: fly directly over the heliport at height, and make the radio call

"Red 850, one Lynx from the overhead, Red 850."

Then it got noisy. The descending spiral was tight, steep, and unmistakable. The distinctive *whomp-whomp* of the BERP blade slap echoed off every wall in the vicinity, reverberating through the old mill buildings like thunder in a barrel.

To the uninitiated, it sounded like a catastrophic failure in progress, but to us, it was beautiful, a display of controlled aggression. Done well, and Chas Challis was the maestro, it was pure theatre. From height to hover in a matter of seconds, spiralling down like a falcon stooping on prey, it was both impressive and unnerving.

Chas was probably the best Lynx pilot I ever flew with, as giant as he was, he merged seamlessly with the machine; his giant bratty fingers swallowing the cyclic.

Night operations were conducted in the same manner.

On this first evening of operations for me, we set off as a pair with Lynx 5. At night we reduced to two aircraft for mutual support. By day we flew a minimum of three, usually with a Puma. We rarely saw the Sea King that was tasked by night; they would be off delivering underslung loads of supplies in, and waste out, as there was no road movement allowed.

We crossed the Newry Road, between Forkhill to the west and Dromintee, to the northeast. Dromintee was a small hamlet, but home to the infamous 'Three Steps Inn', the hostelry where Capt. Robert Nairac had been abducted from in the 70's.

As we did so, the doorgunner saw a car swerve, topple over and cartwheel into a ditch. On NVGs it was obvious, the headlights throwing out a kind of laser light show. Aidan immediately changed the plan. Instead of the routine patrol insertion we were on, he called the air ops controller at Bessbrook, call sign '*Buzzard*' and gave him a sitrep.

Our immediate actions were to secure the vehicle and assist the driver. The patrol we had on board was dropped next to the car and they established a protective cordon around the vehicle with the patrol first aider rendering assistance to the single occupant; a female driver.

Back in R850 the Airborne Reaction Force (ARF) was stood to and a doctor assigned from the Mill.

We made a couple of shuttle runs, replacing the foot patrol from the Welsh Guards with the ARF, and then dropped the patrol at their original assigned drop off location.

Or so we thought.

The international border here follows centuries-old parish lines, hedges, streams and arbitrary quirks. For the first and only time in my NI tour, we dropped a patrol on the wrong side of the border, about fifty metres over, separated by a stream. They called us straight back and ten minutes later they were in the right place.

Aidan was mortified and professionally embarrassed at the navigational error.

The debrief back at R850 was thorough.

That brief, fifty-metre misstep over the border should have been nothing more than a navigational hiccup, but *An Phoblacht*, the IRA's propaganda machine, worked it up into a tale of 'SAS death squads' running amok in Eire.

What had been a ten-minute correction became another opportunity to stir tension and feed suspicion. *The Irish Times* reported it with far less drama, noting it as an embarrassing navigational slip rather than an incursion, which is exactly what it was.

Two very different readings of the same non-event, one designed to agitate, the other to downplay, but in truth it was nothing more than poor map reading, dressed up as news.

Back at the crash site, the doctor had stabilised the driver, apparently the smell of alcohol was overpowering. She was taken to Daisy Hill Hospital in Newry by road ambulance, we picked the ARF up and brought them back to R850 once the cordon was collapsed.

A few days later, a slab of beer was delivered to the gate house entry to the Mill addressed to the pilots.

After being thoroughly checked out, a note was found attached to it. It was from a gentleman called Sean Gerard Hughes, known simply as SGH. SGH was a notorious IRA commander in the South Armagh brigade and a member of the IRA Army Council. The lady we had saved was his wife, Annette.

Years later, his name would be publicly linked to some of the most notorious attacks of the Troubles. At the time, all we knew was that we'd just pulled his wife out of a wreck.

SGH remains a free man, last I heard still running his scrapyard in Jonesborough.

106 Bessbrook Mill Red 850

Chapter 83. Walk Before You Can Fly

Now signed off for duties, I was paired with A Flight line commanders. Sensible enough, until the roster started throwing odd curveballs.

Apart from a checkout on the Chancellor aircraft (it had a couple of non-normal procedures that needed thought) and a couple of trips with Capt. Mark Ketle, the doorgunner flight commander, I was rostered with some nice enough blokes who simply weren't ready to command.

Mark, on the other hand, went on to become a General. No surprise. He had that calm authority people naturally followed. Despite the rank and the titles, he stayed the same decent bloke I'd first flown with. Just Mark: steady, sharp, quietly brilliant. A great pilot too.

One early incident stood out when I was rostered as Lynx 9 pilot, the No. 2 aircraft in the province response pair. There are a few quarries dotted around NI, naturally they need the terrorist's best friend, explosives, to do their job. In our world, that meant extra attention.

Our task was routine: pick up an underslung load of explosives and fly it to the quarry. A police patrol would be there to receive the stuff and make sure it went to the right people.

The aircraft commander was a Sergeant I hadn't flown with before. He took the handling for the whole sortie. Unusual, but I let it go.

On the way back to Aldergrove we descended to low level and punched straight through the threat band. The threat band for us was 300 to 3,000 feet. That assessment was based on tracer ignition and burn-out. If you can't see fall of shot, it's near impossible to hit an aircraft in flight.

The threat wasn't theoretical. The PIRA South Armagh Brigade had recently shown they could coordinate heavy weapons and hit aircraft hard on departure from Crossmaglen. When they came for you, they brought proper anti-aircraft kit, not pea shooters, and they could vanish before anyone could close.

The recent 'Battle of Newry Road' was still sitting in everyone's head.

The technique was simple in principle but unforgiving in practice.

You bled speed, pushed the nose over and dropped hard through the band, then levelled at low level and accelerated away. Done properly it was controlled, smooth, and quick. Done badly it was a mess of over-speed, over-torque and chasing the aircraft.

The aim was always the same: minimise time between 300 and 3,000 feet.

Back on the underslung sortie, this bloke's control was all over the place. I found myself on the controls overriding him to stop us over speeding the rotor head. He stared at me.

"You're not a newbie are you?" he said quietly.

"Nope. Done this before," I said pointedly.

He was '*larging it*', showing off to what he thought was someone that he could easily impress. The cavalier attitude, and the lack of skill behind it, shocked me. I spoke to some of the other guys in the mess that night. It wasn't just me.

He didn't fly again as a commander for a while.

WO2 Grant Better was the opposite of those fuckwits. Tall, balding, polite, and lethal in the air. He didn't raise his voice. He didn't have to. A rare breed. Grant was the C Flight commander and knew exactly what, and who, he wanted in his flight. Flying skills mattered, but just as important was the ability to stay two or three steps ahead of the SF troop commander in the back of Lynx 1.

Anticipating what was coming and making it happen before it was asked was an art form.

No one went straight into 'C' flight; it was a rite of passage that Grant controlled. I did a few Lynx 5 duties with him, then flew my Northern Ireland Combat Ready (CR) check with WO2 Rory Burnett from Regimental Standards. I'd just passed 1,200 flying hours and was signed off as a NI commander.

Some of the fuckwits previously rostered as commanders knew the writing was on the wall. With the characters inbound over the next month or two, a sea change was on the way. Some recognised the errors they were making. One, maybe two, notable exceptions didn't. A posting came their way sharpish.

Some of the inbound commanders' reputations preceded them.

Chas took over rostering, rearranged the flight compositions, and got a grip in about three minutes. By the end of his first morning, the place ran like a watch. The pretenders knew their time was up.

A more relaxed, but far more professional, atmosphere settled across the Squadron.

Flying was still the one place where things made sense. There were rules, limits, and consequences, and if you stayed within them the aircraft looked after you. Skill mattered. Preparation mattered. Ego didn't.

Checks and balances. Torques matched. Limits respected.

I was finding that relationships were the opposite. They were the one place where I had to slow down when every instinct told me to push on. I couldn't fly them harder or lower to make them work. The same decisiveness that kept you alive in the air could wreck things on the ground.

I trusted myself implicitly in a cockpit. Outside it, with Ali, I was learning that knowing when *not* to act mattered just as much.

Between the intense tasking, I'd been over to Tidworth a couple of times for weekends with her. She was living in Tedworth House, the Officers Mess for the garrison. Her room was on the ground floor, which helped.

There was a fire door that made discretion easy.

Her room was huge but had no en-suite, which made it feel oddly primitive. The other women in her corridor clocked me soon enough. They didn't make a thing of it and were friendly with me.

We talked a lot about everything with ease. Ali had been invited to a wedding party on a river boat on the Norfolk Broads. She felt safe enough to invite me, despite it being all officers. I took it as a positive sign. I even floated the idea of leaving the Army to protect what we had. Ali wouldn't hear of it.

Chapter 84. On Ilkla Moor Baht 'at

By the end of July 1997, most of the incoming experience in Northern Ireland had arrived. Everyone went through the full combat ready syllabus and was on the roster. Dean Kavanagh and I had been provisionally selected for C Flight by Grant and sent to the main HQ in Lisburn to *'sign on'*. This was mandatory for anyone working directly with the hangar and amounted to an enhanced version of the Official Secrets Act.

The 12th of August 1997 was my grandmother's 80th birthday. She was still living in Ilkley. I really wanted to see her, so I hatched a cunning plan.

I booked the training aircraft, Lynx 10, and asked Dom if he was up for a long nav to Yorkshire. He was, so I made the bookings. Remembering the Cow & Calf had an HLS, I booked us in to it, along with every Low Flying Area on the route, so we could justify it as proper training.

The groundcrew looked baffled watching Dom and I strap into an operational Lynx with a bouquet of flowers. I strapped the large bouquet into the rear seat like it was royalty.

We went the long way to Ilkley, via Newcastle, where we dropped in for fuel. We approached Wharfedale from the north, skirting the town to the west to avoid overflying anywhere sensitive and generating any low flying complaints.

I landed in the same spot as I had with Chris Keane and Hagar all those years ago as a student pilot in a Gazelle. As we shut the aircraft down, my mother, her friends, and my grandmother, completely oblivious to my arrival, wandered over to say hello. I dropped my helmets' dark visor to hide my face, jumped out, opened the cabin doors, and retrieved the giant bouquet.

Dom and I walked towards the small group. I held the flowers out to my nan and took off my helmet for the great reveal. Hugs, kisses and tears all round. Worth every minute of planning.

After a short pub lunch and a coke, we all walked back to the aircraft and let my mother and Nan climb into the cockpit for some happy snaps. While they poked around, clearly impressed, Dom and I briefed the departure.

I would fly a 'rearwards' departure, transition to forward flight and execute a tight circuit before climbing away to the north. The Lynx could bite the unwary in that manoeuvre. If not properly controlled it would overcorrect and drop the nose further than intended. It did, however, look and sound impressive.

I helped my ol' Nan out of the cockpit and Dom helped my mother. Being the squaddie he was, he helped her a little more than was strictly necessary…

He caught my eye and gave me a wink.

Shameless.

We strapped in, spun the engines up and made the relevant radio calls to Leeds/Bradford ATC.

After the pre-take-off checklist, *"Happy?"* I asked. *"Happy"*

I lifted into a low hover, eased the cyclic rearwards and applied collective. At about 50ft I pushed the nose over to enter forward flight. We rapidly increased speed, I made the '*flyaway*' call and descended to about 10ft, pulling a tight turn, looking up through the window in the roof to maintain depth perception.

'*Flyaway*' was a call the commander would make on take-off, meaning that if we lost an engine, the pilot was to continue flight a single engine; the equivalent of the fixed wing 'V1' call.

The blade slap from the BERP blades was loud and rhythmic. I tightened the turn to roll out pointing directly at my family, all of them with both arms waving wildly over their heads.

I rolled out and pushed the Lynx for a bit more speed. As we flashed overhead, my 'family' hit the ground.

Impressive reaction, I thought.

Then Dom's voice came over calm as ever: *"Isn't that your family over there?"*

In the steep turn, I'd mixed up a group of tourists with my own relatives.

Ah Bugger!

Somewhere in Yorkshire there's still a family telling the story of the day an Army Lynx tried to mow them down.

They must've enjoyed it. Otherwise, I'd have had the usual chit waiting at Aldergrove: *"Termini, call Rudloe Manor."* The RAF's low-flying complaints unit had me on speed-dial.

Low flying complaints were a constant, no matter how carefully we avoided livestock, especially horses, or steered clear of buildings and villages. Annoying someone was inevitable. I covered myself by booking into every Low Flying Area and holding authorisation down to ground level, just in case.

I was delighted to have been able to do that for my Nan. I know she enjoyed seeing me and Dom. On the flight back we agreed that given her service and her own sacrifice to the country it was the least we could do. I just wish my Grandad could have been there.

Who knows, just maybe, he was.

Chapter 85. Know Your Place

I'd managed to get the long weekend pass approved for the trip to Norfolk for the wedding. I rode over to Tidworth, left my bike in the garrison car park, and Ali picked me up. We drove to a hotel in Wroxham and met a couple of Ali's Sandhurst friends. Decent blokes, if a bit plummy.

The next morning, we all gathered at the marina and boarded the party boat. We were greeted with welcome drinks on the lower deck. Most of the guests had drifted upstairs to the tiered upper deck with its ornate wrought iron balustrade.

One of the lads from the night before shot me a look, the kind of look that said, 'you need to see this'. I had no idea what he meant. He flashed his eyebrows again, this time clearly meaning 'look up'.

I glanced up slowly, moving only my eyes, trying not to draw attention. There, directly above us, was a young woman in a flowing summer dress who had, it seemed, decided underwear was optional.

So, there I stood, looking up at 'it', thinking she was very much in full landing configuration, gear down, flaps fully extended.

I gave Ali a gentle nudge and the same *"look up"* signal. The three of us tried not to laugh out loud, but 'Top Deck Girl' instantly became the main talking point for everyone on the lower deck.

Discreet we were not.

The day was going well, full of laughter, until one of the women, from the Military Police, realised I wasn't an officer. It could have been the accent or just the way I carried myself. For reasons best known to her, she decided it was inappropriate for me to be there.

The Army has rules about relationships between ranks in a direct chain of command, but we weren't in the same command, and Ali was not in a command position over me, nor would she ever be. Logic played no part in the Military Police' thought process.

She got louder, then turned on Ali. That crossed the line. I waded in with a few choice words, not my most diplomatic moment, which only succeeded in pissing Ali off. She didn't need rescuing and didn't like me trying.

By the time we moored back at Wroxham, the atmosphere had gone flat. We salvaged the evening in a local pub, minus the Military Policewoman and her coven.

Given we'd all had a few, I naturally assumed that was the end of it.

The awkwardness from the day had thinned a little, replaced by a pull between us that neither of us wanted to ignore. She stood by the bed in her light summer dress, skin still warm from the late sun, her eyes locked on mine in that way that made my world shrink to just us.

When I reached for her, she came willingly, hands in my hair, our mouths meeting hard. We'd been like that since New York, no hesitance, no second guessing, and we knew exactly how to make each other feel wanted. Every touch was familiar.

Later, lying tangled together in the quiet. There was still affection in the way she looked at me, but also something distant, as if part of her was already pulling away.

I rang her from the mess late that night to say I'd got home in one piece. She cut straight to it. The MP had threatened to report her to her Brigadier for misconduct. It was nonsense, but apparently it was her "duty."

I told Ali again I'd leave the Army before I'd see her career damaged. She still wouldn't hear of it.

She ended it there and then.

I couldn't believe it.

When the line went dead, just like that, the future I'd been quietly building in my head was gone.

This time, I didn't have the lonely silence of Gütersloh that Christmas after Bosnia.

I had May. Dom. Dean. The rest of the lads in the mess.

This was my place.

Happy to rip the piss one moment, ready to listen the next. That's an Army mate. Real, genuine, horrible and beautiful, sarcastic and straight-talking. Misunderstood by civvies, only understood by squaddies.

Chapter 86. Options for Change

Flying didn't fix anything, but it gave me somewhere to put the pain. She was gone so suddenly there wasn't even time to argue with it.

The only thing left that still made sense was turning up, flying, and not stopping long enough to think straight. The only cure I knew was to keep moving, and the timing was on my side; Grant had just got clearance from HQ for Dean and me to start training with 'C' Flight.

I had flown the JCOs around Bosnia and got to know a couple of them in Tombstone over a pivo and trout, but this was different, aviation was an integral part of the SF team.

Grant set up a proper training session. Now that we were '*signed on*', he could tell us exactly how 'C' Flight worked; in detail, and with that mix of precision and dry wit that made you pay attention.

'C' flight consisted of 4 Lynx and a Puma, usually supported by at least 1 Gazelle.

Lynx 1 was the command & control aircraft. The SF troop commander would have operational control, the Lynx 1 aircraft commander had aviation control. The two had to be in absolute sync.

Lynxes 2 and 3 were the 'manoeuvre pair', they would execute the orders from Lynx 1, normally carrying the assault force.

Lynx 4 was interesting, it was the support aircraft carrying support weapons, rebroadcast equipment, electronic warfare kit, medics and even body bags. That told you everything you needed to know.

Once rostered as 'C' flight, we would also carry a second pager and on 1 hours' notice to be in the hangar, ready to be briefed.

Grant laid out four standard options, all built around the same ugly truth: if a terrorist vehicle crossed the border, it was effectively gone.

"Option one. Soft stop.

Lynx 2 blocks. Lynx 3 inserts the arrest team.

Option two. Hard stop.

Lynx 2 gets ahead and inserts a stinger team. Lynx 3 follows with the backstop and arrest team.

Option three. Hard stop, failure to stop.

If the driver ignores the stinger or breaks through, Lynx 2 switches from stopping to forcing, escalating to disable the vehicle. Lynx 3 comes in behind to make the arrest.

Option four. Hard kill.

The manoeuvre pair works with Lynx 4, escalating from disabling fire and CR through to lethal force, if authorised."

Using models, Grant walked us through each option in detail, asking questions as he went, pulling us into it, making sure we understood not just the plan, but the thinking behind it.

He closed, as ever, with a simple, *"Any questions?"*

I was good, but Dean piped up in his Nottingham accent.

"Right… these pagers," he said. *"I don't mind being on a pager, but if I carry it, I get the flying, right?"*

Grant looked slightly puzzled. *"Not sure what you mean, mate."*

Dean tried again. *"Right, is there a medals team? In other words, I get the ball ache of being on the pager, it goes off, and someone else swoops in to take the glory."*

It was a good question, one I hadn't thought of, but Grant assured us that wasn't the case.

The next time we were both off the normal roster, we'd start some airborne training up at Ballykelly, an old training area where we could train and fly hard without prying eyes.

We started slowly, practising all four options against a Subaru Impreza driven by the ops team, who took great delight in trying to throw us off.

After a full day of hard flying, we were authorised pilot on all aircraft and commander on Lynx 3 and 4, day and night.

A few sorties later, command on Lynx 2 followed.

All this culminated in a final exercise at Otterburn in Northumberland. This included a full troop of SAS, some from Aldergrove and some from Hereford. Directing Staff from Hereford provided the operational scenarios that both the SF commanders and our aviation component would have to deal with.

I was crewed as Lynx 2 pilot with Darren 'Daz' Porter, an experienced Sgt who'd been with 'C' flight for some time and definitely not one of the fuckwits.

The deployment included WO2 Mel Singer, an old school aircrewman who had become a pilot. I'd known Mel when he was in Lynx Flight 651 whilst I'd been on Gazelles, in fact it was Mel that had picked up me and Bob Scarlet from the Himmelstur after the snap escape & evasion. Mel was balding with a high forehead and glasses.

We flew all the options, and every permutation of them, by day and by night. By night we used various laser pointers to illuminate targets and areas of interest. Under NVGs they stood out like light sabres, incredibly effective, assuming the other side didn't have NVGs as well.

It was a method of target marking that proved useful in normal operations too, a bit like F1 technology filtering down into road cars.

We'd put it to good use soon enough.

On the second day we moved on to perfecting option four. A scrap car was strapped to the deck of a flatbed truck. With mutual support from the manoeuvre pair, Lynx 4 would fly directly above the target, matching its speed. The pilot applied rudder to angle the cabin door over the roof, denying anyone inside the chance to lean out and shoot back, although there was nothing to stop them firing straight up through it.

A trooper carried a pump-action breaching shotgun and a gas round. The breacher punched an entry hole, the follow-up round, usually CR, far harsher than CS, was designed to incapacitate anyone inside. CR was brutal. Burning eyes and skin, it put people out of the fight quickly.

The pilot wore an S10 respirator, just in case any gas made it into the aircraft. Flying in one was bad enough. At night it was worse. NVGs never quite lined up with the eyepieces, so it was like trying to land through a pair of jam jars.

The aircraft settled into position. The trooper, clipped on and leaning out, fired both rounds into the roof of the scrap car.

Then it was back for a debrief, tea, and, no doubt, a few more medals.

Afterwards, we wandered down to the cookhouse for some lunch. On the way, we passed the scrap car, now sporting an involuntary sunroof. Mel decided to take a closer look. I've never understood why.

He came jogging back, wiping his brow. It was a hot day. He must have picked up a molecule of CR on his hands, because it went straight into his eyes.

His world ended.

He screamed like his eyes had been stabbed with a red-hot poker and went down hard, pawing at his face, trying to rip it off like a bad James Bond disguise.

We had tears running down our faces. Ours from laughter. His from his body desperately trying to flush the gas out.

Later, Daz found a lemon in the fruit bowl, drew Mel's high forehead and glasses on it, and left it on his pillow.

When Mel finally managed to focus, he stared at it and stuttered, *"Is… is that supposed to be me?"*

That was us gone again.

"If the cap fits…" Grant added, sealing Mel's humiliation for the day.

That was the last thing Mel needed to hear from Grant, the most respected man in NATO. Even as a joke, it hit harder than it should have.

Poor Mel.

The final exercise on the last night was the hardest yet. Daz and I were flying Lynx 2, running option three, with a landing point to drop the stop team.

It was darker than a witch's tit, almost no ambient light. The NVGs were working hard, and so were we.

The landing point sat on a road along a ridgeline. We came in low and fast. At the last moment I bunted up to land.

As I started the bunt, a roller of wind came over the ridge and rocked us hard.

Without the go-ahead from Daz, the SF commander in the back, thinking we'd already landed, went for it and jumped.

We were still about 50 feet up.

He told us afterwards he'd moved when everything suddenly went quiet. His headset was still plugged in and snapped his head back as he went out.

Then nothing. Just black.

He said it felt like falling into a void. He hit the ground badly on his side, completely disorientated, and rolled onto his back, winded and in pain.

Above him, just a few metres overhead, was the faint glow of the cockpit and the Brightstar IR floodlight.

He rolled clear just in time.

Eight tonnes of Lynx came down where he'd just been.

He carried on with the exercise.

With a badly damaged ankle… and an even more damaged ego.

The last morning, we had the chance to use up any live ammunition that was left over, and the SAS had access to unusual (at the time) stuff. Mk 19 Grenade launchers, RPG 7's, Barrett 12.5mm sniper rifles, MP5-K's and of course the breaching shotguns.

The DS set up a mini range for us, just really to '*have a go*'. I'd never even seen an RPG before so was first in line for that bad boy. The DS explained how to handle it, load it and fire it, but crucially, not how to aim the thing.

I picked it up and looked through the sight, it had a very basic graticule pattern, like a Chieftain TLS. With a graticule, you would put the aiming mark on the target and let rip. Or so I thought.

I put the aiming mark on the target and fired. The thing flew out and landed about 25m in front of us before bouncing off into the distance, luckily, we were firing practice rounds.

Ah Bugger!

It turns out that if you're firing an RPG-7, you need to estimate the range first, and place that range line on your intended victim before firing.

The Mk 19 was outrageous fun; an automatic 40mm grenade launcher should be on everyone's Christmas list.

The final day was debriefing, reviewing what went well and what could be improved. The consensus: jumping out of moving aircraft was best avoided. Noted for next time; it might be for real.

The flight was duly signed off as ready by the DS and Grant.

We departed the training area and went to Carlisle airport, a small airport that we would drop into for fuel and bacon sandwiches whenever possible; the bacon butties were unbelievable. Our final lift was to be a formation engine start, a formation rotor start followed by a formation take off and departure.

Grant made the calls.

"Engines…….. engines go, call in when ready for rotors" a minute or two later:

Two's in, three's in, four's in"

"Rotors….. Rotors Go"

Dave Thompson, a Cpl from the Parachute Regiment was a micro-second late on releasing his rotor brake in Lynx 3 and fucked the whole thing up.

"Thompson, you're an idiot" came from Daz, *"Fool!"* from Grant and that was it, Dave never lived down his inability to follow simple instructions!

On return to regular duties a couple of weeks later, I was Lynx 5, Paul was Lynx 7; we were darting around South Armagh smashing the Buzzard programme. By this stage, me and Paul flying together needed minimal radio chatter, he knew what I was thinking, and vice versa. I could literally point my aircraft in a specific direction, and he'd know what to do next.

This mental connection only ever grew stronger.

We were at G40, a surveillance tower perched high on Slieve Gullion, the highest point in South Armagh and a stone's throw from the border. These towers, along with the 'Romeo' towers, were a chain of fortified surveillance posts monitoring IRA activity and protecting troops.

Landing at any of the towers or bases was fraught with danger, they were choke points.

In March 1994, the IRA shot down a Lynx as it was landing at Crossmaglen. Using a makeshift mortar fashioned from a gas cylinder; a so-called "barrack buster" fired from a tractor hidden behind hay bales only a short distance from the perimeter.

The gas cylinder struck the tail boom, forcing the crew into a desperate crash landing inside the base. Miraculously no one was killed, though several soldiers and a police officer were badly injured.

G10 overlooked Crossmaglen, with G20 and G30 stretching northeast toward Newtownhamilton. Towering over the countryside, these steel and concrete structures housed cameras, sensors, and troops, providing constant observation. Hated by locals but vital for helicopter support and troop safety.

I descended well clear of the G40 HLS, flew nap-of-the-earth for the last couple of miles, then came in fast up the slope, masking my approach.

In mutual support was Paul in Lynx 7.

"5, this is 7. Have you dropped something out of your cab?"

I checked with guns. We hadn't, the door wasn't even open.

"Negative."

"Roger. We have two splash marks just under your approach."

"Splash" meant they'd seen the impact of at least two large calibre rounds close to us.

I called in the immediate contact report.

"Buzzard, Lynx 5. Contact. Wait out."

Paul immediately manoeuvred Lynx 7 between me and the likely firing point to provide cover while I overshot the landing point.

Both aircraft went guns made ready.

We turned back to investigate. Paul took the high slot, I took the low, six pairs of eyes scanning for movement. If any further shots came, they'd be answered with a hail of 1-in-1 ball and tracer from both aircraft.

Nothing.

No further rounds, no identifiable firing point.

I cancelled the contact report, and we debriefed Buzzard and Zero-Charlie Ronan on return to R850. I hadn't seen the fall of shot. Paul had, and as an infantry heavy weapons specialist, he understood exactly what we'd just been on the receiving end of.

108 Lynx at G40

109 Forkhill

110 Lynx at a 'Romeo' Tower

111 A busy R850, note the fast rope frames and ALQ 144

Around South Armagh

Chapter 87. Clerks, Magazines and Hard Landings

Mel Singer wore two hats in 655: line pilot and sergeant major. His office sat next to the boss's, separated by a small sliding window. He had a single Squadron clerk for admin.

Pte Sally Davies was attached to the AAC from the Adjutant General's Corps (AGC). Before 1992, clerks belonged to the units they supported, but Options for Change created specialist clerks posted where needed.

Unusually, Sally was a single mum living in a married quarter, and her mind was often elsewhere, something that regularly tested Mel's patience.

She was shorter than average, dark-eyed, olive skinned and confident. To put it bluntly, she turned heads. Her uniform was immaculately fitted, she clearly knew a magician of a tailor. The queue of young, testosterone-fuelled Air Troopers with spurious admin queries was rarely fewer than three deep.

One morning Mel was in a flap, running around looking for Davies while waving a magazine. It was a copy of *Escort*, one of the harder "soft-core" 'gentleman's publications' in the pre-internet era.

An Air Trooper had been flashing it about in the crew room.

There, in the reader's wives' section, was our very own Pte Davies. No attempt to hide her identity, a couple of the more adventurous shots had her wearing her Cambridge Blue AAC beret perched on the back of her head.

That, oddly enough, was Mel's main gripe. Not that there was a square inch of her that hadn't been on display, but that she'd done it wearing her beret

When Davies eventually arrived at work, she walked straight into the shit storm. Someone had plastered photocopies of her "best bits" around the admin office, clearly expecting her to fold. She didn't even blink. She tore one from the wall, looked it over with a smirk, and pinned it back straighter.

Then, in front of the lot of us, she said it was her; she'd done it of her own free will, and she was proud of it. If anyone, Mel included, in fact especially Mel, had a problem, they knew exactly where they could go and what to do when they got there.

By the time she'd finished, the laughter was on her side, the photocopies were in the bin, and she was walking away with her head held high, untouchable.

Mel toured the camp, buying up every copy he could find and binning them, but the damage was done. Davies' legend was sealed. The entire Squadron had seen more of her than regulations, good taste, or Mel could cope with. None of it landed the way people expected.

Mel's time in 655 was coming to an end anyway, not long after his clerk had bared herself to the world. Grant got his head together with Dick Kalinski (of Aigen fame), and they hatched a plan to give him a send-off befitting an SSM… one he, or anyone else, wouldn't forget.

Grant spoke to his contacts in the hangar; Dick spoke to his in the freefall parachuting world.

At the appointed time, Mel was in his office. We knew what was coming.

He didn't. Not a clue.

Two of Grants contacts, complete with balaclavas were 100% in character and smashed their way half through, half past the office door. Mel was violently thrown to the ground in a hail of invective, phlegm and hard Irish accents.

His hands were zip-tied, a heavy hood pulled over his head, and within seconds he was being dragged by the elbows towards an unmarked car waiting outside, boot already open

We were sworn to silence, I almost felt a twinge of sympathy for him watching this brutal piece of fun happen.

He was in some clip even before he was thrown like a rag doll into the boot, the lid slammed shut.

We could hear him whimpering; *"lads, lads, what's going on?" "I think you've got the wrong bloke"*

The car made smoke and peeled out.

Apparently the 'terrorists' were talking loudly about what fate awaited the *'wee brit bastard'*.

About an hour later, the car arrived back in camp, having done a quick tour of Antrim. They skidded to a stop outside the Royal Navy hangar where Dick Klink was waiting, along with a friendly Sea King crew.

Silently, Mel was dragged out of the boot and unbeknownst to him, was hooked up to a tandem parachute harness that Klink was already set up in.

They were bundled into the Sea King and with clearance from Aldergrove tower, climbed to about 7000ft overhead the field.

Dick slid the door open and simply flopped out of the aircraft. The rest of 655 was waiting for the landing. After some time in freefall, Dick leaned forward and ripped the hood off Mel's head. We could hear the screaming from that point all the way to the ground.

As Dick flared his rig, he expertly lifted his own feet; they landed. Mel was face down in the mud. A short gust of wind smeared the poor bloke across fifteen feet of muddy grass.

Dick picked his passenger up, scared, bloodied and muddied; but clearly the most relieved man in the province to still be alive.

Mel deserved the cheers he received; it was an epic stunt he had just been through. The pair of them were liberally sprayed with champagne.

As for the 'actors', well they were never formally identified, no charges were pressed and the whole stunt went down in 655 diaries as another day on the lines.

The next day, I passed Mel's office. Davies was leaning on the doorframe, arms folded, *"Nice landing, boss,"* she said. Then, tilting her head, she added, *"Pity you weren't wearing your beret, would've made the centrefold."*

Chapter 88. The Festive Season

Approaching Christmas 1997, I had no idea what I was going to do. The rosters came out and unbelievably, I was off duty for both Christmas and New Year, but flying in between, which meant I couldn't really go anywhere or do anything.

I was missing the children, always wondering what they were up to and how they would spend their holidays. Matty and Alex would be going home from boarding school, but I had no real sense of what their life in Germany looked like anymore.

Some nights I'd go round to Paul's house just to sit, watch TV, talk nonsense and drink beer. Once, we decided that one day we'd buy a remote island, with no corners so nothing could sneak up on us, and just be ourselves, doing daft things and upsetting no one except each other.

It was a good plan.

Just before Christmas, 'C' Flight was unexpectedly paged into the hangar. We were to deploy to the Pontrilas Army Training Area (PATA) for an exercise simulating a full-on crisis.

We'd provide aviation assets to assist UKSF in various methods of vehicle stopping.

I was crewed as pilot in Lynx 3, with Rob O'Brian as the commander. Like Daz, Rob was a Sergeant pilot, but unlike Daz he was quiet, deliberate, and Grant's de facto 2i/c. The right man for that job.

The UKSF had several aviation options: their own AAC Flight operating the Agusta A109A helicopters (two of them spoils of the Falklands war, the other two purchased for their adaptability), and 657 Sqn AAC, a full Squadron dedicated to SF support, based at Dishforth, but slated to move to RAF Odiham to be closer to their RAF counterparts.

'C' Flight was also an SF asset, though only for operations in Northern Ireland.

It was an eye-opening few days, and we performed well. It finished with a tour of the infamous "killing house", where SF rehearse close-quarter combat.

This was the place where Princess Diana had famously had her hair set on fire during a demonstration. Prince Charles later signed a letter, now framed on the wall at Stirling Lines:

"Should this demonstration go wrong I, the undersigned Prince of Wales, will not commit B Squadron 22 Special Air Service Regiment to the Tower of London. Charles."

On my last duty at R850 in 1997, coincidentally flying Lynx 7 with the new Squadron commander, I passed the milestone of 1,400 flying hours.

In the end, I had Christmas dinner with Paul and Cath, then made an excuse and left. I wanted them to enjoy their family Christmas, plus I needed space to be myself, without pretending to be merry.

The mess was quiet over the holiday period, only a handful of us around. Living below me was Tina, the Postal & Courier Sergeant with 5 Regiment.

We'd spent plenty of evenings at the bar, swapping stories and laughing. She carried herself with quiet humour, but there was a heaviness behind her eyes I didn't understand at first. She wore her hair long to hide multiple physical scars.

The rest of her scars couldn't be covered.

The year before, she'd been badly injured in a double car bomb attack at the HQ in Lisburn. The first device exploded near a travel arrangement centre, injuring many. As the wounded were moved to the military medical centre, the second bomb went off, hitting both the original casualties and the medics treating them.

The blast flattened the building.

Twelve minutes had passed between the explosions, just enough time for rescuers to be right where the second device would do the most damage. Thirty-one people were injured, four seriously, and one soldier was killed.

Hearing her describe it in a calm, matter-of-fact way, with tears in her eyes, was sobering. I could picture the chaos, the noise, the smell, and her, somewhere in the middle of it, surviving. It made the small gripes in my own life seem trivial.

One person was arrested. No one was ever convicted.

Tina had made it clear she was interested. I liked her, she was attractive, and good company, but I didn't see a relationship on the cards. I'd taken two hits in a year.

I couldn't take another.

Still, in moments of mutual weakness, we slipped into a friends-with-benefits arrangement. It wasn't love, but in those quiet, lonely weeks it was warmth, and neither of us pretended it was more than that.

Chapter 89. Schools, Teachers and Vasco de Gama

1998 started with everyone talking about peace, but on the ground nothing much had changed. We'd reduced mutual support from three aircraft to two, a highly visible compromise, but we still flew as if someone out there was waiting to take a shot.

Because they were.

The politicians, and terrorists dressed as politicians, were edging towards a deal.

But in South Armagh, the threat didn't pause for photo calls.

The Belfast Agreement, or Good Friday Agreement (GFA) was being discussed and for a while the province seemed to breathe out. But the hardliners didn't sign up to anything. Their message was simple enough; they were still in the game.

Before the year was out, they'd prove it in the worst possible ways.

It would be a year of contrasts, hope on paper, risk in the air, romance at home, and 655 were right in the middle of it.

The Apache was coming; I knew that was where my future lay and it was an exciting future at that.

I planned out my path to the rear seat (commander) of an AH, and I saw myself doing what I could to be amongst the first crews to transition. I booked onto a range of courses at the Air Warfare Centre (AWC) at RAF Cranwell to arm myself with skills and knowledge that would be required by the AH force.

The first of these courses, the Joint Air Weapon Systems Course (JAWS) was in mid-January. This was the first course of its kind to be delivered; I was an Army Sgt amongst fighter pilots, bomber crews and intelligence experts.

To back up the JAWS course, I was back in Cranwell the very next month to complete the Air Electronic Warfare Course, which qualified me as an Electronic Warfare Officer (EWO).

Back in NI, I had been invited to a wedding, it was a friend of a friend and although I suspected some matchmaking was going on, it would be a welcome break.

I was correct.

There were only two single people at the wedding, me and a tall, very good-looking air hostess who flew for BMI British Midland. We spent the evening comparing notes on flying, music and anything else that would keep us away from the rest of the wedding gaggle. Annie and I exchanged numbers, nothing more than '*let's keep in touch*'.

We were flying more. Duties were stacking up, with less time for courses. I was lucky to have got mine in when I did.

The Army regularly sent teams to schools and careers events, partly for recruiting, partly to stay visible. A tour had been planned for the Northeast of England and South Yorkshire that would include the school Paul had gone to. A Gazelle crew was also assigned from 665. I should say that we always referred to Gazelles as 'Floppies' because of the way their main rotor blades drooped like they'd given up.

We immediately volunteered for this onerous duty, purely out of self-sacrifice of course. Once approved, we went into overdrive, planning and procuring items for the presentations we would have to give.

The groundcrew element had a range of '*show and tell*' items, including dummy weapons, missiles and a TOW simulator. They set off a day or so before us in a truck.

The plan was simple; the best ones always are. We would base ourselves in Dishforth with 9 Regiment and operate outwards from there.

The first school on the schedule was near Shotton in County Durham. We landed in the playing fields in formation with the Floppy. The ground crew had already set up the stands, cleared the landing areas and ensured no kids ran out Bosnia style.

The presentations were delivered to individual classes, starting with 13-year-olds, and working up to those just about to leave school at 16 or 17.

The Floppy was crewed by Ned O'Leary, a SSgt in the Royal Signals that bore an uncanny resemblance to Lou Ferrigno of '*The Hulk*' fame. Ned had stood up and was giving his last brief of the day to the older class. At the end he gave the usual *"any questions"*. A girl who must have been 15 or 16 had her hand straight in the air, stretching for the sky.

"Yes, what's your question" asked Ned.

"See yer git big boots you've got on mister?"

Ned glanced down to his boots, puzzled as the where this was going. *"Yeah, they're flying boots"* he answered a tad bewildered.

Quick as a flash, and obviously prepared, this kid shot back

"Does that mean you've got a git big cock?"

Paul and I made a very hasty departure, but I know for a fact us laughing drowned out the rest of the questions.

Although we had powers of self-authorisation, we had received pre-clearance from Regimental standards to take some of the kids on a quick flight, a 5-minute circuit of their village. We had a little route ready to go; lift from the school, down a local nature area to the coast, up the beach a little way before looping back to the school.

As we oversaw the loading of our passengers, I couldn't help but notice who they were. Paul's dad, his mother, his sister and his brother. To be fair, his brother Adam, was at the school and went on to join the AAC becoming a highly accomplished Apache pilot and later instructor that flew several tours of Afghanistan. So, the recruiting tour certainly paid its way.

Back to Dishforth and prepare for the next presentations in South Yorkshire two days later. But not before a night in York; my favourite city in the world.

We hit York hard. Me, Paul, Brian 'Cappy' Capstick a Sgt from the Ops and Training Wing who was managing all the ground aspects of the tour and his number 2, Simon 'Si' Burrows.

Having done the decent thing and frequented as many of Yorks' pubs as we could, we were all in a friendly, chatty frame of mind. I was talking to a young lady I had met; it turned out she was a German teacher at a local school and was recently divorced. I shall call her Heidi for the purposes of this tale.

I turned on my best German.

Towards the end of the evening, Heidi asked if I would like to go back to her house to practice my language skills, she would grade my efforts accordingly.

Hoping for an A+, I reluctantly agreed and told the others that I was off, muttering something about Bratwurst, and that I'd see them *'im der morgen, und tchuss'.*

Heidi hadn't been drinking, and her car was parked just around the corner. Driving through the narrow streets of York to get out of the city, we stopped at a set of lights, which, by a cruel twist of fate, happened to be outside a kebab shop.

Timing is everything. And our timing could not have been worse, stumbling out of the shop, each with a giant kebab, dripping garlic sauce everywhere, was the trio of Paul, Cappy and Si. It was pure Pulp Fiction.

They clocked me and Heidi. In their delight, and my horror, all three of them rolled, Starsky and Hutch style over the bonnet of her car, the windscreen ended up covered in garlic sauce, chilli sauce and shredded lettuce.

She glared at me, gave me the nod that says, *"get the fuck out of my car NOW!"*

She sped off with her wipers smearing kebab contents into abstract art. I went inside, bought my own, and the four of us shared a £100 taxi back to Dishforth.

The next morning, we had breakfast in the mess with sore heads. Cappy still had garlic sauce in his hair. Gallons of coffee later, we headed up to the 9 Regiment Ops room to finalise the planning for tomorrow. Cappy and his crew set off that evening to prepare the school and thank goodness they did.

We checked the NOTAMs as usual. A NOTAM is a 'Notice to Airmen' that provides new information such as temporary airspace closures, areas to avoid or any other changes you needed to be aware of. There was nothing out of the ordinary that would affect our flight, and then home to Aldergrove.

Paul was the commander, and I was flying. As the commander, Paul had the maps and operated the radios. Talking to ATC, Paul must have had a 'bad kebab', because he wasn't feeling well and was shitting through the eye of a needle.

Years before, in Gutersloh, I was flying with him, but in a reversed role; on that occasion, every time I had tried to talk on a radio, he would pull power, forcing my head into my chest and making me grunt into the radio.

Karma's a bitch and revenge was mine.

Talking to the newly opened Sheffield Airport, Paul gave a position report, "*Sheffield tower, Army Air 842 approaching the field*" as he did so, I pulled an armful of power of his head went down into his map. Not feeling well, he was on the very edge of chundering.

Sheffield ATC came back *"Army Air 842, confirm your position"*

Paul couldn't talk, so I jumped on the radio *"Army Air 842, we're just passing through your overhead"*

Army Air 842, Sheffield tower, you're not passing through my overhead, confirm you have the correct airfield?"

SHIT! The new Sheffield airport had only recently opened; the maps we had from NI hadn't been updated and we hadn't seen anything in the NOTAMs. We looked like rank amateurs.

"Sheffield Tower, Army Air 842, apologies we've passed the old airfield. I'll be having a conversation with 'Vasco de Gama' to my left here when we land". Lowering the tension worked, ATC laughed and were content we did know where we were. Paul, however, was not impressed that I'd just managed to pass all blame for being 'navigationally embarrassed' on to him

If Paul's school had been rough, Barnsley was openly hostile.

Cappy and Si were on the cusp of binning it when we arrived, they had a female Air Trooper with them who was feeling very unsafe. She was getting unwanted attention from some of the older students. Cappy and Si were more than capable of looking after their team, but were not happy at all.

We had to place a guard on the aircraft and after the presentations we didn't let anyone near them, certainly no flying with these apprentice criminals.

After another friendly chat with Sheffield ATC, we made it home with no further incidents.

Back in the Squadron lines, standard duties were belt-fed. 'C' flight seemed to be getting called out increasingly, which was a welcome break from South Armagh.

The boss called me and Paul into his office. Wondering what we were in trouble for this time, we went in. Had Sheffield ATC done the dirty? It was something nice for a change, we were both told that we had both been selected for promotion to Staff Sergeant.

Chapter 90. Belfast and Romance don't Mix

I started going out with Annie, first in the city around the university area, then closer to her home off the Cavehill Road in North Belfast.

Belfast is carved into distinct areas: West, largely Catholic and nationalist; East, largely Protestant and unionist. There are enclaves in each area, flashpoints that host the infamous '*peace walls*'.

North Belfast was a mix of both, and usually more stable. Annie was Catholic, but only on paper. So am I. It didn't bother me one jot. The first time I met Annie's mother, Agatha, I was wearing a suit; we must have been going somewhere nice.

The whole family had gathered to suss me out. Annie's elder sister Marie, her partner Trevor and Agatha's friends were assembled. Marie was the Base Manager for BMI British Midland in Belfast.

I felt at ease straight away with Agatha and Marie. They were both kind and easy going. Agatha had grown up through the Troubles on the Falls Road, one of the hottest hotspots Belfast ever produced. She had married a widowed Protestant man, Annie's father, who had passed when she was only 13.

Agatha had endured a difficult life, but the woman I met carried no bitterness, no deep-seated hatred or bigotry. I came to deeply respect Agatha. The problem as I saw it would be Trevor.

Trevor was a legend in his own imagination, he tried to be a wise-cracking wannabe gangster, the sort of bloke who thought he was auditioning for Goodfellas every time he opened his mouth. He was never off towards me, but he liked to hint that he was 'well connected' and got protective over Annie.

I was having none of it.

I quietly got him checked out. I made some discreet enquiries, passed his details to the intelligence cell in the hangar. They could run 'P' checks quietly, without lighting up the RUC switchboard.

The report came back before I'd finished my brew: Trevor was indeed 'known', but only as a small-time chancer. His rap sheet read more insurance scammer than gangster.

I kept Trevor onside, but I stayed wary.

Riding the Honda VTR around Northern Ireland was great fun, but

I needed a car. I didn't want to spend much money, the Yorkshireman in me coming out, so I bought a new Peugeot 206.

This meant I could travel more safely between Aldergrove and Annie's house in North Belfast without drawing attention to myself.

After being granted the appropriate clearances, I was able to stay at Annie's house. This meant declaring where I was and what I was doing. Belfast was divided into coloured security zones, each with specific rules. Annie's house was Amber. The rules were clear, and I understood them. Breaking them never crossed my mind.

Eventually Annie and I decided that I should move in with her. More clearances, more 'P' checks on the neighbours and the issuance of a Personal Protection Weapon (PPW). I had contingencies in my head. I never told Annie about the PPW, it would have scared her to death.

Always alert when we were out anywhere, one night I knew we were being monitored because I recognised a pair of operators from the hangar. I later found out it was a routine check to ensure that I was being safe and following the requirements of the clearances I had.

Although my relationship with Annie was progressing, I was deliberately taking it slowly.

There were times when I couldn't get to Annie's for weeks on end, duties in Bessbrook were stacked. I passed 1,500 flying hours during a long stint down south, including four Lynx 5 duties back-to-back. When that happened, I still had my 'apartment' in the mess to fall back on.

I needed those nights to catch up with the blokes and unwind like I couldn't anywhere else.

In April 1998 the Belfast Agreement was signed, but it still required a referendum approval for it to be implemented. That approval came in May, and it was overwhelmingly endorsed. I was amongst those that had voted 'YES'.

One consequence of the Good Friday Agreement was the accelerated release of terrorist prisoners.

There was a party next door to Annie's house. It was a release party for a notorious IRA terrorist who had been inside for a well-known bombing. Collecting my stuff and making sure my PPW was handy,

I packed quickly and went back to the mess in Aldergrove.

I never went back to that house.

By then I was used to living out and the freedoms that came with it. I cashed in a small endowment policy I had and raised enough money to put a deposit on a house. I found a lovely 'doer-upper' in Jordanstown, a small suburb between Belfast and Carrickfergus. Annie let her house out and we moved in whilst it still needed a lot of work.

Chapter 91. Tankers!

On 21 June 1998 my increasingly happy life came to a standstill.

We were in the blast-proof Lynx crew room in Bessbrook on a routine Lynx 5 duty. Lynxes 8 and 9 were due down later for random Eagle vehicle checkpoints. Nothing about the morning felt unusual.

The Lynx crew room was closest to the helipad and was constructed from concrete with very high, small windows for light. Just over a year earlier, L/Bdr Stephen Restorick, aged 23, had been shot dead by an IRA sniper protecting the helipad. His killer was later released under the GFA, along with Annie's neighbour.

If 'crashed out', the pilots sprinted out to the 'cabs', whilst the commanders would receive a briefing from 5. By the time a quick initial brief was given, the pilots would have the aircraft burning and turning; ready to pull pitch the moment the commander was in, often pulling the skids off the ground whilst the commander was still strapping in.

We had a dispensation from Army Flying Standards to leave the aircraft with multiple switches in positions that allowed for these rapid starts.

In the back of the aircraft the doorgunners would be loading the doorgun whilst simultaneously organising the Airborne Reaction Force (ARF). The ARF was a full section of infantrymen from the resident battalion; in this case The Royal Gloucestershire, Berkshire and Wiltshire Regiment (RGBW), spread evenly between Lynx 5 and 7 and were on a minute's notice to move whilst in R850.

The ARF also usually included a policeman from the Royal Ulster Constabulary (RUC) and a dog handler with his search dog.

My pilot for the rotation was Charlie Greene; a Captain attached to the AAC on a flying tour. The Lynx 7 was commanded by Dean Kavanagh and piloted by Sgt Dan Brodie.

Just after 10 in the morning, whilst preparing for the day's routine flying programme, we heard the call from Buzzard *"CRASH-OUT; Lynx 5 and 7 CRASH-OUT"*.

The pilots sprinted out to their waiting aircraft and initiated the abbreviated start sequence, the doorgunners marshalling the ARF into the back and made the doorgun ready. Meanwhile I sprinted up the concrete stairs to the tower where Buzzard was waiting for me.

I was given a rough location to head for, *"1.5K's north of Cross on the Blaney Road, patrol in contact"*. Running out to the cab a million thoughts were going through my mind but was this just another day in South Armagh?

It wasn't.

Climbing into my armoured seat, whilst Charlie gripped the controls between his knees and both hands in case I knocked them, I pulled my flying helmet on, being careful to ensure my ears were laid flat under the earphones; it was going to be a long day, and the worst thing was having your ears fold under the ear cups, a slow burning agony!

My helmet was already connected to the aircraft intercom, so it was a lightning quick brief to the crew, *"Charlie, head for Cross, guns, shots fired, make ready."* The doorgunner communicated with the ARF section commander and let him know what was occurring.

"7 this is 5, lifting"

"Red 850, two Lynx departing up the tube, Red 850".

Pulling the maximum allowable power, the pair of Lynx screamed into the air, Lynx 5 in the lead as our spot was furthest west, so the first to enter the tube. Charlie executed a 'threat band climb out'.

I gave Buzzard a situation report (sitrep) and an ETA at the incident site. I then received a more in-depth report in return.

A joint RUC and Army patrol had stopped a Northern Irish registered heating oil tanker at a random checkpoint on the Blaney Road near Crossmaglen. The driver was questioned by the RUC, and it was believed the police had asked him to accompany them to the station in Crossmaglen when he drove off.

The vehicle was stopped again by the RGBW patrol, when the section commander, Cpl Gary Fenton tried to question the driver, he drove off a second time, fatally injuring him. The cut-off section fired several rounds at the vehicle, but the driver escaped across the Border, abandoning the tanker about one mile from Longfield Cross.

Gary Fenton was married.

He had a four-year-old daughter.

Arriving at the incident I contacted the ground commander on the secure tac frequency and received a sitrep. The Rules of Engagement (ROE) in Northern Ireland, and especially in South Armagh, were very complicated, but, we did have the legal ability to cross the border for up to 4 kilometres if in contact with the terrorists of one of the many branches of Irish extremism.

On the second secure tac comm I contacted callsign Zero Charlie (ØC), the Army Air Corps liaison officer based with the 39 brigade HQ in the main building of the mill.

Capt. Ronan Thorne was an attached AAC officer pilot from the Royal Signals, hailing from Fermanagh in Northern Ireland, Ronan was an extremely well-liked and respected officer that wasn't averse to straight talking and socialising with the warrant officers and senior NCOs of the Squadron.

I knew Ronan was the duty ØC and working hard in the background to support the operation. I requested clearance to cross the border. We could see the truck.

We'd already taken a fatality.

In my view, we were in contact.

Unbelievably Ronan came back and denied permission for the cross-border chase down. Ronan was the voice on the radio, but the decision wasn't his.

Rage is a poor co-pilot, but adrenaline is worse at diplomacy, I was livid and ordered Charlie to head for the border where it crossed the Blaney road, the doorgunner leaning into his sight line, I'd already given the order to *"make ready"* as we'd lifted out of R850, in case it was a 'come on' style ambush, his index finger poised, and I just need to give the orders to engage.

As we approached the invisible border, I relented and gave Charlie the order to return to the incident site. Conscience and training had won, I broke off, orbiting above the scene while the tanker sat there, just over the border. I didn't know yet that decision I'd made in the heat of the moment would haunt me for years.

The rule was simple: Lynx 5 held control until an officer on the ground established a command post.

By then, a Company Commander from the RGBW had arrived in the Lynx 8/9 pair. We exchanged SITREPs, and about an hour after leaving R850, I handed over.

The Lynx was nearly always fuelled to 450 Kgs of fuel from a maximum possible 750Kgs, this trade off gave us the lifting capacity that we needed. Fuel was always close by, but we only had just over an hour of usable fuel.

By now we needed to make the 10 mins back to R850 for a hot refuel, delivered as always by the unfailing fuelling experts from the RAFs Tactical Supply Wing (TSW), known to crews as 'Tiswas'.

Having taken on a suck of fuel and boarded more troops from the RGBW we started a supply run out to the command post on the Blaney Road, but we also still had to maintain the normal days tasking of supply to, and movement of troops around the chain of towers in the area. I handed over the remaining operations to the commander of Lynx 8, Dom Giles, and started the days programme.

On completion of the daily ops programme, tired, frustrated and feeling ashamed, I walked out of the crew room and headed over to mill to talk to Ronan, and then to the brigade staff to provide an in-person debrief. Ronan explained it was the RUC that denied permission for the border crossing.

In the background there was of course the recent signing of the GFA, but the reality was there was only us observing it at the time.

Politically, a British Army helicopter crossing the Irish border and shooting dead a tanker driver would have caused the politicians some headaches.

They would have moved on. I would not.

I had the authority to act as Lynx 5 commander. Instead of requesting clearance, I felt I should have taken it and engaged the driver.

Had I done so, I would most likely, all these years later, be sitting in a courtroom as "Soldier P", my name removed, while those who set the conditions carried on regardless.

In 2000 a jury cleared a Dundalk man of killing Gary. He was convicted at Newry Crown Court of not being covered by insurance to drive the tanker involved in the incident. The judge fined him

£150 and slapped six points on his licence; that was the value placed on a soldier's life.

In August 2019, at the 50th anniversary commemorations of the start of Operation Banner, the operational name for British troops in Northern Ireland, at the national memorial arboretum in Alrewas, near Lichfield, I met some of Gary Fenton's section that were there on that day.

I talked with them and admitted my long-held shame in not taking the decisive action I felt I should have; that I could have brought some justice for Gary's family; the judge certainly didn't.

As true brothers, they simply said that Gaz was gone, nothing anyone did could bring him back and that they too could have done things differently. They'll never know how much that quiet moment lifted a weight I'd carried for two decades.

Back in the mess at Aldergrove that night, I stayed in, I was on duty the next day. Dean, Dom and I lifted a pint to Gary Fenton. A quiet toast, no speeches.

Then we changed the subject, because that's what soldiers do when the alternative is to sit there staring into your glass.

Paul had just been through QHI selection and was nudging me to have a go. I'd thought about it, but with HWI and EWO work, and Apache on the horizon, it wasn't high on my list.

Still, he wore me down.

Over that pint, I told Dom I'd put my name in for QHI selection.

And, without mentioning it to him first, I'd put his in as well.

The look on his face was worth it.

The language even more so.

Army Standards wouldn't send the team out from Netheravon to test just one candidate, so my little recruitment tactic had worked.

On the day, my flying was fine; my ground school prep was… well, less so. When the day came and Maj Norman Ryan put me through my paces, the flying went well, but a couple of basic aircraft limitations slipped my mind.

I failed the assessment on the ground school portion only, which meant I could try again.

Dom passed with flying colours.

I was happy at home in my own house. I had a car and the VTR in the driveway. A beautiful girlfriend and a dog we could walk along the foreshore.

But things with Emily were still poisonous. She went out of her way to make my life difficult. I had been over to Germany to see the kids again. This time when I went to pick them up to go to a nearby safari park, I wasn't allowed anywhere apart from the front porch.

On arrival she made a series of demands that I just couldn't meet; more money on the spot, exact places I would be taking the kids and, incredibly, copies of my medical history *"in case the kids inherited something from me"*.

All four children were crying and screaming at her to *'stop'*. Telling them I'd be back in a moment; I went out to my hire car and called Annie. I couldn't speak; my throat was pounding.

Eventually I managed to blurt out what was going on, she calmly said *"oh Jim, just come home"*.

We went into the annual July 'marching season' not really knowing what to expect that year. Normally we would fly an extraordinary number of extra hours, providing additional Chancellor surveillance aircraft, Lynx fitted with a live-feed camera system to watch the parades and likely flashpoints.

We were hoping for a quiet one. The GFA had been signed, things were on the up.

Until the annual march through the small village of Drumcree.

Chapter 92. To Drumcree and Beyond

By the summer of '98, Drumcree had gone from a yearly Orange Order parade to a full-blown standoff. The march from Portadown to Drumcree Church, and back along the nationalist Garvaghy Road, was tradition for some and provocation for others, and nobody was backing down.

Each July brought roadblocks, riot lines, and petrol bombs, with police and soldiers caught in the middle while both sides claimed the right to stand their ground. 1998 was the apogee for the annual battle.

And battle is the right word.

On Friday 3 July, about 1,000 soldiers and 1,000 police were deployed in Portadown. The Paras and engineers built barricades across all roads leading into the Catholic area. In the fields between Drumcree Church and the Catholic area they dug a wide trench, which was then lined with rows of barbed wire.

Our role was twofold, primarily we deployed at least 1 chancellor cab on station 24/7, which meant 2 or 3 aircraft fitted out and 'on station' handovers. These aircraft became 'must-have' support. Of course, we still had the full range of duties out of R850. Lynxes 8 & 9 were shuttling around doing province wide Eagles to disrupt movements.

About 10,000 Orangemen and loyalists arrived at Drumcree from across Ulster. A loyalist group calling itself "*Portadown Action Command*" issued a statement which read:

"As from midnight on Friday 10 July 1998, any driver of any vehicle supplying any goods of any kind to the Garvaghy Road will be summarily executed"

After Bosnia, I knew exactly what humans were capable of. This wasn't theatre, this was about to get real. And this was the United Kingdom. United indeed!

Over the next 10 days hundreds of attacks occurred, in both directions.

As with Bosnia, we were in the middle between sides that just wanted to massacre the other side. The village of Dunloy was put under siege by over 1000 Orangemen.

The whole of Northern Ireland was a tinderbox starting to go up in flames.

Even our aircraft were targeted. Pete Royle, a Royal Marine pilot on exchange, was flying a night Chancellor sortie over Drumcree on NVGs when someone aimed a high-powered industrial green laser into the cockpit. With the goggles amplifying the light, a direct hit could have blinded the crew instantly.

Pete broke away, reported it, and a few nearby Paras asked the gentleman, very politely, to stop.

Apparently, when he eventually regained consciousness, he was arrested.

The battle of Drumcree came to a head on Sunday 12th July when three little boys, the Quinn brothers, were killed when their home in Ballymoney was firebombed by Loyalist terrorists.

Ian Paisley, a "man of god", went on to claim: "*I think that if the Press waits a few days, they will find there is very little relationship between those deaths and what happened at Drumcree. The IRA carried out far worse murders than we had at Ballymoney over and over again. No one said to them to call off their crimes and now they are received as acceptable democratic politicians*"

Three little boys, miles from Drumcree, burned alive in their beds for something they didn't even know existed.

This is what people are reduced to when religion becomes a weapon.

I thought Bosnia was bad. Three boys burned alive in my own country was worse.

The murder of the Quinn brothers should have ended it. It didn't. All over who could, and who could not parade down the Garvaghy Road.

The following week the police, supported by the Army and a Chancellor aircraft, began a search operation in the fields around Drumcree. Several loyalist weapons were found, including a homemade machine gun, spent and live ammunition, explosive devices, and two crossbows with more than a dozen homemade explosive arrows.

The end of July couldn't come quickly enough.

Drumcree had bled into every corner of the province, roadblocks, raids, threats, and three dead boys whose killers would never face the punishment they deserved.

We'd seen this before, in Bosnia and closer to home, and it was starting to feel like a rule: if you kill in the name of your cause, someone in a suit will find a reason to let you walk away.

Relief came in the form of a very different mission, taking a Chancellor-fitted Lynx to the Royal International Air Tattoo at Fairford. One week I was flying over barbed wire barricades in Portadown, the next I was taxiing past a B-1 bomber and waving at spectators behind the rope line.

Relief is the wrong word.

Nothing had eased. Nothing had resolved. We simply stopped looking at one thing and were told to look at another. The same week that ended with burned children and laser strikes on our aircraft now required a pressed flying suit and a smile for the public.

The theme of the 1998 event was coincidentally *'eyes in the sky',* and as we still had multiple aircraft fitted with Chancellor, I was given clearance to take one of those. I was crewed with Tyrone Selfe, a new, young and excitable Captain.

Arriving at Fairford amidst hundreds of other participants from most countries was awe inspiring. Taxiing between an example of pretty much every fighter, every bomber and aircraft of all sizes, up to and including the C5's of my childhood was an experience I'll always cherish.

First task was to put the aircraft to bed; we were only doing a static demo. The information boards were safely in the back, and we locked her up.

Two days this aircraft had been evading a deliberate attempt to blind its crew. Now it needed an information board and a rope line.

Welcome drinks awaited at the obligatory hangar party. And here my memory fades.

Squadron zap stickers came out and were applied to anything that didn't move, and some things that did. I'd brought a small bottle of 'edge sealer', so once a sticker was in place, it would be absolute bastard to get off.

By the end of the night, most female aircrew were sporting zap-sticker bikinis.

Somehow, the crabs managed to arrange transport for everyone to an allocated hotel in Swindon or its surrounds and then pick us up the next morning. Instead of using the bus, like normal people, Tyrone wanted to *"quickly nip to Bicester shopping village"*. I thought he had something specific in mind and went along with him.

What he really wanted to do was wander around the shopping complex in a flying suit larging it massive in front of Japanese tourists.

When I realised what his plan was, *"Sir! We're going back, right fucking now!"*

I met up with my friends from Oxford and enjoyed showing them the aircraft, opening the tensile barrier and letting them climb in was the least I could do.

The whole Fairford experience had really been most excellent, the flight back to Aldergrove was uneventful and of course, I was on duty the very next day.

From airshow to shitshow overnight.

I took some local leave, the house in Jordanstown needed some TLC. Like a whole new kitchen. When we moved in, the first thing we did was have a kitchen extension built.

Annie had found a fantastic ex-display kitchen; solid Oak with granite surfaces, it was truly a thing of beauty. I needed to get it fitted as Vince & Colleen were coming over for a visit from Savannah; the house was a building site as it was, but we had to have a kitchen.

We had a lovely time showing the yanks around Northern Ireland, despite the troubles it's a great place to visit.

Just before going back to work, a group of us; me, Annie, Trevor, Marie, and a few others had tickets to see U2 play in Belfast's Botanic Gardens. It was more than just a gig; it felt like a ceasefire for the soul. The air was electric, but for once not from tension, just joy.

The only edge in the air that night was 'The Edge', standing on stage with a guitar. There was no alertness, no perceived threat. Just people, thousands of them, laughing, singing and holding each other like the war really was over.

Circling quietly overhead was an Islander from 1 Flight, supporting the RUC, its presence more symbolic than strategic. Each time the searchlights picked it out, the crowd erupted in cheers, not fear, just pure release.

For that one night, under a wide Irish sky, peace felt possible. Tangible. And for a soldier who'd seen the worst of human nature, it was oddly disarming to witness the best.

Annie and I were getting closer, but neither of us were at the point of anything permanent yet.

Chapter 93. The Long Road to Peace

By late July, our daily briefings had narrowed to a single promise: a statement attack, a "spectacular", was imminent. Intelligence was tracking movement south of the border, and the assumption was simple.

They would come north.

The first real indication came with a large explosion near Moyry Castle, right on the border.

The crater was enormous, impossible to miss, with blast markers radiating out every ten metres. Someone had test-fired something serious.

The ARF secured the site, and we brought in EOD specialists.

It was a dress rehearsal.

A few days later, intelligence was still warning of imminent movement north, but this time they had more.

A description of a possible target vehicle.

One early evening at the very beginning of August we were crashed out. I had Paz Hewal as my pilot and Pete Royle was Lynx 7.

A solid team.

One of the Romeo surveillance towers that watched the main A1 Dublin – Belfast highway border crossing point had picked the target vehicle up heading north towards the border.

I got eyes on the target vehicle, a white minivan, heading north towards Newry.

I put Pete on its tail whilst I headed north at max chat to put a blocking action in place. This would involve closing the main north – south arterial highway.

En route up to Newry, I requested Gazelle 1 to move into an overwatch position.

The Gazelle was on station within five minutes, feeding me a calm, constant commentary that gave me total situational awareness.

Just south of Newry is a very large roundabout, which is where I planned to put the stop team down. I had about 11km from the first sighting to where I was going to intercept them.

Less than 8 minutes.

By now it was getting dark, I transitioned to NVG, Paz stayed 'mortal'.

Illuminating the roundabout was a ring of tall, bright streetlights. Paz flew us into the black hole inside the streetlights without needing me breathing down his neck. I didn't have time to babysit and didn't need to; I let him get on with it.

"Tango approaching your position at speed, 500m to run, 200m, 100m," reported the floppy crew

I looked up, I could see the headlights coming towards us at some speed with Pete low and on his tail.

The ARF spilled out of the aircraft, but we were too late.

I got Pete to bound ahead whilst I picked up my patrol. The floppy continued his descriptive commentary.

As I lifted to take up the back stop position, I gave Buzzard a sitrep and requested Lynxes 8 and 9 to be deployed, we were probably going to need them. I also asked for RUC road unit support to close the A1; granted.

Pete found a suitable area on the A1 and put in the front stop. The target blasted through his position seconds before they were ready.

"Tango 1 heading north, speed 90"

I bounded ahead of Pete, who had picked up his troops, but he was also nearing his 'bingo' fuel state, the point at which he would have to return to R850 for fuel whether he wanted to or not.

On NVG I could clearly see the target moving at speed.

I wasn't far behind Pete on bingo fuel.

This next action would either stop them, or we lose them.

I leapt ahead, diverting away from the road to give the impression we'd knocked the chase off. Pete flew back to R850 like a bat out of hell, 'tiswas' were waiting for him, he was on the ground for less than 2 or 3 minutes, getting just enough fuel to continue. Tiswas produced the goods in record time.

I also now had Sea King 1 spinning up with more troops that had been 'stood to' in support for us.

I bounded up the eastern side of the A1, by now closed both ways by the cops. The A1 arced to the west; by flying up the eastern side I could save some time. I told Paz to head for the viaduct, a well-known feature to the south of Banbridge, where I could get a decent stop group in place.

"Guns, make ready"

The words you use when you're trying not to think about consequences.

"READY!" screamed the doorgunner in a heartbeat, the butt firmly in his shoulder.

We dropped the cutoff group and set the trap; they smashed into position like a well-oiled machine. Stinger out. Weapons ready.

I established comms with the patrol commander on the secure tac frequency.

We came into a high hover; gun trained on the approach.

I rebroadcast the overwatch commentary to the patrol commander.

The target vehicle arrived as if it had rolled straight out of a film.

The patrol commander stepped out to activate the trap. The target screeched to a halt and he advanced, covered by his section and our doorgunners.

Gary Fenton was in my head. Not this time. Not again.

History wouldn't repeat itself. GFA or not.

I had eyes on the occupants, the very slightest twitch from either of them and they would have been converted to a pink mist in the blink of an eye.

A couple of RUC cars arrived, blues and twos going, our NVGs flaring green with every red flash in our eyes.

Then everything stopped.

Nobody moved.

Nobody spoke.

I told the patrol commander to "*send sitrep*", what the fuck was going on?

He gave me a full report and asked us to pick his team up; Pete arrived back on scene and went into mutual support mode without being asked.

We landed on, and the patrol climbed back in. *"Seriously, what the fuck is going on?"* I said over the intercom, *"people were close to getting killed!"*.

And that's when reality took its NAAFI break.

I cancelled 8 & 9 and stood the Sea King down with Buzzard, we were on the way back to R850. I thanked Gazelle 1, that night they were simply critical.

Pete and I, together with Buzzard went over to the Mill to Brigade Ops, via Ronan in his office.

During the heat of the chase, everything seemed logical, obvious even. I didn't have to think twice about any of my actions, DODAR was working at light speed. Call it adrenaline, call it whatever you like, but the training, the experience and my own intuition all kicked in.

The pilots' course sold us the dream: clean inputs, clean decisions, clean outcomes. Gutersloh and Bosnia reinforced it. When it worked, it was beautiful. That night, it did work.

What failed came afterwards.

The debriefing was a clusterfuck of epic proportions. What should have triggered alarm instead triggered defensiveness.

The target vehicle was in fact, a separate RUC operation, at first, they claimed, they hadn't seen us. Utter bollocks. A lie so lazy it barely qualified.

They then claimed that when they did see us, they didn't realise it was a stop operation. More bollocks. The Police liaison officer in the Mill went very quiet, he had authorised the closure of the A1, but was allegedly unaware of this vehicle, that matched exactly intelligence reports, just after a trial run of a huge bomb on the border.

We had been sitting in a high hover over a roadblock on a closed A1, with armed soldiers on the ground, a Lynx overhead and police vehicles arriving with blues and twos flashing.

It was hard to imagine what part of that might have suggested a village fête.

I got vocal, still high on adrenaline and Sharps extra strong mints. I had been on the very edge of killing this pair of idiots; how many lives would have been destroyed? Ronan near punched me in the ribs.

The crews had all performed to a standard that Pete and I were proud of.

On return to Aldergrove the next morning, I was signing the aircraft back in at the -700 offices when the phone went. *"Go up to see the CO as soon as you're done here"*

It felt like Bosnia all over again: write it down, file it away, then silence. The near miss stayed with the crew, not the system. I was simply asked for a brief of what had gone down. The CO, Lt Col Nate Mossman was quite rightly boiling and asked for a written report. I knew Nate had my back if anything was to come of this; he was a great CO.

I never had any feedback after that; I don't know what drove the cops to be so stupid and unprofessional. I don't know what happened to them, I just know what very nearly did happen.

I stayed in the mess again that night. I needed time. And that time needed to be with Dean and/or Dom. On this occasion it was Dom, we were on duty together the following day. It was becoming the norm for us to lean heavily on each other, the importance of mess life never more critical.

Surely someone asked if we were OK?

Nope.

Not even a token *"how was your day?"* Nada, never mind, take a day off to recover. My support was Mr Mason, Mr Grant, Mr Kavanagh and Mr Brodie, with many others thrown in for good measure.

They know who they are.

I wasn't even on duty when Omagh inevitably happened, in fact I was on leave. I remember it like it was yesterday though; a 'JFK moment'. I was at home in Jordanstown, working in the garden building a deck.

Annie called me in to look at the news; she was visibly distressed.

A massive car bomb had been detonated in Omagh town centre.

I still see the footage when I don't want to. People moving like they'd been switched off, faces grey with dust. You didn't need to hear the commentary; their expressions told you everything. I felt physically sick.

The huge bomb had exploded in a crowded shopping area. It tore the car into deadly shrapnel and created a fireball and shockwave. People were caught in "a storm" of glass, masonry and metal. A thick cloud of dust and smoke filled the street.

Omagh town centre ate its own people.

The blast was so strong that it tore up concrete and pipes burst; the water, running down the street, turned red from the blood of dead and wounded people. That's a sentence I hate writing, but it's true.

Twenty-one people who had been standing near the bomb were killed outright. Eight more died on the way to or in a hospital.

I immediately called the ops room to see if I was needed; they had enough crews; for now.

I needed to help. To do something. Anything.

Every available helicopter in Northern Ireland was used to ferry people to hospitals across the province, including as far away as Altnagelvin in Londonderry.

Lisanelly Barracks in Omagh was used as a makeshift morgue.

Three days after the bombing, the 'Real' IRA claimed responsibility for planting the bomb but strongly denied intending to kill civilians and apologised to the victims. It also announced that *"all military operations have been suspended"*.

They apologised. Then announced they'd *"suspended operations"*. Brave stuff, once the killing was done.

Just a few weeks earlier, we'd stood in a sea of optimism at the U2 gig, watching a new future trying to be born. Sunday, Bloody Sunday was still ringing in my ears. For a few hours it had felt as though that song belonged to history rather than the present, a reminder of what had been rather than a warning of what still lay ahead.

Now, all that hope lay in ruins. The promises, the power chords, they were no match for home-made explosive and cowardice.

President Bill Clinton was due to visit in early September in a bid to reaffirm US support for the GFA, Omagh had driven a coach and horses through the national confidence that these 'troubles' would ever come to an end.

Having just witnessed the hope and a bright future, I too felt the despondency most of Northern Ireland was feeling. Talking to Agatha about it, she was sanguine, having lived through this shit right from the start.

Prior to the Clinton visit, the 8 and 9 pair were tasked to fly 'special advisors' from the presidential advance security detail around the province, including South Armagh. We knew that the IRA had access to SA-7, a soviet era, shoulder launched, surface to air missile.

These old missile systems were powered by a thermal battery which had a time limited shelf life.

In July 1991, the IRA had fired one at an RAF Wessex operating near Kinawley, on the Fermanagh border. Fortunately, the missile failed to lock on and detonated harmlessly on the ground, but it marked a worrying escalation. Int was saying that the thermal batteries were approaching their shelf life, so why not use them?

Talking to the 'men in black', they were much more concerned about another possibility, the Stinger missile. In the late 80's, the IRA had been trying to acquire these things, but our int was saying they didn't have any; the guy in the sunglasses in the back was saying something else.

We had a better than evens chance against an SA 7; the ALQ 144 'disco ball' would help with that. But the Stinger, well that was a different matter. An all aspect, third generation missile. None were ever launched, but the fact they might be out there, and so soon after Omagh; anything was possible.

We'd trained for small-arms fire and even multiple heavy machine guns, but the idea of man portable, surface to air missile system in the province raised the stakes. It forced a rethink in altitude, routing, and complacency, no longer could we rely solely on terrain and speed to keep us safe.

Seeing Air Force 1 taking up most of the civilian side of Aldergrove was very impressive. Watching Clinton talk to the crowds was a special time.

3 months prior, Paul had departed Aldergrove and moved Cath and his family to Wallop whilst he attended the Qualified Helicopter Instructor (QHI) course at RAF Shawbury. The time had come for his graduation, and we had an invite. Pulling together a plan was easy, largely because we also been tasked to support some fast rope training taking place at RAF Brize Norton at the same time.

I talked Jon 'Adlow' Hadlow into coming with me, it didn't take much persuasion. Cappy and Si, as fast rope instructors, came along for the ride.

RAF Shawbury. It's the home of rotary flight training but more importantly, it's home to every QHI, and the Vatican of the 'sky police'.

As Jon and I arrived at Shawbury it was late afternoon on a Friday, air traffic was about to close and wanted to go home for the weekend. So, we elected to expedite our arrival, just to be kind to the hard-working controllers. I requested an overhead arrival.

I looked at 'Adlow. 'Adlow looked at me.

"Shawbury, one Lynx from the overhead" I announced on the radio.

Jon entered a textbook split 'S' manoeuvre, and we thundered directly to the dispersal, Bessbrook style. The racket we made was a proper symphony, AC/DC would've asked us to turn it down.

Apparently, Paul was already in the mess when the bar went quiet. All eyes turned to the vibrating windows; Paul just put his head in his hands; he knew we'd arrived. And thanks to the unmistakable blade slap, so did every 'trapper' in the station.

After our dramatic arrival, it just went downhill from that point. Knowing we'd caused more than a few senior RAF eyebrows to be raised, and after getting changed into formal mess kit, I put on a battle helmet and a full set of body armour on over the top. There was no way I was going into a mess dinner with a room full of newly minted QHIs without sensible protection!

Cappy, for whatever reason made some decisions that even I couldn't begin to explain. After crawling under the long legs of the dinner table, he thought it was a good idea to tie the station commander's shoe laces together, which would have been very funny, but he was busted before he got anywhere near his target.

Not to be beaten, he wandered out into the foyer, apparently to clean the chocolate pudding from his face that had just been slammed into a bowl of the stuff, and being ever the champion of health & safety, smashed a fire alarm just as the speeches were starting.

I was both horrified for Paul, and proud of Cappy at the same time.

We left the next morning in a more 'compliant' manner. Paul left with a huge bill for his guests' behaviour and actions.

He was not amused, everyone else was.

A few weeks earlier, Omagh had been destroyed. Now we were heading back from a mess dinner.

It looked like piss-taking. It was, partly. Mostly it was a pressure valve. If we didn't let some steam out, it would have come out somewhere worse.

Back in Aldergrove, after successfully delivering the fast rope training at Brize, the fallout from the Omagh attack was about to get real.

If you read official reports from the time, they will tell you that 12 arrests occurred on 22nd September. That's not entirely accurate.

On the 20th September Dean was the Lynx 5 commander. He was tasked to put together a plan that saw the coordinated arrest of suspects at 10 sites in South Armagh.

I was the Lynx 8 commander, my pilot was Capt. Makay-Thompson, lovingly known as 'Micky Thump'. We deployed that night to R850 with Lynx 9. There were multiple other aircraft at R850, including a Puma, unusual for night operations, and a Chinook, 10 in total.

For a theatre where aircraft numbers were normally measured in twos and threes, it was an unmistakable sign that something significant was about to happen.

Seeing all these aircraft converge on Bessbrook for one purpose, knowing what was coming their way brought a sense of satisfaction that for once, we might just come out on top.

Just once.

Dean briefed the mission in a packed Ops Room. We were each given a specific time on our nominated target. We departed R850 en masse, split up as part of a deception plan and went to each target.

Hand-held laser pointers marking targets as we'd rehearsed with C flight.

Within 15 seconds of each other, each target was hit with precision. A superbly planned and executed operation.

The police teams, backed by the Army, moved like well-oiled clockwork. At my target, the suspect was hooded, bound, and on board within sixty seconds of the knock.

Watching this coward dragged into my aircraft, I felt nothing but contempt. He wasn't a freedom fighter; he wasn't a soldier; just a thug who thought planting bombs made him a warrior.

His war was over.

I hated him with everything I had; I hated everything he stood for, and I had no sympathy.

We then took each suspect back to R850 where they were transferred to the Chinook, still hooded and cuffed. From there, the Chinook flew them to a police station near Castlereagh.

As dawn was breaking, after a full Ulster fry in the mess, I led the Lynx pair, the Pumas, Sea Kings and Gazelles back to Aldergrove under the collective callsign of "*Victory Formation*". How naive was I?

It was very far from victory.

No one has ever been convicted in a criminal court for the Omagh bombing. However, in a 2009 civil case, Michael McKevitt, Liam Campbell, Colm Murphy and Seamus Daly were found liable for the attack.

There have been multiple investigations into what the authorities knew and what they did not know. I can only speak to what I saw.

The intelligence briefs we were receiving in the weeks before Omagh left little doubt that a major attack was expected.

I saw what appeared to be a trial run near the border.

I closed the A1 on a vehicle that matched the intelligence picture almost line by line.

A few days later, twenty-nine people were murdered and hundreds more injured.

Perhaps those events were unrelated.

I have never been able to convince myself that they were.

What troubles me most is not what happened that day. It is what happened afterwards.

To the best of my knowledge, nobody ever asked our crews what we had seen, what we had chased or what had almost happened. Nobody sat us down and asked for our recollections. Nobody seemed interested in whether the pieces we had witnessed might fit into a larger picture.

It was as though operational memory itself had become politically inconvenient.

The chase up the A1 and the devastation at Omagh shaped me in ways I only understood years later.

We had been trained to fly in the dark, hunt armed terrorists and make life-or-death decisions in milli-seconds.

What nobody taught us was how to live with the possibility that we had come within touching distance of disaster, seen fragments of the puzzle and then watched the pieces disappear into a drawer.

Nor how to live with the knowledge that, despite everything we did, it still wasn't enough.

Chapter 94. The Apprentice, without Sugar

As we approached the end of what had been an extremely busy year, we had an influx of new pilots, and unusually many of them were Captains.

Officer pilot entry took time: Sandhurst, pilot selection, then the full flying course. By the time many arrived at their first unit they were already wearing Captain's pips. Most understood that rank didn't translate to competence in the cockpit.

A few didn't.

In this latest intake, we had one such Officer, Capt. Roz Parnell. She arrived into a Squadron stacked with experience, and the gap showed quickly.

I'd flown with Roz a couple of times on air tests before we were rostered together in R850. Our new SQHI, Captain Gabe Smitt, had just taken over from Marty Gilbraith. He asked what I thought of her.

I told him I hadn't had a duty with her, but from what I'd seen, she was OK.

When I was eventually on a Lynx 5 duty with her, I had my eyes opened wide! I would estimate that about 30% of our flying hours at that time of the year were NVG.

Map Reading around South Armagh, on NVG when you're new to the province can be difficult. Aidan Melly was a very experienced and highly capable commander that managed to drop a patrol 'sausage side' on my first duty. If it could happen to him, it could happen to anyone.

After we'd run the daytime Buzzard programme, we transitioned into night ops, I gave Roz the job of map reading whilst I would fly the aircraft. This was a routine patrol drop off. The patrol commander leaned forward between the armoured seats and gave me his grid insertion point.

I knew exactly where it was, so I nodded him towards Roz, she spent an age plotting the grid, she took long enough that I could hear the fuel gauge groan.

"Come on, let's go", I was getting impatient, watching the fuel level go

down, I could hear she was getting frustrated from the tone of her reply.

Before she was ready, I made the call *"Red 850 two Lynx over the eastern fence, Red 850"*

Lynx 7 lifted and peeled off to the right; I followed suit. In the climb out I asked her if she was sorted.

"Yes, all good, I've got it thank you." She replied somewhat tersely.

Heading in the right direction and at about 3 mins out, I asked if she had the landing point. She started slowly rotating the map, staring intently at it, a classic sign that a pilot was 'unsure of their position'. Read 'lost'.

This was no time to be pissing around; we were close to the border and getting closer.

On NVG, 'close to the border' is not a geography lesson. If you're lucky, it only ends careers.

I put the pressure on.

"Come on, talk to me, where do you want me?" I already had the drop-off point; Lynx 7 was leading us in and had started the descent through the threat band.

"I need you to talk to me." Silence. *"Roz. Talk."* Silence. *"Talk to me!*

I may not have been that polite.

We dropped the patrol and lifted to head back to R850. *"Guns, unplug for a moment"*, the doorgunners clue that not all was sweetness and light in the front, he disconnected from the intercom.

"What the fuck was that all about? I barked.

"If you're lost, you say it. You do not go quiet here. I'll take 'I'm lost' every time over standing in front of the CO explaining why we crossed the border."

I knew I'd upset her, that genuinely wasn't my intention, that's not my style. But I had to get across how far behind she was, Captain or Corporal – it was irrelevant.

Flying operations in Bandit Country was not the place to be feeding rookie pilots. The enemy had spent twenty years perfecting their trade. We couldn't afford to be learning ours.

To her credit, Roz must have improved significantly, although I didn't see it. In 1999 Roz became part of the first all-female crew on operations with 5 Regiment.

Chapter 95. Leadership v Management

1998 had been a monumental year for me, professionally and personally, and my tolerance for bullshit wasn't just low.

It was gone.

As we approached the end of the year, I had another couple of courses to attend at the Air Warfare Centre in RAF Cranwell. The Targeting & Battle Damage Assessment (BDA) course was all new information for me; and it was riveting.

I finished 1998 back at Cranwell, absorbing everything I could. Targeting, battle damage assessment, electronic warfare. It was all about understanding consequence, not just execution. It was obvious that a lot of the RAF were getting very nervous about the role that the Apache could take away from them, and it was beautiful to watch!

My fourth, and last course was the Advanced EWO course.

I was AH ready.

Back at 655, we had yet another new boss. Major Matt Leggett joined us from 3 Regiment in Suffolk.

One of his first decisions was to change our working environment. We each had a desk, because we needed one. He replaced them all with a single enormous table no one had asked for.

It looked tidy.

Tidy and completely impractical.

I suspect he thought it was leadership.

It set the tone for his tenure. Management over leadership.

Leadership absorbs risk. Management pushes it down.

What really upset the crews was his habit of imposing ideas without the faintest hint of consultation. The worst of which was his decision to enforce a fixed changeover time for Lynx 5 and 7 at Bessbrook.

Until then, Ops varied the timings day to day, giving us at least a fighting chance of avoiding setting predictable patterns, the terrorists pal.

Thanks to this bright idea, we were now running on a timetable the enemy could set their watches by. We all told him it was dangerous. Predictable patterns in South Armagh get people killed. Nothing changed.

We were forced to fly patterns we all knew could get us killed.

When I first went to C Flight, Dean had asked whether there was a medals team. There was now. And it wore rank.

I think we all expected the tactical situation to calm down given that political tensions seemed to be on the wane.

Another standard Lynx 5 duty in South Armagh proved otherwise.

We had a call that a body, albeit the man was still possibly alive, had been located on Doran's Hill, a short distance from the notorious Barcroft housing estate, on the western edge of Newry.

We flew in the ARF with medical teams. The man was however dead. Very, very dead.

It was Eamon Collins, a former senior IRA intelligence officer from nearby Camlough, who became disillusioned with the organisation's internal culture and brutality.

After a period of arrest and interrogation in the late 1980s, he'd agreed to give evidence against his former comrades, but later retracted. In 1997, he published *'Killing Rage',* a powerful and controversial memoir detailing his life inside the IRA and his subsequent rejection of violence.

Believing (incorrectly) that the ceasefire would offer protection, he returned to Newry. He set a pattern of walking his dogs and was ambushed.

He was brutally murdered near his home.

This wasn't just a murder. It was a message. Collins had been placed on the road, his chin on a kerb. A car was used to run over half his face. His eyes were gauged out and his ears cut off. The coroner described it as one of the most "brutal, horrific and grotesque" murders he had encountered. His family could not have an open coffin at his funeral.

Once it was clear that this was a recovery operation, I called in Lynx 8 and 9 to take over at the incident.

No one has ever been convicted of murdering Eamon Collins.

Some things never changed.

Matt appointed himself as a Lynx 1 commander and brought Gabe along for the ride. He removed Grant and the SF troop commander from the loop and decided only he would appoint Lynx 1 commanders and then, the names would have to be published on part 1 orders from Regiment.

I spoke to Gabe; he was just as baffled about these appointments but couldn't argue.

By now, I was regularly being rostered as the Lynx 2 commander and was knocking on the door of Lynx 1 command.

Paul Mason had learned the same lesson years on predictability earlier: that violence doesn't announce its arrival; it waits patiently until you stumble into it.

He learned it on a foot patrol in Carrickrovaddy Wood, near the village of Belleek.

In August 1991, a Coldstream Guards patrol was three days into the ground when a 300lb landmine detonated beneath them, possibly triggered by the very countermeasure equipment meant to keep them alive.

Lance Corporal Simon Ware was killed instantly.

He was twenty-two, married for only five months.

Paul carried that moment like a fracture that never quite set. The noise, the smell of burnt peat and explosives, the sudden absence of someone who should have walked away with them.

It changed him in a single second, the way sudden violence always does.

When his ground tour ended, Paul applied for Army pilot selection. Maybe he wanted distance from the road, maybe he wanted a different view of the fight, but it gave him purpose again.

It also put him on a path that crossed mine.

Had Simon Ware stepped a metre to the left, Paul and I might never have met. That's how close these things run. Careers, friendships, whole lives redirected by a moment that no rulebook can account for.

The men who planned and celebrated these murders walk freely today, some recast as politicians or community leaders. The soldiers who were sent in with rifles and rulebooks grow old under investigation.

The politicians who set the conditions vanish into history, reputations intact.

That imbalance still burns.

Annie and I had set a date to be married towards the end of 1999. The house was coming along nicely and was a haven from life in Bessbrook Mill.

Back in 655 the dysfunction was continuing, this wasn't what I expected. Matt had brought with him a handful of Sgts from his previous Squadron in 3 Regiment. There were a couple of good guys, but largely they consisted of sycophants.

They were immediately placed in positions that would 'enhance' their promotion prospects, much to the chagrin of Sgts who had been with the Squadron and served their time.

Trust in a Squadron is oxygen; once it leaks, everything else suffocates. Not deterred by the fate of Eamonn Collins, at least two of the new Sgts (they shall remain unnamed!) became 'touts', acting as a conduit to Matt for things going on and being said in the mess.

This did not go down well at all; the mess was our space.

After one weekend, I had heard that one of the doorgunners had been in trouble on the Saturday morning. A sound lad that had got drunk on a Friday evening and gobbed off to a chef the next morning, still worse for wear.

Instead of a beasting, or some other suitable punishment, Matt decided to bust him back to Air Trooper and post him out to Germany. A total overreaction, and an object lesson in how not to handle minor discipline.

I went into the SSMs office to talk to Mel's replacement, WO2 Mike Gilliam.

"Mike" I said, *"what's all this crap about Dave Clarkson?", "he's a good bloke, we should be looking after these guys, not shitting all over them."*

Matt was in his office and heard the whole conversation through the sliding window. His head shot through the hatch like a whack-a-mole and he went berserk about it being his Squadron and me needing to wind my neck in before fucking off.

On a positive note, I can say that mutual dislike never interfered with flying operations. Once we climbed into the cockpit, whatever we thought of each other on the ground stopped mattering the moment we plugged in.

We both understood the job and respected what the other brought to it. We didn't have to like each other, but we trusted the aviator in each other, and in that environment that mattered more.

'C' flight training was moving at a fair pace, and I was crewed with Matt for a couple of sorties at Ballykelly, mainly so he could check me out. In the aircraft he was focused, competent, and entirely professional.

I passed the Lynx 1 command check and was duly appointed on Regimental Part 1 orders and started to carry the L1 pager, as well as the standard Bessbrook duties.

We knew there was a major SF operation/exercise in the offing. Part of the exercise was to perform a hostage rescue mission. A local landowner allowed us to use his remote, stately home as the target location.

As duty Lynx 1 commander I was sent to conduct pre-exercise daylight recces.

These recces would mean mapping all possible obstructions; wires, loose items like bins, waste and anything else that could hurt us by day or by night.

The exercise, Op Priest, was called.

My pager went off, and I reported to the sqn lines, met the rest of the flight and signed out the aircraft. Hopping from the main dispersal over the fence into the hangar took less than a minute.

I was rostered Lynx 3 commander, my favourite role: hard manoeuvring with Lynx 2, no pissing around. The Lynx 1 commander?

Yep, the boss.

In the hangar Ops Room we sat through extensive briefings from all aspects of the SF, very much an 'all-informed' briefing. I stood up to give the aviation briefing to the assembled team. I had prepared an overhead projection slide that showed all locations of things that would affect helicopter operations.

I highlighted a set of wires that were 30ft above the ground, the telephone wires to the house. To get the assault troops close enough that they could effect an immediate, explosive entry, the plan, I briefed, was to use fast ropes.

Lynxes 2 and 3 would attack simultaneously from different directions, on NVG, the SF assault teams would deploy the fast ropes and do their thing. We would either ditch the ropes or have them brought back into the cabin, if there was anyone left onboard.

A full stick can be on the ground in seconds, after which the aircraft departs immediately to minimise exposure.

A tactically robust plan. *"Any questions?"* I asked.

"Yeah, we're not doing that, you're landing" said a shortish member of the SAS with a Glaswegian accent.

I explained why fast-roping was quicker and safer. He cut me off.

"We're not fucking fast roping! Alreet!"

Jokingly, I said *"what's up with you, you're a trooper in the SAS!"*

His demeanour changed in a heartbeat.

"I haven't been a trooper for a fucking long time. We're landing – got it?"

In the sequence of events leading up to the final assault, I was just in awe; listening in to the tactical radios and the running commentary coming from the guys, already in observation positions directing the 'trojan' vehicles, and then calling us, the aviation option, in what was part of an extraordinary team.

We landed, as instructed, about 50m from the house, the assault went ahead, and we won the day.

I heard during the extensive debriefs that my man was carrying an injured back; hence his reluctance to make it worse by fast roping in an exercise; entirely understandable.

Chapter 96. Tragedy Strikes Again

The danger now wasn't so much the enemy anymore.

It was complacency.

Over-enthusiastic pilots itching to throw a Lynx around like they'd heard about, but without the finesse South Armagh demanded, even if the kind of multi-weapon ambushes that had defined the place were becoming increasingly unlikely.

An interesting sidebar to 'C' flight was that we were there for full support to SF, not just exercises. SF soldiers that were freefall qualified still had to remain current on those skills.

Lynx 1 was called out, yep, I was on the pager, but as there no possibility of a medal, I got to do the task. It was dropping freefall parachutists up at Ballykelly.

We trundled up to the north coast and were briefed. All the jumpmaster needed from us was altitude and accurate nav. The Lynx ceiling was 12,000 feet pressure altitude, or 10,000 density altitude. We did the sums, 12,000 was possible, but crew limitations capped us at 10,000 to avoid hypoxia.

We checked winds at 8,000 and again at 10,000 with streamers. Conditions were acceptable, if a little marginal. The jump master was satisfied all was in order, even though we were directly above the centre of Lough Foyle, a large body of water that separates the River Foyle from the Irish Sea.

I created a waypoint in the RNS252 navigation computer as the point to drop, as we had no visual markers on the ground.

The next day, Dean came up in Lynx 1, I went up in Lynx 2. Load up, climb up, they jump out, we go and pick them up. Repeat.

On the third day, the margin disappeared. One of the free-fallers lost position in the cloud, the wind shifted, and he was gone.

He ended up in Lough Foyle; a long way out.

Rescue launches were dispatched from Londonderry airport, the parachutist was safely picked up, but was near hypothermic, the Irish Sea is not known for its balmy waters.

Not long after, tragedy stopped being a near miss.

It was mid-May 1999 when a Lynx from 657 Squadron lifted from Dishforth en route to RAF Odiham. Near Tilton-on-the-Hill in Leicestershire, one of the engine oil pressure gauges flicked from green to amber and back, just for a moment.

It resonated uncomfortably with my own incident in Bosnia, under the wires north of Bugojno. But unlike mine, this one ended in tragedy.

The aircraft suffered a catastrophic engine failure. Major Whitely, the Squadron commander, fought to keep it in the air, trying to steer it clear of houses while black smoke filled the cockpit. He leaned out, searching for the ground through the smoke, and brought the aircraft down in a field.

Three crew members; WO2 Andreij "Batco" Prenczek, SSgt Stu Donnan, and SSgt Pete Clyne died instantly. Whitely survived, badly injured.

It was a gut punch.

A stark reminder that all the tactical sharpness and elite-level training in the world meant nothing when pitted against a mechanical failure. You're always just one failure away from the worst day of your life.

This was an elite crew. Experienced. Respected. Gone in seconds. 657 needed replacements. I was called into the boss's office. He asked me if I'd be willing to move across.

It wasn't a straightforward decision. I'd already received my preliminary posting order to become the first EWO on 656 Squadron, 9 Regiment AAC, the unit earmarked to convert to the Apache first. Paul had recently passed his QHI course and looked likely to be posted there as well.

For both of us, this was the dream posting.

Annie and I had already planned our wedding for later that year. I'd assumed we'd marry and then move to Dishforth for the Apache. This just brought things forward. After discussing it, we agreed: we'd either sell the house in Jordanstown or rent it out.

I returned to the boss the next day and accepted, on the condition I'd be granted leave to return to Northern Ireland and get married as planned.

For the first time in a long while, everything lined up. A front-line Squadron. A clear path to Apache. Paul likely alongside me. A wedding on the horizon.

It felt earned. Not lucky. Just the next step after years of doing the job properly.

Tragedy has a way of rearranging priorities without asking permission.

Matt was keen to see the back of me. I was more than ready to leave him. I had a week to hand over and report to 657. We both got what we wanted, and for once it felt like the future had finally arrived.

UK, Dishforth: 9 Regt AAC, 656 Sqn, Jul 99 – Jan 01

Chapter 97. Hero to Zero: Betrayal

I arrived in Dishforth with some trepidation. I took my bike and left Annie with the car. After checking into the mess, I headed over to the 657 hangar and found Lynx Flight.

The crews had no idea I was inbound, and it showed. I knew most of them, but the crew room felt like a place where someone had died, because someone had. To them I wasn't reinforcement. I was a replacement.

The OC wasn't exactly holding court. Some crew members openly questioning his role in the crash. Everyone processed it in their own way, mostly with silence, suspicion, or passive aggression.

After a brew and some awkward silences, I went to find the OC. While the leadership were expecting me, they were clearly unimpressed to learn I'd be posted out again soon to 656 on Apache. They needed someone permanent, and although I was ready to throw in everything I had, they didn't want a stopgap. I got it. Still, it stung.

The next week I was sent to 656, my future home with the Apache. 9 Regiment's CO was Lt Col Ivor 'Thacko' Thackery. Bob Scarlet was now a Maj and the Regimental second-in-command.

The RSM was Len Best, a legend in the AAC, he had been in the Falkland's war and was 'the face of the AAC', along with Dick Kalinski when they were photographed with a Scout aircraft and a severely injured soldier.

The Adjutant was Capt. Patrick Le Man. I already knew the shape of this individual. He was well known in 5 Regiment, for all the wrong reasons. Integrity wasn't a priority in his world.

Settling into 656 was like slipping on an old pair of boots. Familiar faces, familiar bullshit. Before I even flew, I was volunteered as Lynx Flight Commander for a Brigade & Battlegroup Training (BBGT) exercise. I knew the format well from Command Troop days.

The AAC was starting to get its own Battlegroups now, traditionally the armoured corps or infantry would command these formations, and this was the first time an AAC Lt Col would take command of one.

Sadly, for me, the AAC and the Army in general, Thacko was that Lt Col.

In my view, Thacko was loud, aggressive, and about as tactically sound as a brass band navigating a minefield. I tried to offer quiet advice. He took it as mutiny. So, I did what SNCOs do with bad plans. I stepped back and watched it fail.

That night, Sam Regent, a Gazelle pilot and former Paratrooper, and I slipped out of Catterick Garrison and headed to Richmond for a beer. Thacko had gated the staff like we were teenagers.

We got back unseen.

The next day's looks said we weren't as invisible as we thought.

The BBGT was a disaster. We packed up and went back to Dishforth with Thacko visibly seething.

I had a couple of flying check rides with the SQHI, Chris Ewell, a wiry Brummie who I hadn't met before, but got on with immediately. He was a pragmatic and realistic QHI with good skills. Once UK CR checked out, I was tasked to work with 'them' again.

'Them' being the bloody Italians. It seemed I was doomed to an eternity of working with them and couldn't escape. 656 had a long-standing exchange and partnership programme with an Italian Regiment from Bolzano in northern Italy.

I was to escort their two Kiowa's and a Huey from Dishforth to Chelsea Barracks in London, take them out for a night in the town, and then escort them to the channel coast.

I was flying with a young Captain, Fred Mills. Fred was a great guy, but had an immature sense of humour, he'd just say random things to himself and then piss himself laughing regardless of anyone around him.

When I briefed the flight plan to the Italian crews, I was explicit about sticking to the route and close to me, London City Airport, who controlled the heli-lanes, would not tolerate the slightest deviation.

The plan was to enter heli-route H4 at the QEII bridge, then head westbound to Chelsea bridge before landing at the barracks parade square. All the bookings were done and predicated on this routing.

All went well until Stansted refused us entry to its airspace. We had to detour slightly; the Italians panicked and ran themselves almost dry of fuel.

We landed in Chelsea barracks with them on vapours.

That night, thanks to some shameless blagging, we all ended up in the VIP section of Stringfellows. A good night.

Best left at that.

As Fred lifted us into the hover to go home the next day, I could already see the Italian Kiowas racing down the heli-lane, completely ignoring the reporting points, the Huey was well behind them having had fuel delivered by truck from Battersea heliport.

I told ATC we were no longer in formation and that the Italian callsigns were responsible for their own actions. I wasn't going to trash my callsign.

That was the job done, and for once, nothing followed us home.

For the Millennium, we went to Osnabrück, where Ed and his wife Alana were now posted, to spend time with the kids. We dressed from a charity place called Kilo's and looked like a gang of bankrupt pimps. Then we bought enough fireworks to frighten the Germans into another treaty.

We survived the Millennium Bug.

A few weeks into the New Year of 2000, 9 Regiment was on a tactical exercise on Salisbury Plain, I was operating with a timid floppy commander and not much reconnaissance was going on.

I used some initiative. I contacted my old mates in 'A' Sqn, The Light Dragoons, who just happened to be playing enemy.

They gave me their tactical radio frequencies and some of their HQ locations. Good lads!

Listening in to their radio comms, I fed in real sighting reports. RHQ was thrilled. Thacko could barely contain his indifference. I think he knew I'd bent the rules, but I'd made him look good, well, not as bad as he was, so he tolerated it.

Meanwhile my divorce dragged on like an over-budget soap opera. Emily insisted on German family law, which meant paying lawyers to translate Kafka into invoices.

It finally landed in the Royal Courts of Justice, three and a half years in, and still bleeding money.

656 had received a brand-new pilot, Roger Smithfield, a nice lad. On arrival in the sqn, he had done his standard induction flight with Chris, who had immediately decided to prolong his flying probation and shorten the interval between check rides from the standard 6 months, down to 3.

Additionally, Chris would only allow certain commanders to fly with him. Deep joy.

I went into the first sortie with him with an open mind, as with Roz Parnell, and had my eyes opened. I had given him a low-level map reading exercise to plan and fly.

I wish I hadn't.

He was on the very edge of control and simply could not get ahead of the aircraft; he was all over the place. This was one of maybe 1 or 2 occasions where id taken control from a pilot and didn't hand it back!

I briefed Chris on return. The word 'liability' may have been used.

I think that was the last I flew with him, he was RTU'd shortly afterwards, highly unusual for a 'winged' pilot to get chopped from a Squadron.

I rang Paul in Wallop to see who his instructor had been and to see who was going to take responsibility for this lizard.

It was him.

Paul told me he'd tried to chop him at Wallop but was told the same thing everyone was told, the system needed bodies more than it needed competence.

The time came for our wedding, and I went back to Jordanstown. Now being a visitor to Northern Ireland, I had to report to a local barracks for an intelligence briefing and to book in and out of the province every time.

The wedding was at St Gerards Catholic Church on the Antrim Road. Most of 655 had been invited and we had a great time.

Truthfully, I wasn't nervous about marriage. I was nervous about Annie and me. We laughed, we worked, we looked solid from the outside. Underneath I felt a quiet disconnect, like I was playing the part of the man she wanted. I hoped effort would build what I couldn't quite feel.

Even then, I think I knew.

Maybe that's on me, maybe I let the momentum carry us forward.

The girls came over as bridesmaids, but they didn't seem thrilled to be there. I don't blame them. It was a strange day for me too, not cold or unhappy, just… clouded.

Paul was my best man.

After a wedding lunch in Belfast Castle, the party was in our garden, on the huge deck I'd built, in a marquee. We had fireworks and it really was a great night. The next day, as planned, we headed off to Glasgow for a flight to the Caribbean where we had a honeymoon cruise around the islands booked.

On return to Dishforth, as usual I went to the orderly room to give them my updated personal details, date of the wedding etc. I needed to so that I could then apply for a married quarter, although we had been looking at houses in nearby Boroughbridge to buy.

In the meantime, Paul and I had been back in touch with Vince Stella who was now based in Hunter Army Airfield, Savannah, Georgia with a different Apache unit, The 159th Aviation Regiment.

We were both holding posting orders for Apache conversion, Paul would go to the US to complete the Apache training with the US, I would go through conversion in the UK, so we thought that we should be preparing for our future.

Vinces' CO invited us to visit the 159th to observe their operations in the field. Four of us went on this learning mission, Paul, Me and two Captains.

The trip was worth every penny. We learned a lot about how the Americans ran their Apaches in the field, including the bits the brochures never mention.

We also learned how to order food in a breakfast joint in the US without being asked seventeen questions. That took longer.

We got back to the UK late on a Friday evening, I had a weeks leave.

In my absence, 656 had received a new OC, Maj. Alan Taylor. I went in on the Monday morning to say hello.

"Ah, Staff Termini, nice to finally meet you, how was the US, I hear it was a blinder" he calmy said.

"Yes, nice to meet you too sir. The US was a revelation in many ways"

"Well, Staff, I have to tell you that you're confined to the mess with immediate effect" he replied

"What? What do you mean?"

"Allegations have been made; serious allegations about your wedding" "I can't tell you anymore"

At that moment, my life fell apart.

The chain of command abandoned me and made my life a misery.

The only, and I do mean only, support I had was SSgt Barry 'BJ' Jones and Capt. Arvin Williams, the Lynx Flight commander.

So, what the fuck had I done that was so bad?

When I had reported my details, the clerk transposed my wedding and divorce dates. On paper it made me a bigamist.

Instead of asking me to sort it, Le Man went straight to SIB in York. No phone call. No quiet check. Straight to the guillotine.

The charges, as ludicrous as they were, came down to:

1. Bigamy: due to the clerical error transposing my wedding and divorce dates.
2. Entering an "illegal building": St Gerard's Church
3. Being in an out-of-bounds amber area: St Gerards location.
4. Falsifying a document: my leave pass.

Thacko devoted tremendous energy to pursuing these ridiculous allegations against me, seemingly more focused on making them stick than on considering whether they were warranted.

I attended a formal interview at York SIB with an RMP Sergeant who was not as intelligent as he thought. Arvin came with me. By lunchtime it should have been over.

The bigamy charge was easy to make them drop; a couple of certificates and a decree absolute did that.

St Gerards was indeed in an amber area of Belfast. I asked the RMP to produce the out of bounds map, which they promptly did, almost as if it were close to hand in a file...

I then asked them to read the small print regarding amber areas.

I was totally legal.

I challenged their assertion that any church was out of bounds, and they quickly crumbled, soldiers have right of religion.

And finally, the leave pass. I had put my mother's address as a contact, standard practice. I asked them if they would be able to plot the Lat & Long of every turning point for the ship even if I knew them.

Within a matter of hours, I had proven every charge to be false and yet they kept going.

They weren't investigating anymore. They were trying not to look stupid.

For five months they kept going, not to find the truth, but to find something that would justify the noise they'd already made. I was marched into Thacko's office again and again. Same desk, same spittle, same rage.

BJ and Arvin were the only two who treated me like a human being.

My solicitor, a deeply unimpressed reservist Captain in 4 Para, was horrified by the pantomime unfolding in 9 Regiment. Without telling me, he'd hinted to the RMP that if the papers caught wind of this circus, they'd need extra ink and a double page spread.

Paul, as my best man, faced the same charges, apart from the bigamy rap. His CO could see it was nonsense and dismissed it immediately as there was no case to answer.

Meanwhile, I was on first-name terms with the CO's doormat. I was getting marched in so often I could have installed a turnstile. Each time, the same routine, screamed at from about eight inches away, flecks of spittle on the paperwork, the kind of rage usually reserved for courts martial or an unexpected mess bar bill.

Word had got back to RHQ that I'd gone to the papers, which I hadn't. But Chinese whispers being what they are, that didn't matter. BJ lent me a portable dictaphone, just so I had a record of what was being said, or rather, screamed.

One day, it loudly clicked off halfway through another session. Thacko nearly had a seizure.

After 5 months they trumped up another charge. The RMP had been through every leave pass I'd ever submitted whilst in NI and cross referenced them against every possible ferry crossing.

I had once used the Holyhead to Dublin route, technically illegal, but on a motorbike, it was a no-brainer.

I admitted that I had used that route, but told them if I was charged, I would name every other biker in 655 Squadron who had used that route, including Matt Leggett, the person that recommended the route as being the most convenient.

When it eventually came to a head, Arvin had been advising me '*to accept the CO's award*'. A CO's award is the Army's way of letting you plead guilty to something you didn't do, in exchange for them pretending it never happened.

I had zero intention of doing so and was ready to go to courts marshal, in fact, I wanted to. I wanted my time in court to expose this travesty for what it was; personal and vindictive.

I had made the decision to leave the Army; I wasn't going to be treated like this by anyone. I longed to see my Commanding Officer and his Adjutant held accountable for how they treated me.

Arvin, who I trusted, was insistent. *"Just accept his fucking award and all this goes away"*

The day finally arrived.

Dressed in Service Dress (without medals), Len marched me into Thacko's office like a condemned man heading for the scaffold.

I sauntered in.

Thacko was already turning purple and I hadn't even opened my mouth.

"FUCKING SALUTE!"

I threw up the laziest salute I could manage. It was technically a salute, in the same way a garden shed is technically architecture.

That didn't help my case.

If anything, he became even more incandescent.

By now he was well beyond angry and heading rapidly towards a medical emergency.

He read out the charge, the single charge about the ferry into Dublin.

"Do you accept my award?" I looked at Arvin.

Arvin looked at me.

I paused, looked again at Arvin and the people on the other side of the desk.

I stared at my boots and despite my best intentions I quietly said *"Yes, I accept your award"*

"SIR!" He shouted back, beginning to sound like Chewbacca with haemorrhoids.

"Yes, I accept your award. …………sir"

"Case admonished, **GET OUT!"** *"Get out of my office"*

An admonishment is typically recorded but does not count as a disciplinary conviction or criminal charge.

I walked out, only to be shouted at again. Arvin was all smiles. I wasn't. This was the end of my career, one that I had fought hard for, all my adult life. Lost my family over. And this was it.

I was sent immediately to the orderly room; they had heard all the shouting. *"Here, I've got something for you,"* said the chief clerk, without looking up. He literally lobbed the Long Service and Good Conduct Medal across the desk towards me like a takeaway menu. I caught it.

That told me exactly what I needed to know.

That moment ended my Army career, not by discharge, not by choice, but by betrayal. I'd given everything. Lost mates. Lost my family. And in the end, it was a clerical error and a ferry route that did me in.

They threw me a Good Conduct Medal like a dog biscuit and expected silence. I'm still angry. And I always will be because it wasn't just the injustice, it was the abandonment. It was cowardice. And it was personal.

Back at the Squadron I became a recluse, hating everything about 9 Regiment. BJ lived out in Boroughbridge, I'd catch up with him whenever I could, but in the end, I had to use the SSAFA helpline for support. A services mental health charity.

I was especially disappointed with Len. He was my RSM; the most senior soldier in the Regiment and in my opinion, he should have been the first person to defend me.

Many years later Len and I would go on to talk these events through over a beer. We listened to each other's version, had a drink and now wished we'd been closer at the time. I'm pleased to report that Len and I are good drinking buddies.

I feigned happiness that I was leaving the Army; the truth was that I was devastated. Everything I'd worked for, flushed away because one officer couldn't admit he was wrong.

As 656 prepared for another deployment back to Bosnia. I wasn't allowed to go, instead, Thacko appointed me as the training wing clerk, normally a position held by a L/Cpl. I'd been reduced to booking courses for other people because of one man.

I decided I'd go down the commercial pilot route, bypassing civilian helicopters for larger jets. I was told I was dreaming, I'd be lucky to get a turbo-jet position.

After 9 Regiment, dreaming felt like the sensible option.

Life out of Uniform

My first Remembrance Day out of uniform with Capt. P Mason

Chapter 98. The Twilight Zone

My last 9–10 months in the Army were a limbo of contradictions, still in uniform, but no longer trusted to fly; still a soldier but already looking beyond the wire. Being grounded wasn't just embarrassing, it was personal.

I started the process of becoming a civilian pilot in February 2000.

The first thing I needed to do was increase my fixed wing hours from the 46 Chipmunk hours, to over 100. The cheapest place to do that Florida. It meant 6 weeks in Titusville. I booked in with a small aviation training company called Walkwitz Aviation, based at the Arthur Dunn Airpark.

Luckily, the Army resettlement scheme paid for much of the accommodation, a nearby B&B. I started off by gaining a single engine private pilot's licence issued by the FAA, the US aviation authority. Next step, upgrade to a multi-engine aircraft. Once that was achieved, I just had to burn holes in the sky to build flying hours.

At $12 per hour, including fuel, there was plenty of cash left over to enjoy Florida. Annie, Marie & Trevor and Agatha flew out for 10 days in the middle. One of the highlights was getting an invitation to the NASA launch pad viewing area to see the space shuttle Atlantis launch on STS 101.

I stood there, rooted, eyes fixed on the launch pad as the countdown hit zero. A sudden, searing light burst from beneath the Shuttle, brighter than anything I'd ever seen, even through the morning haze. The ground rumbled like distant thunder, then all at once, the noise hit, not a sound, but a physical force that pressed against your chest and rattled your bones.

I watched in absolute awe as that machine, an improbable beast of engineering and courage, tore itself from Earth's grip and climbed into the sky. It was violent. It was beautiful and emotional.

Once I'd passed the US FAA multi-engine tests, I just cruised up and down the Florida coastline in a PA-44, a small twin engine aircraft, with a guy called Joe, a US Blackhawk pilot doing the same thing I was, just building hours.

We had a route. It involved flying to down to West Palm Beach or Boca Raton, park up. Grab some chicken wings at the local Hooters and head back up the coast to Titusville. We took it in turns to get P1 or P2 time logged. One evening as we were approaching Fort Pierce from the south, Joe looked out at the engine on his side and said *"Dude, how's your single engine work going?"*

"No worries, why?" I replied wondering why he'd asked that.

"'Cos this side's about to quit" he'd seen a stream of jet-black oil surging out of the engine coaming.

We called an emergency, shut the engine down and landed in 'Treasure Coast International' at Fort Pierce, called up the owner and told him where his POS aircraft was, reiterating that he was paying for our hotel for the night.

And the Hooters.

The owner, Heimi Walkwitz, wasn't having much luck. At the same I was there with Joe; there were a couple of Tunisian students. Neither could speak much English. They kept having minor accidents; a trip across muddy grass wrecking a nose wheel or infringing Miami's airspace. They left without paying their considerable bill.

I left Titusville having logged just over 100 flying hours, which when coupled with my Chippy hours, was more than enough. Next up, the UK CAA Commercial Pilots ground school exams.

There were 13 exams grouped into either Tech or Nav. You had 3 shots at each exam, fail the third attempt and the whole, expensive group reset. I attended ground school at a Cranfield airport called 'Bonus Aviation'.

Still technically in the Army, I was able to book into the closest Sgts Mess, The Army Intelligence Corps were not far away in Chicksands. This absorbed much of the cost for accommodation and food.

The subjects I had been used to and gone through at Middle Wallop were easy; the others not so much. The CAA curriculum was out of date. The exams themselves were ridiculous. The navigation exams required precision that was not possible without a laser and a micrometre.

After I'd finished at Cranfield, I signed up for revision classes, joining me were Mac and a couple of others I knew from the AAC, they were going for rotary licences though.

I sat the nav group first and failed 3 of the 8 exams. Then passed all the tech exams first time around, which was a massive relief.

On my first resit, I knocked off 2 of the 3 failures, scoring 74% on Nav 1, the hardest exam of them all. Fail again and the whole group was reset.

My second resit was in the CAA HQ at Gatwick airport. This would also be my first attempt at the aural morse code test.

I sat there when the invigilator said, *"you may begin"*. I turned the paper over, read through it, and panicked. I couldn't see a path to any of the answers.

I took a deep breath and gave myself a mental slap.

I eventually put my plotting instruments and pens down with a minute to go.

Straight into the aural morse code exam. In a portacabin at the second busiest airport in the UK. Straining to hear the 'dah's' and the 'dit's' from a clunky old cassette player; they would play 10 groups of 3 letters, replicating the identification signals from old radio navigation aids.

Get a single letter in a group of 3 wrong, that counted as a failure. Get 2 groups wrong, or more than 3 letters, that fails the exam.

I went home unsure of how I did. I knew I'd got a couple of letters wrong, I totally guessed on one. It must have been correct because I passed the morse exam.

Then the letter from CAA arrived in the post. An 80% pass on Nav 1.

I nearly cried with relief. Ok, I actually cried.

The final hurdle was a Commercial Instrument Rating. Whilst I had been good on the dials in helicopters and held a master green rating, this was a different ball game altogether. holding patterns, precision and non-precision approaches and asymmetric approaches (single engine).

I had never had instrument landing systems (ILS), VORs or the range of other ground-based aids I now had to learn, fly and master in very short order.

I signed on with the Professional Pilots Study Centre (PPSC) at Bournemouth International Airport, living in the mess at Bovington this time. My instructor took me straight back to basic fixed wing at Middle Wallop, a retired former RAF pilot called Dickie Snell.

Except this man was good, very, very good at what he was doing. He took no prisoners but could explain all the different entries to various holding patterns like I was 5 years old.

In late November I took my commercial instrument flying test with a CAA examiner. In nearly 2.5 hours I flew from Bournemouth to Exeter, via a zone crossing, onwards to Cardiff, then Bristol and home to Bournemouth without seeing the ground for any length of time. Different approaches at each airport all the way down to minimum decision height (DH). DH is the point at which you decide whether to land or go around.

Anyway, as we taxied back to the school at Bournemouth having done the final approach and landing, I was handed a pass certificate. I bought Dickie a bottle of bubbles and went home to Jordanstown.

Throughout all of this I had been attending job fairs and had done an interview techniques course. I had had a couple of bites from recruiters; one from Aurigny, an operator that flies turbo-prop Trislanders, a 3-engine version of the Islander, out of the Channel Islands.

The other was with Maersk Air. Maersk flew the Canadair CRJ regional jet and 737s out of Birmingham.

I knew which I wanted.

On passing my IRT, I now held a frozen Air Transport Pilots Licence. This meant I could only exercise the rights of a commercial pilot until I reached 1500 eligible hours, only 100 rotary P1 hours could be counted. I called Stan Lewis, the Maersk recruiter. Stan was a South African expat; I had got on with him immediately at the jobs fair.

Stan invited me for an interview with the Chief Pilot, Chris Ward. I was offered a position as First Officer on the CRJ fleet starting in Mar 2001.

In less than a year, I'd gone from booking training courses for other people to sitting in the right-hand seat of a commercial jet.

Chapter 99. The Aftermath

Nearly 18 years in. Four years from pension. Close enough to smell it. Just three hours short of 2,000. The sort of number you notice when you're not allowed to fly anymore. A string of outstanding reports. I had gone from a Junior Leader on the YTS, to Special Forces aircraft commander, an Advanced Electronic Warfare Officer and Helicopter Weapons Instructor.

And yet my final commanding officer couldn't even be bothered to sign my annual logbook summary.

That was the level of care.

My last summary was "*PP'd*" by someone I'd never met. That's how it ended.

I'd worked through stables troop and dog section, driven an ambulance and operated a Chieftain tank. I'd made my way through to command troop through not taking no for an answer.

Sure, I had a couple of lucky breaks and was helped by some outstanding officers, for it all to come to a spectacular end because of one appalling individual.

Looking back, I don't miss the Army.

What I miss are the people.

I miss the friendships forged through shared misery and bad decisions as well as flashes of complete professionalism and brilliance in the very face of adversity.

I miss the laughter. I can't ever imagine serving a blend of kitchen leftovers to anyone else.

I miss the sense of belonging.

I miss being surrounded by people who would, and did, risk their lives for each other without a second thought.

I wish my children could experience that bond, that sense of purpose, to walk in my boots and feel what it meant. The Army can make or break a man. On balance, it made me. And for that, I'm grateful.

I do have regrets. I regret not staying on the career path that had been laid out in the posting to the Apache Fielding Team. I regret not fighting the Apache in Afghanistan. But that's about it.

There was one small moment of satisfaction. When I returned from the States, the CO of 159th sent a letter of commendation. He misspelled "Thackery", missing a 'K.' Thacko lost it and scribbled it out with angry exclamation marks. It was petty.

I still have the letter as a reminder.

He even banned ketchup in the mess. That tells you everything you need to know about his relationship with joy.

That detail, and a few others, made their way, anonymously, onto a pilot chat forum, PPRUNE. I ended the post with: *"Would you go to war with this man?"* It caught fire. The Daily Mail, The Sun and The Guardian picked it up. Steve Wright mentioned it on his BBC radio show. The Sun sent topless models to the gates of Dishforth waving ketchup bottles.

Col. Ketchup, they called him. I like to imagine the volcanic rage that followed.

And I smiled, hoping that no innocents were caught in the crossfire.

I was ready for the next chapter, self-respect, stability, a predictable roster and time with my kids. Maersk offered all that. More than that, it gave me back my pride. I still think of Gary Fenton and that day in Bessbrook, that unresolved feeling of being stuck in the middle, as I closed this chapter of my life.

If Thacko was the baddie, there were many other officers that were the opposite, that were truly inspirational. Simon Edwards, Chris Collett, Gary Coward and Nate Mossman.

I'd rebuild myself on my terms. I'd work hard, earn command, and stand tall again. I'd stay close to the people who mattered. The rest, those who tore me down or did nothing to help, when they should have, could watch me climb from a distance.

Part 5: A Civilian in Uniform

Canada, Montreal: Feb 01 – Mar 01

Chapter 100. Settling in

My first experience in civilian life was very, well, civilised. Having accepted the job offer from Maersk I received a pre-course study guide and tickets to Montreal. Business class tickets, which, after 18 years of military travel via Crabair (RAF) felt like flying in luxury.

My Canadair Regional Jet (CRJ) type training was being conducted at the Bombardier facility in Montreal, and it would last 6 weeks.

My journey started from Belfast and met up with my two fellow trainees at Heathrow. Dave Bird and Richard Woodward. Dave was also a helicopter pilot having come through the civilian route. He had worked for Bristow's helicopters and was on the Westland EH101 programme before that. Richard was on his first piloting job having previously been cabin crew for British Airways.

I liked them both, but I had much more in common with Dave in outlook and sense of humour.

We arrived in a snowbound Montreal and settled into a serviced apartment just off St Catherine Street, the city's main drag, which transforms into a buried underground shopping precinct when winter dumps snow across the pavements. Useful for food, shops, and dodging frostbite.

A minivan from the training centre collected us daily. We dove straight into computer-based training (CBT), with an instructor hovering nearby in case we had questions. It was efficient, but soul-sapping. No Jon Slade winding up the instructor, no Pim pretending not to understand lift vectors. No jeopardy. As long as you didn't completely cock it up, or die, you were going to pass.

After 2 weeks, we were introduced to our sim instructor, Philippe, a young French-Canadian lad with a heavy accent. He'd never held a commercial job but was a reasonable instructor. In the simulator we went through every possible non-normal procedure, normal procedures and reversionary modes of flying.

The emergency handling was much more relaxed than in the Lynx. We only had to react immediately to a handful of extreme situations, most of the time we would have time, and were expected, to get the Quick Reference Handbook (QRH) out and follow it in a very deliberate manner.

One person would read aloud the drill, the other would then execute the actions whilst being monitored: very steady, specific and controlled.

The basic tenets of airmanship apply regardless of what you're flying though;

Aviate – Navigate – Communicate.

Translated, that means fly the aircraft safely, know where you're going, and talk to people.

The sim sessions were scheduled throughout the night, so occasionally we would have a full day off and work by night. Compared to the Army's old Lynx sim, which looked like someone had strapped a television to a tumble dryer, the CRJ simulator was a revelation.

The graphics were exceptional, the motion realistic, and the experience immersive enough that I occasionally forgot we hadn't left the ground.

Weekends involved a nearby ski resort; Trois-Rivières, not to ski, but for snow tubing and beers. A perfect balance of thrills and hangovers.

We were signed off and heading home. The next time we met it would be for 'base training', flying a real jet, but with no passengers.

I got home to Jordanstown in good spirits, so far, civilian flying was ticking all the boxes: clean hotels, decent food, and no one barking orders before 6am. I also got the news that Annie was pregnant and due mid-August. That gave me a jolt of nerves.

Things with the girls had thawed since the wedding, but there was still a coolness. I wondered if, like I had as a kid, they'd quietly held out hope their parents might reconcile. The wedding might have felt like the full stop they hadn't wanted.

Maersk Air Ltd was the UK-based subsidiary of the Danish airline Maersk Air; part of the Maersk Group empire more commonly associated with ships and containers. Based at Birmingham Airport (BHX), Maersk Air UK operated under a full British Airways franchise.

The cabin crew wore BA uniforms; we wore Maersk uniforms. The aircraft were in BA livery, served BA's catering, and used BA's flight numbers and booking system. It was all very British, on the surface.

The fleet at that time consisted mostly of the CRJs with a few ageing Boeing 737s still hanging on. Routes spanned the UK and Europe, including regular runs to Belfast, Edinburgh, Copenhagen, Gothenburg, Stuttgart, Milan and Zurich; basically, joining the car manufacturing hubs of Europe.

It was busy, sharp turnarounds, but civilised compared to military life.

Before the CRJ, the company had operated the BAe Jetstream, a small twin-turboprop that required a crew of two. This legacy meant that Maersk had historically recruited heavily from the Royal Navy, where multi-crew cockpit discipline and small aircraft experience was the norm.

As a result, many of the senior captains were ex-Naval aviators, which added an interesting dynamic to life on the line. It wasn't unusual to hear nautical expressions on the flight deck or to feel like you'd signed up to the Fleet Air Arm by mistake.

In the air, we were BA. On paper, we were Maersk. In the crew room, we were something else entirely, a hybrid of Danish logistics, naval nostalgia, and British Airways polish. It worked well, so long as you didn't poke it too hard.

I was provided with staff tickets to get to Birmingham for the start of my training. The operating crew saw me as staff on the passenger list and could not have been more welcoming towards me. I had arranged a B&B nearby to stay for a couple of weeks whilst I completed my base training.

On reporting to the Maersk operations centre at Diamond House at the airport, I was issued an airport security pass and my uniform as a First Officer. I met up with Stan and Chris. As it happened, Chris knew Marie, Annies's sister, from his days in BMI before coming to Maersk. Small world.

Chris briefed us on what to expect for the day.

We would depart BHX and just '*hop over*' to Shannon in Ireland for circuits and some non-normal procedures. We'd change crews over, do more circuits and then come home. This was amazing to me. A trip like that would have taken hours, if not days, to plan in the Army. That was easily a 2 – 3 day trip, not just a couple of hours as we would do.

Chris had also ensured the aircraft was catered for just the 4 of us too, plenty of coffee, sandwiches and snacks; could things be better?

On the 10th March 2001, I was first up in the right-hand seat, grinning like a Cheshire cat, and took off from Birmingham International Airport on my first flight in a civilian airliner. Chris must have thought I was high.

The main difference I found between rotary and fixed wing was the crosswind. There's no such thing in rotary; you always land into wind, so in light aircraft it took a while to get used to the techniques involved.

The CRJ was no different as I found out in Shannon. I had to work bloody hard on a landing that was enjoying a stiff 15 knots across the runway.

The CRJ also had a slightly unusual landing attitude. Most airliners descend the glidepath with the nose pitched a few degrees above the horizon. In the CRJ, the attitude is almost level; sometimes even a shade nose-down, which felt odd at first.

Flying the approach was very much a numbers game: airspeed, power setting and flap schedule are precise, with little reliance on "feel." At about 50 feet, Chris told us to bring the thrust levers smoothly to idle and ease the nose up into a gentle flare. The aim was to arrest the rate of descent just enough for touchdown, without over-rotating or floating.

Another adjustment was getting used to keeping my eyes *inside* the cockpit. In the Lynx, we were constantly scanning for conflicts. In the CRJ, we were encouraged to trust the instruments and the traffic systems.

Below 10,000 feet, Maersk ran a silent cockpit policy, a concept I initially found absurd, since I'd rarely been above 10,000 feet, let alone been silent while flying.

All in all, I was finding the training, the people and the aircraft a lot of fun. And the coffee was much better.

The next time I flew, it would be with full fare paying passengers, who rightfully expected an on-time departure, a smooth ride and excellent service. There would be a full crew, on the CRJ that was 2 or 3 cabin crew.

I had my roster for the next 6 weeks; a handful of Vienna's, various French and Scandinavian airports and a Belfast.

Welcome to life as a civilian in uniform.

UK, Birmingham, Maersk Air: Mar 01 – Apr 04

Chapter 101. On the Line

The next week I was into 'line training'; flying with a Type Rating Instructor, or TRI, the civilian equivalent of a QHI: more relaxed, more coffee and better conversation. I now checked in for duties a minimum of an hour before the scheduled time of departure, but for the line training phase we always made it earlier so the TRI could deliver any additional briefs above and beyond normal planning.

We had to successfully complete 50 line-sectors before undergoing a line check and release to the fleet.

In the crew room were a gaggle of crews, all getting ready for the first wave of departures from BHX. Sitting there was Pete Carre, with a brew, briefing a flight. I had no idea he'd left the Army; the last time I saw him was in 651 Lynx flight all the way back in Gutersloh.

I felt like a stalker. I'd followed him from 15/19H to the AAC, and now to Maersk.

It was good to see that there were other former Army pilots there, Chris Morley, another Lynx pilot who's path I never crossed across the deployments, but many of the captains were ex-Royal Navy or civilian, with a couple of RAF fast jet pilots thrown in to keep us honest.

I liked all the TRIs I was rostered with, they had their differences, but they were limited to personality, not opinions on how to fly. The standardisation across the captains was impressive; there was only one way to do things. The grandaddy of the TRIs was undoubtedly the inimitable Mr. Robin d' Erlanger.

Robin was one of those quietly remarkable aviators whose presence lifted every room. A decorated air racer in his younger years, he once ditched a Turbulent in the Channel as casually as most people stall a car and later won the King's Cup Air Race. By the time I met him he carried himself with a calm, almost courtly manner.

Robin had a gentle stammer on the ground, but the moment he pressed the transmit button on the radio it disappeared; smooth, clipped, utterly composed. Flying seemed to tune him perfectly. He was universally respected, universally liked, and for the younger or more nervous, he had a way of making the cockpit feel like the safest place in the world.

The only time I ever really, properly fucked up a landing happened to be with Robin, I must have lulled him into a false sense of security. We were landing in Vienna with a full passenger load, the aircraft was heavy and there was a strong headwind, straight down the tarmac.

Instead of thinking about the situation, I just followed the numbers, and at 50 ft I closed the throttles. We dropped like a Russian being pushed out of a third-floor window. Robin did what he could, he took control and flared as much as possible without striking the tail.

We smashed into the runway. I heard gasps and maybe a scream or two from the cabin. Then came the PA from the purser, cool as a cucumber:

"Ladies and gentlemen, welcome to Vienna where the local time is 5 to 10. I'm pleased to confirm that we did actually land and weren't shot down…."

I looked at Robin. Robin looked at me and very calmly said *"Well, that was a learning experience, wasn't it. What could we do better next time?"*

Ah Bugger!

The next time I flew with Robin, he was conducting my final line check. He simply said, *"how are the landings coming on?"* Having demonstrated I could land the CRJ, and not in Chipmunk style, I passed the line check and was now released to the full roster with line Captains.

Time now to settle down and get into the rhythm of the roster, a more varied set of destinations that now included night stops down route.

Pete was living in a shared house when he was over in Birmingham, he lived in Norfolk. This place was known as *'The Dutch House'*. Before my time, Maersk had recruited a batch of Dutch pilots, and they collectively rented the house in Shirley. A room became available in there, so I took it on.

Off-duty, Annie and I were house-hunting around Birmingham. We must have seen dozens: too big, too small, too expensive, or soulless. Nothing compared with Jordanstown, which we both loved.

In the days prior to 9/11, we were still permitted to have cockpit visits and even let friends and family use the jump seat. I loved having kids come up to the flight deck, usually followed very closely by a nonchalant dad, just accompanying his kids of course. It made life interesting.

The purser would come into the flight deck and ask permission for a visit, naturally we always said yes. There followed a flightdeck sport called '*question cricket*'.

We would take it in turns to nominate a question and assign it a number of '*runs*'. The person with the most runs of course won.

I tried to get *"can you get Eastenders on that mate"* as a six, and if I was lucky, I'd get *"can I try your hat on?"* as a four.

On one Milan Malpensa night stop I invited BJ along. I was under military law for a while yet, and BJ was keeping an ear to the grapevine on my behalf, ready to give me a heads-up if Thacko decided to stir the pot.

A free trip to Italy seemed a fair way to repay him.

He crashed at mine the night before and we spent the evening just shooting the shit like the days in the mess at Dishforth. The next morning he came with me to the briefing, where I introduced him to my captain, Jim Hodges; an ex-civilian S-61(civilian Sea King) pilot who'd flown the BA 'Airlink' Gatwick–Heathrow shuttle.

A rare beast.

I would fly the first sector to Milan, so the preflight walkaround was my job. Walking around the high 'T' tail with BJ, he just stopped, looked at me. Looked up at the size of the aircraft (which was small in real terms) and said, *"I can't believe you're flying this thing, this time last year we were flying around in a Lynx"*

The crew were excellent, made BJ feel welcome and treated him with kindness and to whatever he wanted from the bar. I was proud, they were making my friend, someone they didn't know from Adam feel welcome. Once in the crew hotel, affectionately known as 'Tracey Island' because it resembled the place of the same name in Thunderbirds, we got on the wine.

I had previously discovered they stocked Travaglini Gattinara, one of the wines I'd discovered on exercise Roman Hussar.

It was a perfect midsummer evening, old friends and new colleagues.

Eventually, I found a house that looked like it had potential. It was in a village to the north of Lichfield called Barton-under-Needwood. I did a drive by on it. I was driving an old convertible Triumph Spitfire I'd bought from a bloke on one of my ground school courses after selling my bike.

The car was cold, leaky and rusty, but I really loved it. It had character and never let me down. I also really liked this house, it didn't have much of a garden, but was also coming down with character, beamed ceilings and not a straight line in the place.

Annie came over for a weekend, looked at the house and although she liked it, it wasn't as much as I did; I could see potential in every crevice. The couple selling the house, Marcus and Kara, were staying in the village, just moving up to a larger house.

We put an offer in and after the standard negotiations moved in at the start of June. It needed some work, as with Jordanstown, which was now on the market, it needed a new kitchen!

The house was the old servants' quarters to the much larger house it was still attached to. In fact, there was still a sealed-up doorway that would have led to the main quarters.

Between the living and dining rooms was a large original red brick chimney breast that had a false wall between it and the main wall. I took great joy in ripping it out and taking the chimney back to brick. I fitted a double-sided woodburning stove in between the rooms.

The smell of the wood smoke and the heat the thing kicked out made the house very homely and brought it to life.

I kitted out the small room at the front of the house as a nursery for the baby, we didn't know what flavour we were going to get, so the room ended up a very neutral magnolia/pale yellow.

Annie had stayed in Belfast; she wanted to have the baby there, surrounded by her mum and sister for support, which was 100% understandable, I had planned my leave around her due date in August and was working hard to get the house ready.

At the end of July, I was on a Milan night stop. This rotation had us landing back in BHX early, which was fantastic as we got the rest of the day off. As soon as I landed, my Nokia was going off.

It was Marie; a very excited Marie, telling me Annie had gone into labour and was at the Crumlin Road Hospital.

I ran over to Diamond House and told crewing what was going on. In a flurry of *'oh my gods'* and *'congratulations',* they had me ticketed as staff on the next flight to Belfast. Luckily, I still had my night stop kit with me.

I arrived at the hospital, still in uniform, with my flight case and overnight bag in tow, just in time to be there when our son was born.

Chapter 102. Baboons, Babies and Twin Towers

I wanted our son to be aware of his roots, both Annies and mine. Annie was keen on calling him James Michael. We settled on Patrick James, and I instantly fell in love with the little man. He looked like me, according to literally everyone apart from me.

His face was full of expression from his first moment in this world; we knew, and still know, exactly what's going on behind those big blue eyes. He reminded me of the girls; he had Alex's looks and Jessica's mischievousness.

Annie and Patrick moved over to Barton a week or so after he was born. Being cabin crew, Annie knew exactly what the score was in terms of me being away and the planning required to get time off when you wanted/needed it.

Once the rosters were out, getting a change took effort or a swap with another First Officer. We did, however, have a 10-day holiday planned.

Annie had a friend that had a holiday apartment in Simonstown, just south of Cape Town that we could use. Using staff concession tickets, along with Agatha, we headed off with our baby for a holiday.

Patrick was only a few months old, so he mostly slept. I'd converted a suitcase into a makeshift cot with pillows and blankets. The trip was relaxed, the scenery stunning; until the diarrhoea started. Something didn't agree with him.

We'd boiled his formula water properly, but whatever it was, he began to fire near-clear liquid from his backside and showed signs of dehydration.

We took him to a local doctor, who promptly went through us for not breastfeeding him, he was quite angry that we had the boy on formula milk. We were both quite embarrassed and left with a load of rehydration powder and a decent helping of parental guilt.

Outside our apartment block in Simonstown stood what looked like a high-security enclosure, a full steel cage built around the bins, complete with padlock and reinforced mesh. It wouldn't have looked out of place around a weapons cache in South Armagh.

I assumed it was South African overengineering until a local explained why.

Baboons. Specifically, local troops that were smart, aggressive, and terrifyingly organised.

"They're not monkeys," he warned. *"They're a paramilitary with brake lights for arses."*

I laughed, of course. I mean, how bad could it be? Just a troop of fluffy delinquents with a gland problem. I'd be bad tempered with an arse like that too.

On our final morning, I found out.

I went to take the rubbish out, bin bags in hand, walking like a man with no situational awareness. The second I turned the corner, I saw them, a *troop* in every sense. Fifteen of them spread out in a textbook L-shaped ambush, and I was in the kill zone.

The alpha was front and centre, perched on a ledge, arms folded, wearing a helmet and Ray-Bans in my head, with the thousand-yard stare of someone who'd seen too many bin runs go bad.

He was flanked by a younger male with a limp and what I can only describe as the troop's psy-ops officer, a baby baboon deliberately placed front and centre, looking cute enough to make me underestimate the threat.

Tactically speaking, I was already screwed.

As I reached for the cage door, the alpha gave the signal, a sharp scream like an assault team leader issuing *"Go! Go! Go!"*.

They advanced as one.

It was a coordinated offensive: pincer movement from both flanks, distraction unit coming in low, and the big lad going straight for the gate like he had breaching charges strapped to his hairy chest.

All that was missing was a couple of smoke grenades.

I did what any squaddie would do, I went on the offensive, my ammo a half-eaten apple and a cold baked potato. The spud hit him square in the face.

He stopped. Blinked. Then screamed like an air raid siren.

That clearly pissed him off.

The others charged.

I executed a textbook tactical withdrawal, backwards into the cage, bin bags flying, and slammed the gate shut from the inside like I was defending Rorke's Drift with nothing but leftovers and shitty nappies for protection.

Now I was trapped in the bin cage. Alone. Surrounded. Waiting for a giant Marine called 'Camouflage' to appear and save the day. The baboons formed a perimeter and started a siege, pacing, watching, rattling the cage like inmates checking the welds. One of them licked the padlock. I'm not sure why, but I didn't like the message.

Negotiations began.

I chucked out a stale bread roll and the remains of the baked potato as peace offerings. It bought me some time. But they knew they'd won.

Just when I thought I'd have to tunnel out using a broken clothes peg, salvation arrived. Our neighbour, previously nowhere to be seen, burst from his kitchen wielding a mop like a lance and shouting words in Afrikaans that I suspect would make a rugby player blush.

The baboons scattered like trained insurgents under mortar fire. He'd clearly faced this troop before. And won.

Safely home in Barton, Agatha stayed with us for a night or two before returning to Belfast. I was back on the roster.

On the 10th September I was back on another Milan night stop. I got home around 10:00. We'd booked Patrick in for his injections deliberately that day, as I would be home.

Patrick got all jabbed up, He took them like a champ, and we went home for some lunch. I went upstairs and got changed out of my uniform I'd been in since getting back from Milan. Annie made a brew, and I sat down, quite tired and turned the TV on.

What happened next doesn't need narration.

Everyone remembers where they were when the world changed. For us in the UK, the news broke around half past two in the afternoon. I sat, dumbstruck as smoke billowed from the North Tower. When the second aircraft struck, it was clear, this wasn't an accident.

It was war.

I was still subject to recall to the Army and half wondering, half hoping it would happen.

Had I been reactivated, I would have gone back with a smile on my face. If they needed me, it wouldn't be back anywhere near Thacko.

I could have no idea at the time of the impact those attacks would have on my life for the next 25 years.

The only immediate effect on us was the imposition of the locked flight deck door policy. No more flight deck visits or jump seat rides ever again. No more grinning kids and nervy dads popping their heads in. No more flight deck cricket. That little window into the magic of flying was gone forever.

Then came the annual refresher training. Every flight crew had to retake annual security training under the watchful eye of our company security officer. She was a solid operator, knew her stuff, and did what the Department for Transport asked of her: teach us about Bin Laden, box cutters, and what to do if someone tried to hijack a CRJ and demand to be flown to Milan.

The trouble was, I'd seen the other side of it. I'd been trained by people who knew what real training looked like. Compared to that, this was a PowerPoint on a budget and a panic button. But we sat through it, nodded in the right places, ticked the boxes.

She did was she was supposed to, but having come through the military system, and taught within it, I knew this wasn't training. It was theatre and not especially good at that.

Chapter 103. Back to Bedlam

On the line, the subject of most flight deck conversations would inevitably arc back to 9/11. The whys, the what if's and the so what's. All our aircraft were retrofitted with the now-required ballistic-proof flight deck doors.

New procedures were put into place for entry and exit, including what to do if a member of crew were forced under duress to try and gain entry; that was a difficult one to deal with, a friend, wife or girlfriend with a broken bottle against their jugular vein with an *"open the door or the blonde gets it"* became a disturbingly plausible scenario.

Life as a First Officer was becoming second nature. Maersk was small and tightly knit; we spent an absurd amount of time together and inevitably knew each other better than most extended families.

It's fair to say that no one was more surprised than me, when after 8 months or so, I realised one of my best friends in the company was a 6ft 4" fashion model, very proudly gay, and black.

Liam Wilson was an extraordinary man in many ways. He took me for what I was, an ex-Army white man with an ego problem and without really doing very much, made me look at the way I had previously judged people, albeit subconsciously.

He introduced me to gay culture, and I got to know it, and him, well in the nightclubs of Stuttgart and Copenhagen on many an eye-opening night stop. I learned, through immersion, to accept people for they really are.

I would, of course, take the piss out of Liam for being gay, and he would take the piss out of me for being straight. Exactly how it should be, just two mates with differing tastes in partners and fashion.

Before the airline world, I'd never understood the broad spectrum within the gay community. To me, it was either quiet and understated or loud, flamboyant and 'accented'. I wanted to understand; so, I asked.

Flying a regular line sector with Chris, who was also gay and had zero time for bullshit, I asked him: *"What's with the accent and the limp wrists? I genuinely don't get it."* Chris, ever the patient educator, replied, *"It's a badge. Not something you're born with. It's learned. It says, 'I'm gay and*

I'm proud.'" As usual, Chris made sense. I got it. It helped me drop whatever subconscious sorting mechanism I'd been carrying around.

Night stops in Belfast were as common as the Milans and Stuttgart's, except it was only one night. We stayed at a hotel called 'The Dunadry', a stone's throw from Aldergrove and directly opposite a pub we would go to from Aldergrove called 'Ellie Mays'.

Most night stops there, the blokes would come down from Aldergrove for a beer and a flirt with the cabin crew.

Dean Kavanagh was a regular.

Eventually, Chris Morley and I were rostered for a Belfast, I primed the blokes, and a good handful rocked up for beers on Termini. They genuinely seemed to be pleased that I'd turned things around.

On the flight out of Aldergrove the next morning, I dialled up the Bessbrook VHF frequency on the second radio, listened for a moment to the radio traffic from Lynx 5 and 7.

Just for a moment I felt the urge to go back, to forget Thacko, Le Man and the other idiots and just go back where I belonged.

The moment lasted, and, in a pause, I pressed the mic.

"Red 850, one CRJ from the overhead, Red 850"

"TERMINI!" came back from a voice that sounded remarkably like Dan Brodie.

I did still belong down there in the weeds; the pull was stronger than I thought.

Patrick was growing fast, and we had the kids over from Germany as often as possible. The availability of BA staff travel made it easy. I wanted Matty and the girls to accept Patrick as a younger brother, with a different mum.

Emily still made life as hard as she could, sending me regular demands for money for spurious reasons and making contact as difficult as possible.

When I started at Maersk, we had been using the CRJ simulator in Amsterdam Hoofddorp, but as we approached the addition of CRJ 700s into the fleet, the company changed to using simulators in Berlin.

Of course, the company would get us there on staff tickets and we stayed in the famous Westin Grand Hotel in what I remembered as East Berlin. Given the kids were still in Germany, I requested tickets to Hannover, I would then get a train to Berlin via Gutersloh so I could grab a night with the kids.

I arrived in Gutersloh and took the kids out for dinner; I only had one night before I needed to be in Berlin.

Scott was there.

We didn't talk about it, and he bought me a beer. It was a small gesture, but it opened the conversation around the direction my life had taken. He suddenly came across as a half descent bloke; he'd always been kind and generous with the kids. We got quietly pissed on Jack Daniels. A very painful chapter in my life seemed to be receding into something closer to understanding.

No apologies were offered either way. He did indeed know about the hammer and brake fluid and had assumed it was me. We would never be mates, but an understanding developed that would grow over the next couple of years as the bitterness faded.

I rocked up in Berlin the next day, slightly hung over, and met up with Tom Lawson, my stick buddy for the sim. The sim sessions generally lasted 4 hours with a break halfway through, we would always be scheduled at least two sessions.

Tom was an ex-Navy pilot and a great guy. During the next day we walked down from the Grand Westin to the Checkpoint Charlie Museum. Although the actual checkpoint was long gone, there on the wall of the museum was a picture of Daisy, with me stood behind her.

It all came flooding back, poor Tom couldn't shut me up.

We walked down Friedrich Strasse and Unter den Linden, all sausage side as far as I was concerned, and arrived at the Brandenburg Gate.

There is a now brass strip embedded in the road to show the line of the wall; we crossed it and went to the exact place I had executed that neutral turn in Daisy back in 1989. There, on the stone kerb were the deep scratches I'd made.

Unbelievable, I'd made my mark on Berlin, if not history.

The Maersk crew room was growing all the time with new faces, and some not so new.

Darren 'Daz' Robson joined Maersk. Daz was a good friend from my time in Aldergrove, he was a local bloke and lived in Jordanstown with his wife, Amelie, and his 4 kids. It was Daz that had suggested I look in Jordanstown for a house in the first place. I'd flown a lot with him, he was one of the pilots that had been thrown in at the deep end, but had accepted that change was required.

He'd had an unfortunate incident just before I'd arrived.

Daz had been flying with Regimental standards on a Lynx 7 duty. He had landed at G10, a remote tower just outside of Crossmaglen, close to Cullyhanna. Protecting the site was a large earth bund, designed to protect the installation from sight as well as the Mk 11 bunker buster improvised mortar that PIRA had perfected; the HLS was just outside the earthworks.

To get to an aircraft, a cut out in the bund provided access to and from it.

Unfortunately, a Guardsman instead ran over the top of the bund to help unload the aircraft. In doing so, he ran headfirst into a main rotor blade tip travelling at just under the speed of sound. The blade impacted his helmet, which saved the lad's life, but Daz, I think, could still hear the impact in his head.

He didn't talk about it, and it was just a hunch.

We introduced Daz and Amelie to Barton-under-Needwood, and they found a nice house after some searching. It was nice to have a drinking buddy.

Paddy Roche, from 652 Bosnia days got in touch, him and another former AAC OC, Richard 'Mitch' Mitchell were on the hunt for jobs, and could I help?

I introduced them both to Stan Lewis, gave them some tips and an insight, but it was up to them. They were both interviewed and offered jobs, but Paddy took an offer from somewhere else. Mitch joined us as an FO. 2001 rolled into 2002 and we had a new aircraft to look forward to – the CRJ 700.

Chapter 104. Big Plans

Maersk had ordered the larger CRJ. Up until then, we'd been flying the 50-seat variant. With a full passenger and fuel load, she didn't climb well and she rarely made the higher cruise levels, which pushed the fuel burn the wrong way.

The -700 was a 70-seater, more comfortable and had adequate power for its size, all in all, a better aircraft.

The conversion was just a differences course, followed by a final check to make sure you wouldn't bend it. These aircraft were then put on our more lucrative routes, Copenhagen, Stockholm and Stuttgart to start with. Passengers and crews alike thought they were a better experience all round.

We had started to notice that loads were occasionally getting very low. Four or five passengers wasn't unheard of, but AP Moller Maersk had deep pockets, so we weren't overly concerned.

Then a new MD arrived with a serious CV. Out went Gert Kristensen. In came Peter Spencer.

Peter had been BA Director of the Americas and an advisor to the CEO of Qatar Airways before joining Maersk. We were comfortable we were in good hands and well placed to weather the post-9/11 downturn in air travel.

Throughout 2002, the slow, but steady decline continued, but the company kept on expanding, bringing new routes and accepting new aircraft.

Life was good. We had friends in the village, Marcus and Kara were always up for a party, Daz and Amelie were also usually up for it depending on babysitters and I went full civvy, starting to play golf.

Flying by now was just a job. Not the fix it used to be. Not quite 9 – 5, the roster ensured that and the night stops made up for some of the lack of excitement, but it wasn't what it was.

I needed some excitement.

One of the new routes introduced was Helsinki. I cooked up an audacious idea.

I got in touch with Ed and asked him if he fancied a snowmobiling expedition across Lapland. Over a few beers the idea, well I hate to say it, but I'm going to, it snowballed. He roped in a friend of his, Ewen 'Cam' Cameron, a human wardrobe of a Scotsman, and together we went about some serious planning.

In the lead up to departure I was checking the aviation forecasts on a near daily basis, the daytime temperature in Lapland was normally around the -10 mark, and we were getting nervously excited.

I arranged for the three of us to get staff tickets to Helsinki. I'd already spoken to Mitch about the plan. He was close friends with the explorer David Hempleman-Adams, who conveniently had a warehouse full of proper Arctic kit.

We raided it like teenagers in a locked drinks cupboard. Mukluks, those red James Bond-style snowsuits, mitts, hats and proper tents. Not the canvas monstrosities we were used to.

Outdoor shops were scoured for thermal underwear, gas cookers and anything else that "looked right".

We'd also acquired sleds, known as pulkas, from David. The plan was to tow them behind the snowmobiles, loaded with all our kit.

Annie had gone back to Belfast for a break with Patrick, and so the 3 of us went to a pub in Barton the night before we flew out. We met up with Marcus and his brother-in-law, Richard.

After we'd been kicked out of the 3 Horseshoes, we went back to my house and cracked on late into the night.

I vaguely recall naked dancing involving only hats and mukluks.

I wished I'd wiped that from my memory.

We checked in our pulkas at the airport, fully loaded like Ernest Shackleton's. Then, unlike Shackleton, we went straight to the bar.

The cabin crew kept things lively for the whole flight. As we got off in Helsinki, they handed me a carrier bag with the contents of the bar cart.

Dozens of miniatures.

From Helsinki we flew up to Kuusamo, just south of the Arctic Circle. From there, a bus took us on to Ruka, where we hired three snowmobiles.

The guy in the hire shop took one look at our pulkas and told us to leave them behind. Instead, he lent us a proper snowmobile trailer on skis. Much better.

We had a hotel room that night and cross-loaded everything from the pulkas into the trailer. We'd take it in turns to haul it.

Cam was up first.

We'd planned on getting proper maps from the hire shop in Ruka. But they were only a backup to the GPS we'd brought.

We should have got the proper maps.

We set off early the next morning having woken up early, like kids at Christmas. We set off looking for the first of the tracks we would be on for the week. The route was simple: Ruka east to Posio, north to Salla, then a straight-line home.

The hiking guides we were using were shite.

We barely noticed that the temperature was steadily dropping all the time; the skies were clear and a deep azure blue. Absolutely beautiful. The fir trees were each covered in a deep coat of snow causing them to bend at a crazy angle and there was about 1.5m of snow lying on the ground.

We made reasonable progress on the first two days but learned some critical lessons.

First up, like Maverick and Goose, snowmobiles need speed. This is to stay on top of icing sugar like powder snow; stop and down you go, literally straight down.

Secondly, never, and I do mean ***NEVER,*** attempt to use the reverse gear.

Try it, and down you go.

Picture this: the rear end of a snowmobile burying itself into 1.5 m of snow, now you're at a 45-degree angle. Engage forward gear to drive out of your self-made trench. You'll watch the nose disappear. I say "watch" because by then you're no longer driving.

You're just a passenger in *The Mole* from Thunderbirds.

You now require recovery.

And we did.

Regularly.

The third thing we learned was that anything battery-operated in those temperatures will curl up and die in seconds. Head torches would fade away from brand new batteries to a dying firefly in about 30 seconds, you could see their death throes.

The GPS wouldn't function at all; the LCD screen froze solid.

Ed learnt the hardest lesson of all. He had brought a pair of hard looking combat boots. He wore those instead of the uber warm, but ugly Mukluks that Cam and I were wearing. After 3 days or so, of course he couldn't feel his toes.

He chipped off the frozen combat boots, writhing in agony. We asked him why he hadn't worn the real deal. Because, and this is a direct quote:

"Even when you're in the shit, you still need to look good."

We were in the shit. Not just your everyday squaddie shit, oh no, this was Pte Pyle's "*world of shit*," the deluxe, all-inclusive package.

The temperature wasn't lifting above -30. The snowmobile batteries were dying on us, and we needed heat. There were stacks of firewood everywhere, put out for that very purpose. Finding wood wasn't the issue.

Getting it to burn was.

Three squaddies, plenty of petrol, and we couldn't light a fire. Not just dangerous. Professionally embarrassing.

By now, we were crashing every couple of kilometres, the front skis were curled up like a leprechaun's shoes. We had a single wing mirror between us, and the trailer was falling apart.

We really were in trouble. We decided to head for a small village we could see in the distance; we could just about make out the glow of a supermarket sign.

We needed fuel too; we had used our reserve jerry can trying to light the fucking fire!

Ed clearly had a bad non-freezing cold injury to his feet, and my nose had turned a vile custard shade of yellow.

Riding the snowmobiles on ploughed tarmac roads was ridiculous enough, the skis screeching like nails down a blackboard. Cam and I pulled into the petrol station forecourt and turned to look for Ed.

He arrived from stage left, not upright, not even close.

The snowmobile was on its side, sliding gracefully across the forecourt like a drunk figure skater. Ed clung on, rigid, as the machine ground to a halt in a shower of sparks and scraped plastic.

When it stopped, he stood up, brushed himself down and said, completely deadpan, *"Is it open?"*

He'd seen where we were at the last moment and hauled the bars over. The machine had simply rolled onto its side and carried on without him having much say in the matter.

We were in tatters.

We righted his battered machine, filled all three up and headed across to the supermarket. We needed warmth, and probably a priest. Pulling ice from our eyelashes turned out to be a quick way to remove them entirely.

The large neon signs we'd seen alternated between showing the time and the temperature. It was 16:00.

Then it flashed again.

-42°C.

A serious cold snap even for this far north. The supermarket had heat curtains in an airlock style entrance to the shop. We stood in that airlock for about 30 minutes whilst old Finnish ladies shuffled past with their shopping, staring at these grinning idiots.

On our last night under the stars, our sleeping bags froze stiff where our breath had soaked in. Cam wandered off into the darkness, Captain Oates style, for a piss.

Then the sky came alive.

Ed and I joined him as the northern lights rolled overhead. Bands of light formed, green with hints of red, sliding across the sky. Not fast, not slow, just steady. They pulsed, faded, then returned somewhere else, stronger, pulling your eye with them whether you wanted it or not. It was impossible to look away.

The cold disappeared. The discomfort didn't matter.

It was all worth it.

I dug out the carrier bag of booze from the trailer. Vodka for Ed, whisky for Cam, Jack for me. We cracked the caps, ready to toast the moment.

The lot had frozen solid.

Ah bugger!

A fitting tribute to our naivety.

We limped back into Ruka to return three wrecked snowmobiles and a thoroughly abused trailer. The hire guy just stared at us, then at the machines, and shrugged. We got home in good spirits.

Probably how Shackleton felt. Although he didn't have Ed.

So, Cam and I may have had the tougher trip.

Chapter 105. The Writing on the Wall

Throughout 2002 and well into 2003, the load factors slid, slowly but unmistakably. Some of the captains could see the end coming and started to drift quietly away to other companies. I could see it too, but if I rode it out, command would come sooner rather than later.

I was edging up the seniority list. This was timely because we found out in February that Patrick would have a sibling come October; Annie was pregnant. A promotion wasn't just desirable now.

It really mattered.

By July 2003, the game was up. The AP Moller group had put us up for sale.

There were no takers.

Peter Spencer assembled a management buyout and kept the lights on. From September we would be known as 'Duo' airways.

The tails were resprayed in shimmering, purple and gold iridescent paint, which promptly bubbled and peeled the first-time engine oil touched it.

A poor omen.

In mid 2003 I was told that I would be rostered a pre-command assessment check with Robin. I was given 2 trips with a senior TRI that would act as preparation. These went well and I took a lot on board. After these formal trips I enjoyed a couple of line sectors, one of which went with a bang.

The -700 was having problems with its windscreens, we'd had a couple shatter, scaring the crews to death. My turn came on a return sector from Copenhagen, we were at Flight Level 410, in other words 41,000 ft, as high as we could go.

We'd literally just had a coffee brought in when there was one almighty bang on my side. It was so loud, I nearly spilt my brew.

My head snapped right.

The side window was now a mosaic of a million tiny shards; shattered like a car windscreen.

An aircraft windscreen is a multi-layered structure designed to withstand pressure, temperature and impact. It consists of an outer ply of toughened glass or acrylic, a middle interlayer of vinyl or resin to absorb shock and prevent shattering, and an inner layer of glass or acrylic for cabin-facing protection.

We knew there was little danger of it giving way completely, but the thought was clearly in my mind. Out came the QRH; confirm no depressurisation; there wasn't.

We eased the aircraft gently down to ten thousand, slowly depressurising so as not to shock load the damaged windscreen. The cracked pane stayed intact, but the sight of it, just a few centimetres from my face was a sobering reminder that at this altitude, it wasn't scenery keeping you alive. It was millimetres of laminated engineering.

ATC gave us a straight-in routing from way out over the North Sea. We held a slow, steady descent at -500ft per minute all the way to Birmingham. An uneventful landing and the passengers were none the wiser.

The following week I flew my command assessment. It wasn't perfect, but I passed. I was now on the waiting list for a command course.

The month after my command check, Amelia Rose arrived at Burton Hospital. She was the spitting image of her mum, big hazel eyes and blonde hair, delicate as porcelain. Amelia was such a quiet, timid baby we barely knew she was there most of the time. Patrick, on the other hand, adored her with a kind of clumsy devotion.

He'd scoop her up and squeeze her in his arms with all the subtlety of Lennie from *'Of Mice and Men'*. We had to be careful, he was already a unit of a lad, and though his hugs came with brute force, they also came with nothing but love.

On one particularly memorable occasion, Marcus and I decided it was time to test our boys for the right stuff, not courage under fire, but whether they carried the all-important aviation gene. A carefully planned father-and-son bonding exercise, we chose Duxford Air Museum as the proving ground.

The plan was simple: bombard them with Spitfires, Lancasters, and a Concorde until either their eyes lit up with the same sparkle ours had, or they confessed to preferring football and doomed us to years of shame.

We set off early, determined to give our boys a full day immersed in the wonders of Duxford. What more could a boy want?

On arrival, the priority was toilets; Eddie and Patrick both needed a wee. With that emergency sorted, we made for the aircraft.

Unfortunately, between us and the aircraft sat a children's playpark. Halt! Sixty minutes of mandatory climbing, falling, and general self-inflicted injury was now required before we could move another yard.

Then the heavens opened and we ducked into the hangar that housed Concorde. There was a long queue to shuffle up the front steps, down the narrow aisle, and out the back. I looked around; Patrick was gone. Twenty seconds earlier he'd been right beside me.

Now he'd vanished.

Panic hit like a wire strike. I started shouting his name, scanning the crowd, asking random strangers if they'd seen a kid in a fake leather bomber jacket.

Nothing.

Not a flicker.

Marcus grabbed Eddie and swept off one way, while I tore off the other, heart pounding, the hangar suddenly the size of Heathrow Terminal 5.

Then something caught my eye. The slow-moving line of people shuffling down Concorde's staircase had started to ripple and split, like the Red Sea parting under protest.

And there he was, Patrick charging straight up the middle, elbows flying, knocking aside bewildered tourists as if they were bowling pins. His bomber jacket flapping, his face set with determination, he was storming the stairs in the wrong direction, utterly oblivious to the chaos he was causing.

I raced over and the chase was on. "*Sorry, sorry, sorry,*" was all I managed trying to catch him. He looked over his shoulder and saw me. Now it was a game of tig as far as he was concerned. He made it into the very narrow aisle of the Concorde and set off against the flow with me in hot pursuit.

I grabbed hold of the collar on his jacket just as he was about to start his descent down the upstairs, swept him up to a fit of laughter from him and tuts from the other visitors. The looks ranged from sympathy to outright condemnation.

By then it was clear: our boys had absolutely no interest in aircraft. None. Zero. The aviation gene had skipped a generation. Probably for the best. All those dreams of passing down hangar talk and warbird trivia evaporated somewhere between the Concorde stairwell and the gift shop.

We drove home in silence, two crushed fathers and two perfectly normal sons, the air thick with the tragic knowledge that our bloodlines had been hijacked by indifference.

With the disappointment of Duxford behind us, Duo was about to catch up.

Part of the Duo plan was to open a second base in Edinburgh, crewed from Birmingham. In hindsight, it was an overreach the company couldn't afford.

I started looking at other companies. I genuinely didn't want to, but flying near empty passenger aircraft around is a clue no man can afford to miss.

Especially with six children to support.

One company stood out as meeting my spec: larger aircraft, interesting routes and not too big. British Mediterranean, or BMED, another BA franchise looked like a good fit, and they were recruiting.

I went through their selection process in slow time with an interview at Heathrow followed by a check ride in an Airbus 320 sim at Gatwick. I was offered a position in BMED just before Christmas and reluctantly gave Chris my 3 months' notice.

I also gave up my captain position, which hurt more than I expected, it wouldn't have been fair to do the course and then leave.

My slot went to Mitch; he and Dave Bird did the course together; It was bittersweet flying with them as Captains just before I left in March 2004.

2004. Another round of changes, another uniform, and a new

aircraft to master. This time, it was the Airbus A320 and its bigger sibling, the A321.

BMED operated a mixed fleet to some truly remarkable destinations across Africa, the Middle East, and Central Asia; no more four-sector days or quick hops to Europe. This was long-haul flying, departing from Europe's busiest hub: Terminal 4 at London Heathrow.

The stakes were higher, the flying more demanding, and the routes far more exotic. I couldn't wait.

PART 6: Began Married, Ended Divorced: BMED

UK, Donegore Hill: Apr 04 – Feb 05

Chapter 106. On the Busses

My Airbus type rating course was at a training facility near Heathrow. There were four of us, I was paired with Martin Goodwill, a former C-130 pilot. Martin and I were classed as Zero Flight Time trainees because we were coming off jets, which meant our line training would be with passengers from day one.

The other pairing was Paul Wigham, a Zimbabwean whose family had been displaced to the UK in the 70s, and another lad who'd joined BMED from Aurigny.

Three of us got on well, me and Paul would go on to become firm friends. The fourth kept to himself, never quite relaxed, and seemed uncomfortable from the start. I don't think he was enjoying it.

In the end, the lad from Aurigny didn't make it. He took himself off the course halfway through and went back to flying Trislanders, where he was happier.

The CBT technical training was as dull as it could possibly be. It felt like it had been designed by someone hellbent on weaponising boredom, possibly a retired tax man with a grudge against colour and joy. Truly awful.

Having survived the CBT without too much self-harming, we went into 40 hours of simulator time. As with the CRJ sim, this thing was identical to the aircraft. The Airbus is easily the most heavily automated aircraft in the world, and it is the automation that takes time to get used to, not the flying itself.

Airbus and Boeing come from very different philosophies. Airbus puts computers in charge and protects the aircraft from you; Boeing gives you more authority and trusts you not to hurt it. In 'normal law', an Airbus won't let you stall it.

In essence, the pilot makes requests, and the aircraft determines how to execute them safely, prioritising system logic over manual input where necessary.

Boeing, by contrast, gives the pilot greater direct authority, maintaining more tactile control through conventional yokes and feedback systems.

These differences shape not only how the aircraft fly, but also how pilots interact with and trust their machines.

Airbus also dispensed with the traditional '*handlebar*' style control yoke. The Side Stick Controller (SSC) is a small, intuitive joystick located on the side console of each pilot. Designed to be immediately natural to use, it offers smooth, fingertip control, more akin to guiding a cursor rather than pushing or pulling a large mechanical control.

Inputs are transmitted electronically via fly-by-wire, with no physical linkage between the sticks.

Best of all, where the yoke would normally be, there's a pull-out table; perfect for enjoying dinner or doing the flight paperwork. The flightdeck was spacious, very comfortable and an easy office to work from.

At home, we'd decided that it was time to leave Barton and head back to Belfast, Annie wanted to be closer to her family for support. It was the right thing to do. The roster pattern meant I would be away from home on a 3-4 day trip at the short end, or 4 -5 days at the longer end.

Commuting from Belfast into Heathrow was easy; the tickets were cheaper than petrol from Barton.

Once signed off from the sim, Annie and I could get stuck into finding somewhere new to live.

Eventually we agreed, after a few '*debates*' on a remote farmhouse near Muckamore, just outside of Templepatrick. Donegore Hill had just over an acre of land that included a stream and some woodland; I loved it. Importantly, it also had a stable block and a small riding school. I hoped it would give the girls something to engage with, something that might make them want to come and stay.

I had a battered old Isuzu Trooper; I taught Matty how to drive it around the paddock, trying to find something that was ours. Patrick just enjoyed running around in circles 'helping me' do gardeny type things. I'd get him to punch holes in the mud with a big stick while I dropped flower bulbs in.

Donegore was a dream. It reminded me of long summer days at Eastern Hill, and of the farm at Bolton Abbey with Dinny. Life should have felt perfect, but in the quiet, I knew something was missing.

The house was full of potential, and so was the job. If anything could tie it all together, this new chapter was it, I believed that completely. I was as giddy as a kipper when I got my first roster. I was going to Amman, Jordan, then onwards to Khartoum in Sudan; then back to London via Amman, a quick 3-day trip. My line training on the bus was underway, and for the first time in a long while, I felt like I was moving forward again.

Chapter 107. War Zones and Potential War Zones

The BMED route network consisted of three kinds of places: those that had already been blown up, those that probably would be, and those quietly queuing.

Given that, I fully expected the security training to be several levels better than Maersk. It wasn't. In fact, it was arguably worse.

The BMED Chief Pilot was former RAF and ex 47 Squadron, flying C-130s.

As often happens when someone senior gets the chance, old friends and colleagues followed. BMED was no different and former 47 Sqn pilots largely ruled the roost. There were of course pilots of all backgrounds, but the RAF had the lions share.

One notable First Officer was Nick Richardson, the Sea Harrier pilot shot down in Bosnia and subsequently rescued by the SAS and a French Puma.

I had a copy of his book, and I left a note in his pigeonhole with a suggested redraft recognising UK Lynx pilots as the crème-de-la-crème, and a prepared apology for daring to suggest French Puma pilots were in any way good. Strangely, the correction was never printed.

Publishing can be fickle…

From day one, we were encouraged to disconnect the automation and hand-fly approaches. Some airlines ban pilots from disconnecting automation altogether. The BMED way, I think, was the correct approach, it made us better pilots less reliant on computers as well as making the flying much more interesting.

One of the aspects of automation that was new to me was a full Cat 3 auto-land.

We would practice these in good weather conditions so that when it was needed for real, we had something to think about. In poor weather, with near zero visibility It was both eerie and impressive. After briefing, configuring the aircraft, and engaging both autopilots, you become more monitor than manipulator.

The aircraft eased onto the glideslope like an old master. Outside, nothing but thick grey; no horizon, no lights, just the faint sense of motion. Inside, the cockpit became its own world of ritual and rhythm.

The callouts began: *"One hundred above… minimums… one hundred*". The cadence was calm, mechanical, oddly reassuring. Then the sharper ones: '*Fifty… thirty… twenty… ten.*' At '*retard, retard,*' my hand moved instinctively, easing the thrust levers to idle, ready to deploy the reversers.

The nose lifted, just slightly. A soft thump followed as the wheels kissed the runway centreline, then the choreography continued: spoilers sprang up, autobrake engages, and reverse thrust roared. Out of the fog, the blue edge lights strobed past, the only proof that the world outside still existed.

You're alert, but essentially a passenger.

Oddly, one of the biggest buzzes I got wasn't in the air. It was taxiing around LHR from Terminal 4. The controllers at LHR are the best in the world; calm, relaxed, and utterly trustworthy. You never felt alone out there. We would receive a list, sometimes a long list, of markers and taxiways that would be our route either to or from the runway in use.

At night it was simple, *"Follow the greens"* or "*follow the blues*". The taxiway lighting would spring into life in front of us and flow in the correct direction, all we had to do was follow the assigned colour.

A handful of our destinations were strong enough to support a daily flight from LHR.

One of those destinations was Beirut. We overnighted at the Commodore Hotel in Hamra, not far from the Hard Rock Café; yes, Beirut really did have one. The nightlife was eclectic, vibrant, and oddly surreal.

We'd have beers surrounded by buildings still deeply scarred from years of conflict; bullet holes and blast marks formed the backdrop to rooftop bars and crowded terraces. Monot Street stood out as a perfect snapshot of Beirut's character: defiant, stylish, and alive, a microcosm of everything good the city had to offer.

One of my last line training sectors was to Baku in Azerbaijan. I'd been pre-warned about Baku. Nothing to do with the flying, the warnings were all around the social life there. We stayed in the Hyatt in the city centre, centrally located and near some very popular bars and restaurants.

There would normally be at least 2, sometime 3 full crews there. The inbound and outbound, and then another crew that would operate from Baku onwards to Bishkek and back to Baku. These 'shuttles' were a pain in the backside.

We'd take over the aircraft with its Bishkek passengers still on board, fly them out, then wait five or six hours on the same jet for the London-bound passengers.

My first Baku experience did not disappoint. It was an unofficial SOP that either the Captain or First Officer would host 'drinks' in their room. In my case it was the Training Captain, and it was his last trip to Baku. Within an hour, there was a pillow fight that resulted in two ruptured bags of feathers being liberally emptied around the room, a human pyramid that killed the TV and a barrage of complaints from the hotel management.

My final line check was a trip to Tehran, Iran. I really looked forward to this one. I was keen on seeing Tehran with all its culture and heritage. We didn't get long on the ground, pretty much a trip to a hotel that could take hours in a bus, a couple of hours sleep, and the back again.

I couldn't have been more wrong.

Once we had disembarked the aircraft, the girls had to completely cover their hair with a BA issued Hijab. We had to keep the curtains closed on the bus so that we reduced the risk of a kidnap attempt of a BA crew.

But from the little I could see, Tehran, a wealthy city, was filthy. Litter, especially plastic bags and water bottles, strewn everywhere with a lot of obvious extreme poverty.

Once at the hotel, we had a full floor dedicated to us. It was closed off with a set of doors; not until the girls were past these doors, could they relax and remove the Hijab.

We had a couple of rooms that were adjoined and had been converted to a crew room. BMED provided some DVDs, games and sports equipment, but we mainly just watched movies before an early night for the flight back to London the next morning.

Tehran was a huge disappointment, not because of where it was, but because of what daily life there looked like.

Once I was on the line roster, the destinations broadened, and the trips grew longer. As we were paid an hourly rate on top of our basic salary, the longer trips could be very worthwhile.

Back at Donegore, though, something didn't sit right. On the surface, everything looked fine; we had the land, the kids visited, the work-life rhythm was stable. But arguments would bubble up from nowhere, then fizzle out again.

Annie had a way of picking at things until they bled, stirring tension until I reacted. Loudly.

And once I'd reacted, I was always the problem.

"It was just to get a reaction out of you," she'd say.

"Well, it worked," I'd snap back.

Chapter 108. The Blonde Bombshell

With its route network, many destinations were either unsafe for crews to leave the hotel, or there was simply nothing there even if you did. These were business destinations with clearly defined passenger demographics; Central Asia was miners and oil & gas workers, the Middle East mostly business travellers. Certainly, none were well beaten by holidaymakers.

All our crew hotels had the obligatory crew room, meant for quiet downtime but more often used as a party venue. There was always time to recover and freshen up before the flight back to London. Some locations, where there was a shuttle operating to a third destination, there could be several crews down route at the same time: as with Baku. Another was Amman in Jordan.

We enjoyed the Amman Intercontinental when staying in the city, it had a few themed restaurants that we would use. Some crew members were known as "Delsey Diners". Delsey was the suitcase brand issued to us; "diner" meaning the case contained more food than clothes.

Some of the places were very expensive, but Amman was one of those places where you could expect a good turnout from the crew.

On one occasion in Amman, in September 2004, there were three full crews. One arriving from London, one preparing to depart, and a third operating a shuttle to Khartoum.

We all met at a pre-booked table in a Mexican restaurant. It was a long, cantina-style bench. The empty space beside me was quickly taken by a crew member from another flight.

She was beautiful. Blonde, loud, full of life, with a Teesside accent that cut through the piped mariachi. I pegged her as Middlesbrough, but it turned out she was from near Stockton.

Quick, sharp, and funny in a way that caught me off guard.

We bounced through half a dozen topics in the time it took for my fajitas to go cold. Smoggies, the Boro, the ICI plant outside Billingham.

It wasn't just the accent. She had an energy about her, like she knew all the tricks but only played when it suited her. There was an ease to her, and sitting next to her made everything else fade slightly into the background.

There was no flirting as such, no loaded glances. Just that rare, easy connection that doesn't come along often, especially not in a cheap hotel Mexican restaurant two thousand miles from home.

And then, too soon, she stood up, smiled, said goodbye, and was gone. Her crew was flying back to London first thing.

I thought that was that – it usually is.

I didn't even catch her name.

Chapter 109. Blackhawks and Cowboys

About 4 weeks later, in the middle of November, I was rostered to Addis Ababa, Ethiopia. I'd been a few times, and it was always an enjoyable trip. The flight from London went via Alexandria in Egypt. The airport, Borg El Arab was located to the southwest of the city with roughly a north - south runway orientation. This meant we often flew low past the city with its historic harbour and old town.

I have a photo of my grandfather, Harry, sitting on the steps of the old town in central Alexandria.

The turnaround in Alex was quick, about an hour. Just enough time for fuel and any additional catering before heading south down the Nile valley to Addis.

On a clear day, the views were stunning. The pyramids of Giza, then the ribbon of green the Nile cuts through the desert, past Luxor and Abu Simbel. It always sparked conversation about the history we were flying over.

Approaching Addis, the scenery changed again as we entered the Ethiopian highlands. It was nothing like the images fixed in our Western minds from Live Aid, Band Aid and the rest. Not brown, dry or barren.

Northern Ethiopia is green, lush, almost alpine. Addis Ababa Bole International Airport sits at 2,300 metres, and with high ground to the north, the approach could be technically demanding.

The previous year a BMED aircraft narrowly avoided disaster landing at Bole. The aircraft was flying the VOR/DME non-precision approach when, unknown to the crew, the ground-based VOR beacon was transmitting erroneous positional data due to water damage.

As the aircraft descended, warnings were delayed, and the Enhanced Ground Proximity Warning System (EGPWS) alerted the crew too late to be of use. A go-around was executed just seconds before potential impact; at one point, the aircraft came within 56 feet of terrain. The AAIB investigation highlighted how the faulty VOR signal disrupted multiple safety systems, representing a serious single-point failure.

It shook a lot of people.

After reporting at the BMED crew centre, the captain and I were running a little late, nothing serious, but when that happened, we would introduce ourselves to the crew on the bus taking us to the aircraft stand.

Still relatively new and not knowing all the crew members, I always made a point to say hello to every crew member, not just the purser. As I climbed into the bus, I walked down the aisle, pausing at each row to say hello.

I hadn't met the first crew member, an attractive girl with carefully groomed black hair; she was very Mediterranean looking. I took the opportunity to look at her name badge so that I could use her name when I said hello. I saw that Mireya wore a small Spanish Flag on her badge, indicating she was fluent in that language.

Ever being the comedian, I said "*Ah hi Mireya, how are you, I'm Jim*",

"I'm a fine thank you, nice to meet you" she replied very politely,

"And Mireya, what part of Mexico do you come from?" she went ballistic, as only a proud Spaniard can.

"I NO MEXICAN" she blasted, *"I'm e Spanish"*, so, that went about as well as landing with the parking brake on.

"I'm just joking, lovely to meet you".

Note to self simply read: stay away from the Mexican.

Despite our slightly rocky introduction, Mireya and we would go on to become the closest of friends for the next 20 years.

As I continued up the aisle with Oscar, the Captain, close behind me wondering what kind of lunatic he'd been crewed with. I saw her. Towards the back of the bus, the blonde from Amman was there.

That stopped me cold.

I didn't even know her name. I used the same trick, looked at her name badge and spoke.

"Hi Jax, I know you...."

We arrived in Addis after an uneventful flight, just how it's supposed to be. Jax was working in the front, business class cabin; Oscar and

I were well looked after with belt-fed coffee, snacks and meals.

As usual, the crew met in the crew room at the Addis Hilton, an ageing hotel that sat between UNHCR camps and patchy rows of shanty housing. Just up the road stood the Sheraton, a gold-plated offering to the gods of marble, chandeliers, and diplomatic convoys.

It always struck me as wildly out of place. I was glad we were where we were.

There wasn't much to do in Addis. Maybe a trip to the market during the day, or one of a few reliable bars in the evening. The crew room became our default gathering point, with tired uniforms swapped for jeans and sweatshirts or pyjamas and a fridge full of St George beer doing its best to smooth off the edges of a long day.

After an hour and a couple of beers later in the crew room, there was only me and Jax left.

We put a DVD on to watch, the only one we could find that wasn't scratched and/or sticky was Blackhawk Down, that classic love story, perfectly appropriate for East Africa!

We talked. About home, family, careers. About what we'd seen and what we wanted next. It felt effortless; words came easy, and there was a comfort in the quiet spaces between them.

The conversation went well into the night. We both expressed growing frustrations with our relationships, how we'd both been evaluating our lives looking for something more. Jax described the way her boyfriend treated her.

She told me that he would rarely go out of his way to do simple things with her. I couldn't believe this person wouldn't move mountains for her. I knew I would. And in that moment, I realised the truth: I was going to.

We sat opposite each other on a pair of old, battered couches that had seen more room parties than they had any right to survive. She wore a faded firefighter sweatshirt, sleeves rolled up, hair tied back, relaxed, effortlessly sexy. But it wasn't just that, there was a calm confidence in the way she sat, like she wasn't trying to impress anyone, especially me, and that only pulled me in more.

The conversation flowed easily, and the eye contact lingered longer than it should have. I went to bed knowing we hadn't crossed any lines, but we'd taken the first step towards something more.

I knew with absolute clarity that my marriage was no longer what I'd pretended it to be. I wasn't storming out in a fit or chasing novelty. I was looking at someone who made me feel like myself again and that's what made it dangerous.

The hardest part wouldn't be making a choice; it was admitting that I already had, deep down. I just wasn't ready yet to say it out loud.

The next day a few of us piled into a taxi held together with tape and stickers and we hit the markets, this place sold everything from tyres to catfish, from lingerie to carrots. Jax was beside me the whole time, laughing, bartering and wiping oily mud off her shoes; just like Teesside I suppose.

Eventually, we persuaded the others to head back to the hotel, get changed, and regroup for a night out.

We started at a crew favourite; a restaurant called 'The Cottage'. The food was excellent, but something about eating well in a city haunted by famine always unsettled me. Leaving food on the plate felt offensive.

There were two conversations happening that evening: one among the group, and the one between me and Jax.

Ours ran quietly beneath the surface, effortless and unforced. At one point, someone cut across the table chatter with, *"Will you two either join the rest of us or just go get a room?"* I laughed it off, but I already knew which option I'd have chosen if I could.

After dinner, me, Jax and a couple of others went to a bar called 'Dukes', a pseudo-cowboy style dive bar that had a pool table. The pool room was so small, the cues had been cut down into sticks the size of chop sticks.

The Italian guy running the joint always looked after us well, it was good for his business to have the BA crew as favoured patrons. We played a few games, sank a few St Georges, and eventually ended up at a club called Memo's.

Memo's defied description. Mirrors lined the walls, each with a dancer, though 'dancer' was generous. They moved with theatrical effort under flickering blue lights, their eyes scanning the room for opportunity.

When I walked in with two or three white girls in tow, every set of eyes followed me like I was some kind of international pimp. We made straight for the safety of the bar. It was my round; of course it was.

Jax said she was going to the ladies, The other girls formed a three-ship mutual support unit, South Armagh style.

When she came back, she leaned in close to speak. Her voice lost in the thump of bass and neon. Her perfume mixed with mint and cider. I couldn't hear a word, but it didn't matter, I wasn't listening. I was already somewhere else.

I looked at her, having not heard a word she'd said, nor did I care, I leaned back in and said in her ear *"you're hard work!"*

Slightly taken aback, she gave me a puzzled look, leant back and said, *"what do you mean?"*

"It's taking everything I have not to kiss you"

She leaned back into me, and I kissed her.

It was electric. Quiet. Certain. Utterly wrong.

In that kiss was everything I'd ignored; every crack I'd papered over, every moment of absence dressed up as routine. She had a man back home. I had a wife and kids. A history I'd fought to keep intact.

When we finally broke apart, we both smiled. Eventually we dragged ourselves away from the madness of Memo's and all four of us headed back to the hotel. In the lift going up to our floor we prepared ourselves for it to stop, bracing against the walls because the lift wires were more like rubber bungies than steel cables.

It bounced to a halt, the doors slid open, and we looked at each other, said goodnight and headed to our respective rooms.

As soon as the other two were safely in their rooms, she turned and headed back my way; the look in her eye mirroring my own. We went into my room wrapped up in each other. There were no words needed, no second-guessing; just the unmistakable sense that something had started, and neither of us wanted to hold it back any longer.

We hadn't crossed lines before that night, but there was no denying it now. It wasn't lust in isolation; it was something bigger. The intensity came not from what we did, but from what it meant. I knew, with absolute clarity, that this felt like love. Real love.

Chapter 110. Love is in the Air

I flew home with my head full of cartoons. Nothing had been said out loud, but everything felt different; Annie greeted me like nothing had changed. Something had, and it wasn't jet lag or fatigue.

I had to consider the way forward, what was best for the children first and foremost. I'd seen first-hand how relationships decay into sharp words and lingering animosity, and I wasn't going to drag them through that.

I'd been on the wrong side of a break-up before. Emily and Ali both left marks I still didn't have the language for. Annie didn't deserve that, and I didn't want to be the one inflicting that kind of mental trauma on her.

I would go down into the woods and the stream at Donegore to think, away from everything.

Whatever I did, it was going to be wrong.

My last trip before Christmas 2004 was a 3-day Baku. I reported as usual and collected the flight paperwork from Ops. We were a crew member short; someone had called in sick at the last moment. The CAA dictate minimum crewing levels; this would have taken us below that level and meant a delay while a standby was called in.

Once we'd completed the flight deck briefing, I went back to Ops to see what the crewing situation was.

They'd found a standby not far away and she was on her way into the centre.

Jax.

Christmas had come early.

This was a simple trip, no shuttle, just Baku and back with Capt. Margaret 'Mags' Butler.

We did a lot of talking, each of us from our own perspective. What we were positive about was that we felt the same about each other, that was love, without qualification. People might say, in fact they did say, that it wasn't possible; it was just lust. It wasn't. I knew in Addis how I felt, I was sure Jax was the same back then, and she was.

I was madly in love with her, and I told her.

We couldn't take our eyes off each other for the duration of that trip.

I arrived back at LHR around lunchtime, went over to terminal 1 and waited for the next flight to Belfast. Driving from Aldergrove home, the news broke on the radio of the horrendous tsunami that had struck many parts of the far east.

My decision loop was tightening. DODAR kicked in. I knew my options, and I knew which one I was going to action. Christmas complicated the timing. Leaving then would have detonated everything; it's a time for family and we had a big one planned, including going to a black-tie gala ball with some of Annie's family friends on New Years Eve. As I had another trip that saw me arrive back on New Years eve itself, Annie was picking up a rental dinner suit for me.

Leaving would have to be after that.

I must have arrived home late afternoon/early evening. Annie had bought herself a stunning dress that made her look a million dollars. She hadn't collected my dinner suit, instead opting to borrow one from a friend.

When I clapped eyes on this 30-year-old clown suit that was 2 sizes too large hanging up, I hit the roof. I was angry that while I'd been away, Annie had clearly gone to a lot of effort in ensuring she looked amazing; and she did, she looked stunning.

I looked like I'd crawled out of a skip. It wasn't just about the suit. She was showing up for a version of our life that didn't fit anymore. I was finally realising I couldn't wear that version of myself, either.

It felt like a perfect metaphor. She had put every effort into her appearance, and I was left to look like a clown. Thoughtless. Or at least careless in a way that had become familiar.

The bad atmosphere and arguments continued on the drive to the Dunadry Hotel, where the gala was being held. The barbed comments continued both ways all night; I was embarrassed and felt humiliated sitting there in that suit trying to hide how stupid I looked in it.

On the way home it came to a head with Annie exclaiming *"I want a divorce".*

I knew she didn't mean it. I did. And I said so.

It could have been poor timing on her behalf, but it was convenient for me.

In February of 2005 I'd had a couple of trips with Mags, who after Baku, unlike the rest of the company, was very aware that Jax and I were together. We had long conversations about the perils of intracompany relationships, the pitfalls and benefits. She offered sage advice for me professionally and personally.

Things at home had calmed after the New Year's eruption. On the surface, anyway. But beneath it, there was a tension that couldn't be smoothed out. I'd already made my decision. I wanted out. I wanted the kids to have love in their lives, not the cold, bitter version that Annie and I were moving towards. I know now how naïve that sounds, but it was the best reasoning I had.

When Annie found the burner phone, that was it. There was no shouting at first, just a look that said everything had finally caught up with me.

I was embarrassed, she was furious, and rightly so. Under it all, I didn't want to hurt her, no one deserves that. I packed a couple of suitcases and left.

I had calls from Marie asking me to reconsider and to turn around. She would help us sort everything out. *"Marie"* I said, *"I'm sorry, but I don't want to sort it out".*

I called Jax and told her what had happened. She told me she'd just been through similar with her now ex, boyfriend. I was on my way to Heathrow.

Jax met me outside T1 in her little silver sports car. I threw my bags in the boot and climbed in beside her. As I did, the radio came alive with '*Run*' by Snow Patrol. The timing was uncanny.

We didn't need to say anything; the words did it for us. We just looked at each other, knowing this was it: us, together, with a lot of hurt in the rear-view mirror and plenty of people willing and expecting us to fail.

And that song said everything we couldn't.

113 one of the very few pictures of us flying

UK, Feltham: Feb 05 – Apr 05

Chapter 111. Fanning the Flames

Jax lived in a small, two-bedroom apartment in an estate in Feltham; right between the Young Offenders Institute and the airport. You could measure the place by how often you heard jet noise or sirens. The difference between there and Donegore could not have been starker, and yet I barely noticed.

I moved my suitcases into the bedroom; they pretty much filled it. Her brother and his girlfriend had the second room temporarily.

The first night Jax made an extraordinary effort to cook us a meal. She'd been shopping and bought frozen versions of every vegetable available. Together with frozen Yorkshire puddings and a pre-cooked chicken, our plates resembled an active volcano, with gravy lava. Part overcooked and part frozen, it was a disaster.

"Mmm. That was delicious."

Her brother, an aspiring actor, moved out within a week to be closer to the acting community.

Prior to BMED, Jax had flown in Japan with ANA. A couple of friends from those days were now back in London. Fabian later moved in with us; he was quiet, unassuming, and kind.

When we met friends for a night out in Windsor, they were accepting and non-judgemental about our relationship.

Life at home was messy but settling. At work, though, things never stood still.

I reached my 6-month anniversary point in March and went through the standard Line Check with no issues. Jax and I had requested a 'buddy roster' from crewing. This meant that they would endeavour to roster us together.

Exactly the opposite occurred, we started to fly together less than before the request

Word was out, most people were accepting, some pretended not to care, and some were openly vocal with their thoughts. One captain I had flown with previously, and got along with well, was now openly hostile, for reasons that said more about him.

Another FO, who I'd never met, had openly expressed his own feelings for her.

On more than one occasion, I would take over an aircraft she was getting off, or vice versa. We did have one trip together, to Addis of all places.

On this occasion, the runway in use had no instrument landing system and a complex arrival procedure.

This was usually flown from the right seat because the visual aspects and cues were better from that seat. Doug, the Captain, and I, went through the pre-arrival briefing and started down the ILS as normal.

The weather was marginal but improving. Approaching the decision height, it was still gloopy. *"Go around"* called Doug.

The go-around in an Airbus is one of those things you practise endlessly in the sim. When it happens for real, it still has your full attention.

You're on short final, eyes flicking between the PFD and the grey smear outside that's supposed to be a runway.

My left hand went forward, thumb finding the Take-Off/Go-Around (TOGA) buttons, thrust levers all the way up.

"Go-around. Flaps."

The engines surged, the flight directors snapped into life, and we climbed away into grey.

Acceleration altitude comes and you retract the flaps back, stage by stage. Thrust levers back to climb, thrust settles, speed comes alive again.

The radios are full of your monitoring pilot's calm voice telling ATC you're going around, climbing to whatever altitude is in the box. And for a moment, with the workload easing and the aircraft stable in the climb, you realise the adrenaline's only just catching up.

The procedure in Addis involves a long climb due to the high terrain around. Once at the correct altitude we entered the hold and replanned the next approach. We had enough fuel for one more approach, or we were going to divert, and no one wants that.

The tower said the weather was improving. It had, and we landed with some relief that we didn't have to tell the passengers they were in Djibouti – the diversion.

It wasn't lost on me that this job mirrored my life at the time; sometimes stable, sometimes lurching into a go-around with no guarantee the second approach would be smoother. You flew what was in front of you, trusted the training.

UK, Antrim: Apr 05 – Apr 06

Chapter 112. Belfast Child

I was missing the kids badly. Patrick, with his cheeky, bullish confidence, and Amelia, quiet and watchful, were never far from my thoughts. I needed to be closer to them; none of this was their fault and they shouldn't suffer. I'd told the girls in Germany what was going on, and they were not happy with me either. Jessica said some things that took us a while to get over.

Jax and I discussed it at length and agreed that distance from the kids was a bad thing. We'd caused this, so from that point on they came first. Every decision flowed from that, starting with moving closer so I could be part of their lives and Jax could build a relationship with them.

If the kids were to be at the centre of everything, we had to be closer. It felt like the right call, even if it stretched us thin.

We took a couple of weeks leave and moved into a large, old Victorian house between Antrim and Templepatrick, near Aldergrove and Donegore. The house and garden were huge, but it was fun.

Not by design, we moved in on the 1st of April with everything we had between us, which wasn't very much at all. We went shopping for the bare basics of furniture. We came home with a massive sofa; the most comfortable thing I'd ever sat on, a bed for us and some beds with bedside drawers shaped like jigsaw puzzle pieces for the kids. On top of the furniture shops, we visited the Big W in town.

With '*Dakota*' by the Stereophonics in the background, we pranced around the shop like a pair of big kids buying super soakers, electric shock guns, and other toys; some for the kids, and some very clearly for ourselves.

The first time I went up to Donegore to pick up the kids was a disaster. Annie was not happy to see me; and I understood why. I waited for Patrick to get something he wanted to bring with him, excited and happy to see me.

Nearly as excited as I was.

Through an abundance of respect for Annie, the mother of my children, I won't go into any of my memories of these few weeks, it would only serve to resurrect painful memories for us all. Memories which I'm sure would be contradictory with those of others; to quote Her Majesty the Queen "*recollections may vary*"

We had a lovely couple of days with the children. It felt like a new beginning rather than a visit.

Our first visitors to Antrim Road were my Mom and her partner, John. They had met at a cycling event not long before, but talking to my mother, she said straightaway that she saw in Jax and me the same spark that she had with John.

Since I'd left home to join the Army, life hadn't been especially kind to her. She'd carried on regardless, quietly getting on with things in that way she always had, without drama or self-pity. She'd earned every scrap of happiness that came her way, and seeing her finally relaxed, genuinely content, felt like a small correction to the universe.

Cycling had become her escape, her routine, and occasionally her religion. Rain, hills, flat tyres, none of it mattered. And somewhere between Lycra, energy gels, and suspiciously early starts, it delivered John. A decent man, kind, patient, and robust enough to tolerate both her pace on a bike and her son's complicated life choices.

It felt right. Long overdue. Proof, if nothing else, that persistence eventually pays off, and that sometimes the long road does actually lead somewhere worth stopping for.

Chapter 113. The Russians Attack

Jax and I took a 10-day break, and, with Annie's agreement, we took Patrick and Amelia over to Gutersloh to see the girls during the summer holiday. We arrived at Dusseldorf airport and hired a Mercedes minivan; the only thing big enough for us all to get around in.

Patrick bounced around in the back of the van like he'd mainlined Haribo, while Amelia sat quietly beside him, her big eyes fixed on everything outside as if she was filing it away for later.

The hotel was the usual one just over the road from Scott's house, but by the time we arrived it was late, we were all tired, and there was no food to be had. I drove down the road to a service station to stock up on German snacks.

I drove past a young German bloke, absolutely leathered, weaving about on a bicycle. I had to take avoiding action to miss him.

Driving back to the hotel, just as I rounded around a corner, a man jumped off the pavement and straight into the front of the minivan. He slid slowly off the front, like some kind of jellyfish dressed in an AC/DC shirt and jeans, leaving a smear of saliva and skin down the windscreen.

Miraculously, he stood up, stared at me with wide eyes and gave me the international hand signal for '*give me money*'.

'*Ah. I get this*', I thought. He throws himself into the traffic and then demand cash to avoid the Polizei! A classic scam. If he was doing international hand signals, so was I, he got the same hand signals Keith had way back in Split; the '*wanker*' and the '*do one*'.

I could see his mates, 3 or 4 skinheads advancing in my general direction, and it didn't look like it was for a cultural exchange. As I was thinking about just driving through them, the cops arrived on the scene.

The Polizei didn't even break sweat. One minute these guys were strutting like hyped-up Mike Tysons, the next they were face down on the tarmac wondering how their night had gone so wrong.

I stepped out of the van and went to talk to the feds. Between my bierkeller, squaddie German and their abrupt English, the story came out in fits and starts; it just got weirder. I asked if I could let

Jax know where I was and then come back to give a statement. I returned slightly nervous.

So, what had really happened?

The guy I'd taken avoiding action on was just some guy making his way home after an evening singing '*Thunderstruck*' to an oompah band. Probably. As he tried to pass a group of blokes, equally full on Pils, he hit one of them, knocking him over and tearing his victims knock-off Armani jeans in the process.

Enraged, and apparently devoted to counterfeit fashion, the dude wearing the Armani jeans turned the wobbly cyclist into a guided missile targeting my van. His bike was then hurled over a nearby hedge, which is why I didn't see it.

What I'd witnessed wasn't a scam at all, but some poor sod being used as a human Exocet by three Russians who'd had too many vodkas and one too few brain cells. After he'd bounced off the bonnet, he wasn't after money. He was begging for help.

And what did he get? Abuse, hand signals, and a lecture in Anglo-Saxon from the Brit tourist behind the wheel.

Even as Patrick and Amelia laughed with the girls in Gutersloh, I felt Jess's absence. She'd chosen not to be there, and that choice stung more than any words she might have said. The rest of the week was lovely; we all went shopping and spent time just doing normal things.

We handed the van back in Dusseldorf, and the rental company did what car rental companies do best. They emptied my credit card. It took nearly a year to get it back. Financially, that was the last straw, the thing that finally tipped us into rock bottom.

Chapter 114. The Other Side of Divorce

One thing was certain: Annie wanted the divorce, and she didn't waste time in making it happen. To her credit, she didn't make it bitter or drawn out. I kept it simple too; I gave her all the proceeds from the sale of Donegore.

She needed that money more than I did; it gave her and the kids stability near her family. That was what mattered.

When it was just me and Jax, we regressed into oversized kids. We'd tear around that creaky old Victorian house with super soakers or electric shock guns, shrieking like ten-year-olds on Red Bull.

Furniture was overturned; wardrobes turned inside out for hiding in. It was utterly ridiculous; two supposedly responsible adults with bills, professional jobs, and complicated personal lives, staging military-grade ambushes. For all the mess and madness, it was the best fun we'd had in years. It was what made us, us.

Evenings were no calmer. We'd wander down to Ellie Mays opposite the Dunadry for quiz night, where our team's name probably should have been The Outsiders, but was J2 (for Jax & Jim).

The quizmaster was also the local horse dentist, interrogated me like I was applying for citizenship before he let me join in. English accent, short hair; he clearly suspected I was either an undercover cop or planning to buy his horse dentist practice.

We won that night, and more often than not after that, to the growing irritation of locals who'd been trying for years. Between the water wars at home and humiliating the pub regulars with our trivia, it was hard to know what was more absurd, that life had become this ridiculous, or that I was loving every second of it.

Once the rent was paid, we were skint. Pretty much all my salary was going straight to Annie in line with our philosophy of kids first.

We spent a fortune on staff tickets to commute to LHR. It was getting to the point where we had no money, no access to money and no way to get petrol in the car to get to the airport.

We were properly broke. For one week we lived almost entirely on potatoes. Chips, mash, jackets, roasted. If it could be done to a spud, we tried it. By Thursday, I was eyeing up the wallpaper paste as a potential side dish.

Work meant crew meals and flying pay, so we weren't starving, but at home it was starch or nothing. It was so bleak it was funny, like starring in our own low-budget famine re-enactment, only with a microwave. Just as we were learning how small our lives could get, the world reminded us how big and ugly it could be.

Chapter 115. Tehran and Other Great Ideas

It was easy to think our lives were uniquely messy; divorce papers, weeks living on spuds, and water-gun wars in a draughty Victorian house. Then the news from London hit. Chaos wasn't just ours; the wider world was tearing itself apart too. In early July, I was down route in Tbilisi, the capital of Georgia.

Tbilisi was one of our favourite destinations, the Georgian Saperavi wine was excellent and the local cheese bread, khachapuri, the perfect accompaniment. Sitting at a riverside restaurant with its wrought iron architecture, we could have been in New Orleans.

Back at the hotel, the Captain gathered us together and simply said that there was news from London. It looked like the city was under sustained attacks from suicide bombers. We weren't due to fly back for a couple of days.

A flurry of calls home to check on loved ones, but we went to bed hoping it would all be over by the time we got home; and that LHR wouldn't be attacked.

On one of the few times that we were rostered to fly together, we were in the queue for security at Belfast City airport when Jax spotted someone she knew in the queue ahead. It was Ciara, a former housemate of Jax's from her time working in aviation out of Teesside airport.

Ciara was now the FO on corporate and executive jets. A quick exchange of numbers and a conversation that said, *"I'm back next week on another job, see you then"*.

Sure enough, Ciara came back and having relieved her aircraft of all the unwanted catering and champagne, came up to our big 'ol house. Jax and Ciara caught up on everyone and everything, but the subject of my Finland expedition came up.

A drunken plan was formulated. Ciara and her boyfriend, Richard. Me and Jax, along with Ed and his girlfriend would go to Finland for New Year. We had a few months for the planning, but this was a great idea. No; it was.

In September I was on yet another Tehran, and yet again Jax wasn't on the trip. The only good thing about Tehran was that it was the simplest trip; two long sectors and then home.

Unfortunately, whilst we were down route, the catering company, Gate Gourmet experienced a mass strike. That was bad enough. Then it escalated when BA ground staff joined the protest in "sympathy strikes". Within hours, the entire network had seized up.

Of all the places in all the world, I was in Tehran with only the uniform on my back to wear, that or a pair of pyjama bottoms and a tee shirt, neither of which the Ayatollahs' fun police would have found acceptable.

What was a one-night away turned into three before we were allocated a return slot into LHR. On the second day, a couple of us fashioned some half decent outfits and got a taxi to take us to the British Embassy, closely followed of course by our Iranian 'minders'.

At the embassy 'Cockatoo Club' we met with the diplomats who looked after us with great hospitality, that included beer. We were fortunate that on the way back to the hotel, the taxi driver didn't smell the alcohol on us and take us straight to a police station.

Realising we'd been lucky, we stayed in the crew room for the third night….it just wasn't worth the risk of life in prison for the sake of a cold one.

No sooner had we shaken the dust of Iran off than I found myself on another test altogether, meeting the in-laws. Jax took me to meet her dad and step mum. She wanted to show off just how much of a catch I was. Nearly 40, twice divorced with 5 kids; quite the prize package.

All I was missing was a caravan and a warning label.

They lived on a smallholding near Stamford, and all was good. I'd braced myself for suspicion, but what I got instead was warmth and common ground. It meant more to me than I let on at the time.

Patrick and Amelia would play with the chicks and ducklings and enjoyed seeing the pigs. For me, I got on well with her dad and liked him very much, we had a lot in common. He felt like a potential father figure.

Christmas 2005 was a very odd affair. Mags, Jax and I requested the same trip, which we got; unfortunately, it was in Tashkent, the capital city of Uzbekistan, a place that redefined strange. The bars were half dive, half circus, with 'entertainment' that included an electric drill.

It was like someone had spliced Blackpool with a scrapyard.

We had collaborated with the rest of the crew so that each person would bring something to make Christmas day a bit more enjoyable; Mags brought cheese and crackers, someone brought a bottle of Port, and Jax and I took a box of expensive Christmas crackers from Marks & Spencer's.

Going through crew security, the officer tried to claim that we couldn't take them through to the aircraft; which was nonsense. There are limitations on crackers due to the snap, but that was at the discretion of the airline; not security.

I educated the officer the way a SNCO should, out of habit rather than choice.

The officer didn't like being educated. The officer called his manager who also didn't appreciate being educated. Bracing them up to 'attention' was probably a step too far.

The security manager then threatened to remove my security pass. Mags stepped in and made a Christmas compromise, but we kept the crackers. They were the perfect accompaniment to our dinner, boiled chicken and rice with a glace cherry – just because it was Christmas.

BMED was full of characters, too many to list here; and certainly, too many tales to tell about the down route room parties and the antics some people got up to.

But one couple do deserve a special mention.

Krista was a tall, blonde girl from Bolton and was the life and soul of any party; in fact, pretty much all parties. Her partner in crime was Matt, one of our engineers, a top bloke and full-on cockney rebel.

Whenever they could, they'd be down route instigating parties, and always had a spare suitcase stuffed full of fancy dress, just in case anyone thought about being boring

One of the most memorable, was a trip to Amman; the whole crew had invited family and/or friends along because the Hilton hotel in the city would allow us to change to the Hilton Dead Sea Resort & Spa, at no cost to us.

As usual, Matt had the dressy up case with him, and we got stuck into it the first night there. First night madness was a real thing because it gave plenty of time to recover.

The next day was spent caked in mud and then doing the whole floating in the Dead Sea piece. The night before, Matt and I may have persuaded one of the new girls on the crew, who'd never been to the Dead Sea before, that it was like a miracle spa treatment, most effective if you had a shave before. Mean, I know, but very funny.

The poor lass must have got her orange bic razor out and 'tidied the garden'. She hadn't noticed that Matt and I had 12 o'clock shadows. As we all gingerly entered the water, knowing how painful it was in the eyes, she slowly sank on to her knees to wash the mud off.

The water came up to her tummy, slowly being absorbed by her bikini bottoms. Then it hit, like an electric shock the water engaged with her freshly shaved areas. She exited the water like a Polaris missile calling us some very naughty words.

Then someone's mum decided the best way into the Dead Sea was a running bomb. The splash hit all of us like tear gas. Eyes on fire, people staggering blind out of the water, it was one step short of a mass-casualty event. If she hadn't been someone's mum, I'd have called her worse than naughty words.

That night, we calmed it all down, but only a couple of notches. There was a steak house serving the very best Wagyu beef, so we headed in there to tuck into steak & chips. After dinner, we all went to the bar for a nightcap; but then unexpectedly, Mags, who was the Captain binned it and went to bed – very unlike her.

This just wasn't good enough. Jax went to reception and, pretending to be 'Captain Butler' asked for a new room key. The staff handed one over with no questions asked. The remaining 14 of us went upstairs and formed a conga line outside her room, with Jax at the head.

The key was silently tapped and the door eased open; it was half lit. *"Da da da da dat da, Da da da da dat da"* we burst in, the lights went on. Jax leading us around the corner. "*Whoaaaaaa!!*" Jax hit the brakes so hard the line collapsed in on itself.

There was Mags, flat out, stark naked, book across her chest like some bizarre fig leaf. We reversed out like a comedy sketch, laughing so hard we could barely breathe.

Our friends, Matt and Krista married in 2007 and had two boys. In 2017, tragedy blindsided the Schofield family. Matt was diagnosed with bile duct cancer and was given limited time.

Krista, training for a charity run in support of his cancer fund, injured her foot. To manage the pain, she took some of Matt's prescribed morphine, just enough to take the edge off, she thought.

The dosage proved fatal.

Krista died in October, leaving two boys without their mother.

Matt followed Krista six months later, taken by the cancer he never had a chance to beat.

Figure 114 The Addis Crew Room - Krista in the middle

Chapter 116. Finners

Six would-be adventurers, most of us strangers until that morning, met at a grim service station outside Stansted airport, populated almost entirely by truckers who looked like personal hygiene was something that happened to other people. Within minutes it felt less like a travel plan and more like the beginning of a bad sitcom.

We flew to Helsinki with Ryanair, wearing most of our outdoor kit like overstuffed scarecrows to dodge Ryanair's miserly baggage allowance, before heading north to Kuusamo. From there we picked up hire cars for the final push to the stunning, but remote log cabin we'd rented.

That's when it started going wrong, but in the best possible way. We had two cars between us: I drove the lead with passengers; Richard followed with Ciara and the baggage. Finland at that time of year meant 22 hours of darkness out of every 24. The locals had floodlit ski trails so you could glide through the woods in style. But roads? Not a single light. Apparently, headlights and faith were considered sufficient.

Faced with a choice of direction, naturally, I chose the lit option, which turned out not to be a road. The light was a lie. Ten seconds later we were slaloming down a cross-country ski trail in a Ford Fiesta before burying ourselves neatly in a ditch. No medals awarded, but plenty of wheel-spinning and grunting got us out.

The cabin, when we found it, was remote but every bit as stunning as we'd imagined. The smell of the pine structure, warmed by a log fire was just heaven. We ran in like kids on a school outing and chose our rooms before exploring the cabin. There was a sauna amongst many other features, but our first mission was to find a booze shop.

We piled back into the cars and headed for Ruka, found the booze shop and set about emptying it. Ciara found a drink called Gloggi and had to buy it. We had duty free winter Pimm's, Bacardi and vodka already with us.

We also found a shop next door that sold fireworks, so we also set about emptying that too. We bought dozens of tiny rockets; a proper rocket, but about the size of AAA battery mounted on a cocktail stick. There were also some interesting little numbers called '*Dragon Eggs*', a small but powerful explosive package with digit removal capability.

As soon as we got back, Ciara set about making some kind of brew; Gloggi, Bacardi and vodka heated up in a saucepan. And my god it was delicious. Delicious and deadly. Someone had the brilliant idea of firing up the sauna. Truly a brilliant idea.

We all went and got changed into swimmers, collected a fresh Gloggi and piled into the sauna. After ten minutes in temperatures of 50–60 degrees Celsius, someone suggested using the snow instead of a plunge pool. This suggestion survived exactly five minutes of reflection. The flooring between the sauna and the back door was covered in a plastic lino. We missed that.

The lino was doing what lino does best: waiting.

We ran outside and suffered instant regret. The temperature change of around 80 degrees in 5 seconds was enough to kill a polar bear. We sprinted back inside to stay the onset of frostbite. I had snow stuck to the sole of my foot and slipped wildly on the way back in, nearly crashing. We settled back into the sauna with words of wisdom that said *"yeah, that was dangerous, let's not do that again!"*

5 minutes later.

"Everyone out!" same deal, sprint outside. Only this time I was a true hero and gave Jax a shove in the small of her back. She '*starfished*' into a very deep snowbank and melted into it, a wisp of steam emanating from her bikini bottoms. So that she couldn't get me back, and she was very capable of doing so, I ran back inside.

As soon as my lead foot touched the lino, I executed a perfect Arabesque with pirouette, but the landing was less graceful. The bridge of my nose impacting squarely on the edge of a small changing bench. I ended up in a heap with what I thought was a broken wrist. Richard looked at me.

"It's my wrist" I winced, *"I think I've done it"*

"Don't worry about your wrist mate, it's your nose you want to be worried about!"

Just then, I realised I was covered in blood and that I could see my own nose spread across my face.

Ah Bugger!

Everyone in the room sobered up instantly. Ciara decided she would drive me to the hospital; she couldn't even get her salopettes on.

Ed. Well he decided to take my mind off the situation. Lighting sparklers and waving them over my head like some budget shaman, hot magnesium raining down, while I lay there bleeding, looking up at my own brothers' swingers was a distraction I didn't need!

Jax was running around shouting something about hot sweet tea, but at least she did the one good thing. Wrapped some snow in kitchen roll and applied it to my nose to stem the bleeding.

I was holding the kitchen roll/snow/mush swab on my face and realised that I could move my nose without any pain. I grasped it and pulled hard, then moved it back to where it should have been.

There was a loud click.

Five people retched in unison.

The job was done. My nose was reset. So naturally, we toasted my DIY surgery with another round.

The next day, after a lot of fresh coffee, we went to the ski slopes. Neither me or Jax had ever skied before, I was still technically concussed and absolutely shouldn't have been on skis, which in hindsight explained my confidence.

However, I was still in 'impress the girl' mode and made out I would be great. I was. I adopted the standard ski Sunday racing position and pointed down the hill. I flew down the hill, looking like Clark Griswald with flames coming from behind me, on the very edge of control.

Luckily, I didn't need to turn. Just stop. I threw the anchors out and fluked a near perfect stop before nonchalantly joining Jax for a hot chocolate. Breathing in all that freezing air was perfect in reducing the swelling in my nose, but I still had a fantastic pair of shiners.

We had no internet, TV or radio for the week, which meant no distractions and nowhere to hide. So, we improvised: each couple took a turn hosting some form of "entertainment." Ciara and Richard ran a quiz night, complete with arguments over obscure answers and a scoring system that defied logic.

Jax and I went old school, armed only with a set of oversized playing cards and a strictly enforced set of made-up rules for 'higher or lower'. It started out friendly enough, by round three it had stopped being a card game and evolved into a contact sport involving Gloggi penalties and increasingly unhinged forfeits that included full streaks around the outside of the cabin.

We had the best time just doing old fashioned things.

New Years Eve 2005/2006 came. The boys built a bonfire whilst the girls changed into long, elegant cocktail dresses. We then changed into dinner suits, proper James Bond style, not antique clown suits.

That night, we drank cocktails, fully loaded with those tiny rockets; naturally they became '*Rocktails'*. Standing around that bonfire, sipping Rocktails under the northern lights was a once-in-a-lifetime experience, never to be repeated.

After a week of blood, fireworks, and frostbite, it felt like the universe had given us one perfect week to make up for the chaos waiting back in Belfast.

Later, nursing a 'dobra Pepsi' and puffing on a Romeo y Julieta No. 6, I let my guard down.

Big mistake.

Ed, whose definition of "helping" had always been flexible, dropped something small and fizzing into my glass. I'd forgotten the "dragon eggs." My drink didn't fizzle. It evacuated the glass in one violent cough. One second Jack and Coke, the next, a puff of smoke mixed with steam and the faint aroma of burnt oak and sugar.

The chaos at home hadn't gone away. It was merely waiting for us. Our attempt to help the situation and relationships had backfired. We knew that staying long term in Belfast would simply make the situation worse; and that wasn't good for anyone.

The lease on the house was due to expire in March; we wouldn't be renewing it.

UK, King's Worthy, Hampshire: Apr 06 – Oct 06

Chapter 117. Headaches and Brainwaves

We managed one last hurrah in Antrim: a big, messy party with friends from BMED, Aldergrove, and a couple of locals, including Graham from Ellie Mays. We did a hog roast, the cheapest way to feed a crowd, and drank like people pretending they weren't about to leave.

Eventually the time came when it was the last time we had the kids. We told them that the next time they saw us, it would be in a new house in England, that they'd have their own bedrooms. They nodded, processed it in their own way, and carried on being kids.

We packed up again, the van looking like the Clampetts had packed it after a yard sale and headed back to England.

With no actual house to go to in England, BJ, once again stepped in to help me out. He was living in a married quarter in Bulford, part of Tidworth Garrison, but was based at Middle Wallop, now a QHI and lead display pilot for the AAC Blue Eagles.

It was a much-needed respite and stress relief; it meant we could just get on and find somewhere closer to Heathrow. It was weird though living back in a barracks.

Jax soon found a rental house in King's Worthy, a small village not far from Winchester. It was well located for Heathrow being just off the M3 corridor. The cottage was advertised as having 3 bedrooms with a garden. It did have those features, but all were tiny.

As we first approached the cottage with our stuff, I took one look at it and felt something give way. This wasn't failure exactly, but it didn't feel like progress.

Jax calmed me down, *"it didn't matter"* she said, we were together. We no longer had Ellie Mays, but the saving grace was that we did have 'The Horse and Cart' pub almost next door. The first night we had to check it out. It wasn't very busy, but we got roped into having a game of darts with the locals; this resulted in it being christened 'The Dart".

The Dart also had a new addiction: the quiz machine. We developed a habit with it quickly, cheap, competitive, and just distracting enough.

Back at work, things momentarily got interesting. BMED had a lot of shared history with Beirut, the first owners were Lebanese and Beirut was the longest standing destination we had. Unfortunately, it's also long been an area of rising tension and conflict.

On 13th July Israel attacked Beirut airport, expertly placing a large bomb on the runway intersection, effectively closing the airport.

The BMED crew in Beirut were evacuated overland to Damascus. I was called into operate an unscheduled trip out to Damascus, via Aleppo, to recover the crews and bring them home. We flew via Aleppo every time we went into Damascus, but we had a night stop there this time.

Years later, seeing what Syria became was heartbreaking. At the time, Damascus felt timeless. Politics and religion had done what they always do.

Damascus itself was an incredible city before it was torn apart. Religious divides were only apparent if you looked closely. The centre of the city was home to the central Omayyad Mosque, with various Christian cathedrals not far away. We had crew favourite restaurants, usually in the Muslim quarter, but then we'd walk half a block to a piano bar in the Christian quarter for a beer.

Dominating the city centre was the Al-Hamidiyah souk, a sprawling market under cover of rusty, bullet holed corrugated tin. Anything you wanted was available, from perfume and cologne mixers to spices; from burkas to lingerie, all under one roof.

The jewellery section of the souk was well visited, the diamonds and gold were all good quality, but a fraction of the cost in the UK. Damascus was in my top two destinations.

Professionally, things were going exactly to plan, I'd passed 3,500 hours, about 1,500 on the Airbus, and my flying career was moving exactly as I'd hoped.

After moving into the cottage, we promptly went out in Andover, some of the blokes were over on a course at Wallop. In one of the pubs, we bumped into Cappy. The last time I'd seen him was in Aldergrove, just before I'd gone to Dishforth. He was also out of the Army and was running his own security company in Iraq.

We talked a lot about aviation security, most of it me ranting about how poor the system was. I thought Cappy might pick up the idea, but he wasn't interested. He smirked and told me that if I thought it was such a brilliant idea, I should put my money where my mouth was.

I might have laughed it off, but Jax didn't. She leaned in, told me I should do it, simple as that. Coming from her, it wasn't encouragement, it was an instruction. In that moment, a throwaway conversation hardened into something else. I didn't know it then, but it was the first brick in what would become Redline.

I talked to Paul about the concept; he was now a Captain, deep in the weeds of bringing the Apache into UK service but struggling with the toll that years of Lynx vibrations had taken on his back; he was in constant pain.

We decided to give it a crack.

For the first time in years, we were back together again. That alone was a spark, we'd always been better as a team, two halves of the same circuit. Paul had the Apache and military training systems in his blood; I had the airline world in mine. Between us, we saw the full picture: how security was supposed to work, and how it played out in civilian airports.

We spoke for hours, mapped out ideas, and in August 2006, two major events occurred in the world of aviation security.

Jax and I were down route in Ankara, Turkey when the liquid bomb plot hit the headlines. At first it was just a murmur on CNN in the hotel lobby, then the reality sank in; Heathrow had gone into meltdown.

The authorities had foiled a plan to blow up multiple transatlantic flights using liquid explosives disguised as Tang, a revolting orange drink that tasted more like sulphuric acid, and overnight the global aviation system had been turned upside down.

Nothing was allowed through security, not even bottled water or baby milk, and Heathrow was gridlocked.

For us, it meant we weren't going anywhere. We sat in Ankara for days, watching the chaos unfold back home. Passengers queued for hours, flights were cancelled left and right, and people broke down in tears as their possessions were binned at security checkpoints. It was surreal: we were in Turkey, ready to operate back, yet completely stranded because London was locked down.

At first it felt like an unexpected holiday extension, lounging around, exploring a bit of Ankara, sharing a meze and bottles of Efes, but the atmosphere shifted quickly. We were glued to the news, phones buzzing with updates from colleagues trapped elsewhere.

It was a strange mix of boredom and unease: on the one hand, a forced pause with Jax in a place we'd otherwise never have lingered; on the other, the creeping sense that our entire profession had just changed.

When we finally flew back into London, the mood was heavy. Security had become theatre overnight: shoes off because of that idiot in 2001, liquids bagged, queues snaking into the terminal car parks. For the first time in my flying career, I felt the industry was running on fear as much as procedure.

Watching Heathrow buckle under the weight of fear made something crystal clear: aviation security wasn't built to cope. Procedures were reactive, not proactive. If the system couldn't protect itself, maybe we could help redesign it.

Chapter 118. A Star is Born

The second major event was the registration of Redline Aviation Security Ltd. The name carried weight for us: high performance, precision, a line you don't cross.

The damp little cottage in King's Worthy became our ops and planning room. Post-it notes and sketches spread across the wall like a mission board; every late-night scribble felt like strategy.

But what were we selling? After hours of debating it, the answer landed with a thud. It wasn't training. It wasn't compliance. It was trust. Trust was the only product that mattered. We even reserved the domain. "Trustredline" was a bit of a mouthful, but it was honest.

That word on the Post-it board – TRUST – became the centre of everything.

We weren't just going to build a company; we were going to build the company we wished we could work for. Every idea on that Post-it board felt like a promise of something better. The idea that two former soldiers turned pilots might reshape aviation security was bold.

For now, though, it was pure lift, the kind you can't help but ride.

Turns out excitement doesn't fix leaking roofs, and Post-it notes don't stop damp or make a shoebox house feel like home. After the adrenaline of Redline's launch, the reality of King's Worthy crept back in.

The cottage was really getting me down. I hated it; it was too small when the kids were over. The landlords weren't interested in upgrading anything and repairs never happened. Ed was now a commissioned officer, based at nearby Worthy Down; we saw a lot of each other, especially after Finland and he had a girlfriend who Jax really liked.

We spoke and thought it would work if we got a larger house but shared it to spread costs.

Jax and Naomi (Nomsie) found a house on a housing estate in Andover that was big enough for us all.

UK, Andover, Hampshire: Oct 06 – Feb 07

Chapter 119. African Promise

Life in Andover was good; we finally had the space to have Patrick and Amelia over every school holiday, even if it meant flying back and forth to Belfast to make it happen. Nomsie had a little girl too, and the kids got on well. For the first time in a long while, it felt like we weren't constantly firefighting.

Jax and I were only getting closer; if that were possible. Her relationship with all the kids was great, apart from Jessica who was still not talking to us.

My divorce from Annie was made absolute and at last, I could pop the question I'd wanted to for so long. It wasn't the down-on-one-knee, fireworks-and-violins moment Jax deserved. I promised her I would ask again in a much more appropriate manner, when the time was right.

On our next trip to Damascus, we strolled around the jewellery quarter behind the central mosque. The shops smelled of cardamom coffee, sweat, and Oud aftershave. I'd saved up enough, I hoped, to buy her a beautiful ring. After visits to several of these small, boutique shops and looking at every ring they had, we eventually found one that Jax liked, and I loved buying it for her.

We headed off to a very popular restaurant called '*The Jabri House*' to have a celebratory meal with the crew, followed by a few glasses of wine in the piano bar.

That engagement ring wasn't just silver and diamond; it carried everything we'd survived to get there. She was gutted, though, that I kept hold of it. She wasn't even allowed to look at it until it was presented properly.

Where was the perfect place to ask the real question? Memo's back in Addis of course. Ed and Nomsie came with us on family tickets and, as Jax and I shared a room, they even got a free hotel.

The next day we lazed in the naturally heated pools at the hotel. In the evening, we retraced mine and Jax's first trip together: dinner in the Cottage and then headed over to Dukes for a game of chop sticks pool.

With some excitement and eager anticipation, we all, including the whole crew, went over to Memo's. It was my round; somethings NEVER change, but this time I was happy to buy everyone a drink.

The crew gathered in a loose semi-circle, knowing what was about to happen.

I dropped to one knee, ring in hand. Jax gasped, cried, laughed, and said yes; all in the same thirty seconds. There were more tears, and not just from us. Memo's had that effect on people. When I slid the Damascus ring onto her finger, it felt like every mile travelled and every battle fought had led right here.

And that was that. We were official.

I stood up and turned around to get our drinks to propose a toast. They'd been stolen.

Perfect.

Memo's did a lot of things for a price, romance wasn't one of them.

Ed replaced the drinks, and we raised our glasses to Jax.

We would have one last Christmas in Andover. Alex was coming over from Germany and had asked us if she could bring a friend. We were just delighted that she was coming over, she had grown into a beautiful, responsible and intelligent young lady, her bringing a friend was a bonus. Her friend was 'Jo'. Or so she told us. Jax went out to buy Jo a Christmas present so that she didn't feel left out.

When Alex arrived, she was with some bloke. That some bloke was her boyfriend; she was too shy to tell us she had a boyfriend, too shy to tell us he was in the Army and too shy to tell us his name was really Barry.

Jax made an excuse and dashed out again to buy a boyfriend appropriate Christmas present. To break the ice, I just said *"Ah, you're a fucking squaddie then"* imitating Bobby as best I could.

We liked Barry straightaway. He was kind, genuine, respectful, and clearly cared about Alex. That was enough for me.

Chapter 120. Genesis

Whenever we had a spare moment, Paul, Jax and I were digging into what actually needed to be done. We were determined that Redline would set off on the right foot. Our original business plan saw us just as a security training provider for operating crews.

All we had to do, in theory, was replace the awful training Jax and I had endured. What we offered instead was pedigree: the kind Paul and I had received and delivered throughout our military careers.

The quality was there. The trust would follow. What we didn't have was the time or money to commit fully without breaking our families in the process.

To provide these services, we had to achieve certification from the UK Department for Transport (DfT). Between us we had to complete an aircrew instructors' course and an aviation security managers course.

Paul took the 3-day course, then delivered at Thames House in London; I took the 7-day managers course, delivered by a major UK security company, one of only three approved providers in the UK.

Paul went first and was not impressed with the standard of the course. Mine was no better, in my view it was death by PowerPoint, boring, lacklustre and poorly delivered, seven days of droning slides delivered with all the energy of a damp sponge.

At the end of my course, I met a senior representative from DfT who had delivered a module. In the coffee break I chatted with him; he asked for my thoughts.

"The truth?" I questioned,

"….the truth" he replied.

I didn't hold back. At all.

Like Cappy, he said *"so what are you going to do about it?"*

"Let Redline deliver it" I instantly said back.

The problem, he explained, was that only authorised providers listed in the National Civil Aviation Security Programme could deliver it.

"So, list us," I said. If the problem was that we weren't listed, the solution seemed obvious.

He blinked like no one had ever suggested it before. He was learning fast that we wouldn't be taking no for an answer.

The path to listing was not defined, but he agreed, after some insistence, that if we presented a manager's course syllabus to his department, they would assess it and evaluate Redline for listing in the NCASP, a huge, restricted publication that most people only ever complained about.

There started a long and complex relationship with the government that neither of us had predicted.

I'd been talking to another Senior First Officer in BMED and former police officer about Redline and what we were planning. He was keen to help where he could and told me he knew of a possible location that would be ideal for us to establish ourselves.

The newly opened airport at Doncaster, formerly RAF Finningley, had a brand-new terminal and old wartime hangars once home to Vulcan bombers, were being transformed into an aviation training academy called 'Directions'.

Inside were classrooms, cheap office space, and even a full Boeing 727 fuselage, available for training. As a Community Interest Company (CIC), Directions needed credibility and content. We needed classrooms, credibility, and somewhere to start. It was the first time the pieces actually fit.

At home, the lease was up on the house in Andover, and despite having had a great time there, it was time to look elsewhere again. Ed had split up with Nomsie, so we really had no reason to stay.

Paul had resigned his commission by then, moving straight into Apache pilot training in a civvy company. It was a natural step, the soldier pilot becoming the instructor, but it also underlined that both of us had crossed a line we weren't going to step back over.

Jax and I thought about and discussed our domestic situation.

Where we were heading as a couple.

Our commitment to the kids and to Redline.

Jax was on the hunt for another house that met what we needed, space for the kids (and Paul), a garden of some sort and not a difficult commute to Heathrow.

We packed up Andover the way we'd packed up everything else by then: efficiently, without ceremony, already looking forward to a quiet village in South Yorkshire.

Ranskill wasn't glamorous, but it had what we needed. Including a pub with a quizzy; which in our priorities ranked only just below Heathrow access.

Part 7: Grounded

UK, Ranskill, North Notts: Feb 07 – Jun 09

Chapter 121. Pulling Together

Paul and I had visited Directions a couple of times to watch the hangar conversion take shape. We'd been in the deal early enough to specify the dimensions of our new office. By night we were doing our best to raise funds. We'd drafted dozens of business plans and were attending 'Dragon's Den' style pitches to so-called 'angel' investors who showed more interest in themselves than our business.

It was a very frustrating time.

Borrowing £3,000 from my grandmother's funeral fund just to attend the manager's course wasn't a proud moment, but it was necessary. Our pot wasn't just empty; it was leaking.

We tried government grant funding and considered a bank loan, but we would have been unable to service the repayments. We had an office to equip.

Our first employee, Joy was starting imminently as our admin manager; a critical role, especially when you're known as the admin 'Vortex'. We knew that as Paul and I were still maintaining our day jobs, we also needed continuity.

Drafting the syllabus for the managers course took considerable effort. Paul brought in a friend of ours from 651/652 days.

Mike Rhodes was a former bomb disposal and ammunition technician. His skills were invaluable in writing a very niche and technical element of the syllabus; the Recognition of Firearms and Explosives (RFX).

The course build was going well, but we knew that there was a much larger and potentially lucrative course about to become a mandated requirement.

The European Commission had also introduced a new regulation that mandated specific General Security Awareness Training (GSAT) for anyone that required an airport security pass. The estimate was that nearly 300,000 additional training hours per year would be required.

That number changed everything.

We simply had to seize this opportunity while it was hot, the requirement would become mandatory in October of 2007.

We had a plan, as you would expect. The first phase was to develop a classroom based GSAT followed by a computer-based version. This was a quick way to deliver quality training that was so desperately needed.

While Redline was fighting for recognition, BMED was fighting for survival. While we were drafting syllabi late into the night, I was back in the flightdeck by day, watching another airline circle the plug hole.

It was clear that BMED was going the same way as Maersk, load factors were well down, and the company wasn't responding as nimbly as it needed to. We were flying empty aircraft from Heathrow to Cardiff just to maintain 'use it or lose it' slots.

Gossip was rife across the company; we expected BA to acquire BMED, it made total sense. They apparently weren't interested.

Jax and I discussed what our own options were, if we were both made redundant; and that was a real possibility, we were in deep trouble.

Jax reluctantly resigned from BMED, swallowing a lot of pride in the process. She went job hunting in Bawtry, landing a job as a head-hunter for a recruitment agency in the catering sector.

The news broke in February that BMI British Midland were going to acquire BMED. The transition and integration were to occur rapidly, with BMED continuing to operate as a British Airways franchise until October 2007.

In the intervening period there was a swapping of crews, Captains from BMI would come across to BMED to learn our very different route network, whilst SFO's went to BMI to backfill. I was in the first tranche to go to BMI and requested to be based at Leeds/Bradford.

BMI had a perception that BMED crews were somehow inferior. Instead of just an Airbus differences course to learn BMI SOPs, on the SAME aircraft, they made us start from the beginning, as if we had never flown. We were back in cardboard mock-ups of a flight deck.

BMI banned SFO's from doing a take-off in the A321 because one of their Captains had tail scraped one of their aircraft some time before.

Stupidity dressed up as flight safety policy.

On one day at Stockley Park, the BMI training centre, I had just collected my BMI uniform and was accused of upsetting a cabin crew trainer.

I hadn't (for once), it was a case of mistaken identity; but that didn't prevent the BMI Chief Trainer from calling me in to his office and shouting at me, he wasn't listening. I shouted back, demanding an apology; when he refused, I left with two words:

"Fucking coward."

Thacko had trained me well for this sort of nonsense. Not my finest career move, but one of my most honest.

Being based at Leeds/Bradford was a nostalgia trip for me though. The turning point for runway 14 was pretty much over the HLS at the Cow & Calf pub, which stood out clearly. But the flying itself was soul destroying in a way I hadn't expected. Double Brussels, Amsterdam's and then the worst sectors possible.

Charter flights to the holiday destinations.

The passengers were usually pissed before the 06:00 departure and the atmosphere, camaraderie and friendship amongst the crew was just not there like I'd been used to.

Paul continued to suffer badly with his back. He had significant spinal damage and was in a legal battle with the MoD that was absorbing a lot of his attention. He had been head hunted himself by a new, up and coming air superiority consultancy and was building an attack helicopter division with them.

There were no arguments, late-night silences, or moments where it felt like it might all collapse. We still knew what each other was thinking.

We were spinning plates on every front; personal, domestic, professional. Any one of them could have crashed. Looking back, the only reason they didn't was because we were utterly dedicated to each other. Things weren't about to get easier any time soon.

Chapter 122. Wobbles and Wins

At home, Jax was easing into her new job. She didn't really enjoy it, but she was giving it her best shot; we needed the money. As she was now at home every night, we did something we'd wanted to do for a long time; we got a dog. Dogs had always been a big part of both our lives, and it was a sign that we were settling down together into something that looked and felt like normality.

There was only one possible name for the dog: Addis.

A reminder of where everything had changed.

Normality at home didn't soften everything. I was hating BMI with a growing passion.

But Redline was starting to bear some fruit. We had all spent hours on the phones calling every company, every person and every security pass office in the country that we could think of to promote the Redline GSAT course. We launched at around £100 a head, but the market immediately collapsed into a race to the bottom. Within weeks it was nearer £25.

It would go even lower.

In March we finally presented the managers course to DfT over a 2-day grilling. We had to walk them through every slide, every set of instructor's notes, and demonstrate our instructional capability.

We were successful and listed in the UK as the fourth, and only private, company approved. A major milestone. In recognition, a letter was issued by DfT to all industry representatives notifying them of this change and, crucially, the dates of our first course.

In the end, we didn't quite make the minimum number required to deliver the first course; So we did what good SNCOs had always done and cheated. We placed a couple of ringers on the course to bump the numbers up to the minimum level.

That first course was a triumph. A major aspect was the required live explosives demonstration; we'd roped in an expert who was licensed to blow things up, which in our circles was rarer than you'd think.

We used the nearby South Yorkshire police range with a range of props to vividly demonstrate the effects even a small bomb would have on an aircraft; we had a section from a fuselage and recreated the shoe bomb, a liquid explosive device and other actual attacks against aircraft.

The culmination was blowing up a catering cart; from a safe distance, with only 50 grammes of explosive.

It reverted to kit form rapidly and spectacularly.

The delegates loved the course and word spread faster than a Lynx in a rapid descent. From 9 paying delegates on course one, we went to nearly 20 on the second; after that we were delivering more courses to cope with demand; this was THE course to attend. People wanted to be trained by Redline.

GSAT had also blown up, but in a different way. We were inundated with delivering half day courses all over the country. We'd put a couple of adverts in the Heathrow staff newspaper; The Skyport Heathrow and it had paid dividends.

The revenue meant I could now go full time for Redline, albeit with a very reduced, token salary. Paul was living with us in Ranskill throughout the week, only going home at weekends. His family were enduring everything me and Jax were; only from a distance.

I resigned from BMI and flew my last commercial flight on October 6th, 2007; a Leeds–Alicante charter. A planeload of Brits already shit-faced at six in the morning, a fitting send-off if ever there was one. As we disembarked in Spain, a young lad wanted to step into the flightdeck for a looksee. Of course, I said yes.

The poor bugger stepped in to the flightdeck with his lagered-up mother staggering behind him, her top doing nothing to disguise the absence of a bra. The lad changed his mind and reversed out crying because his mum was embarrassing him, his first trip to the cockpit and he'd already had enough of aviation. I felt so sorry for him.

The flight had been subdued, but when I crewed out for the last time, I wasn't sad or nostalgic. Just relieved. Relieved to walk away from BMI's pompous culture and cardboard-cutout training, and from people who thought arrogance was leadership. Most of all, I felt the weight lift, the certainty that my airline career was probably finished for good.

And that was fine because now it was Jax, Paul and Redline or nothing, and that felt exactly right.

Paul was to continue working his twin roles a little while longer. His back pain was having a devastating effect on him; this wasn't the same Paul that Jax had heard about, it was a twisted version of a man doing his best in constant pain.

It came to a head, it had to. He'd been on a work trip to Australia, a hideous flight in any class of travel, but being in economy all the way was too much. When he got back to the office, I was at home sick. Joy called me from the office telling me there was an issue between Paul and Mike. It was an issue alright; Paul had given Mike a right hook.

I called Paul, talked him down from Flight Level 420, and suggested he go home to recover. I knew it was his back doing the swinging, not him. His body had become his enemy, and sometimes we were collateral damage. His daily diet was a tramadol and diazepam salad.

Mike, clearly, was equally upset and decided that it would be better if he left the company. Mike had been our RFX guru and a source of great credibility, his loss was something we had to work around. Turning negatives into positives is a key to success in business, so that's what we did.

The DfT RFX instructors' course was only delivered by one approved entity, the Royal Military College of Science (RMCS) in Shrivenham. It was not outstanding; it was quite stale. If we could get listed to deliver the managers course, surely, we could get listed for this one too.

I wrote to DfT seeking the requirements to deliver the RFX course. Again, this hadn't been done before. Shrivenham wouldn't release a syllabus, so we made one up that was better suited to aviation.

Same drill, build the course, present it to DfT. And lo! After an assessed course, we were listed as only the second entity to deliver the RFX Instructors' course. Not taking no for an acceptable answer was becoming a way of life and the Redline credo.

Flying had been my constant. From the wide-eyed kid at Schenectady airshows to the soldier strapping into the Lynx, it was always there. Bosnia, Northern Ireland, the endless exercises; the cockpit was my comfort zone.

Leaving the Army wasn't just a change of uniform, it was stepping out of the Squadron, leaving behind a brotherhood that had defined me.

Civil aviation became the next campaign. Maersk, then BMED, even BMI, each kept me in the air, but at a cost. Every roster, every night away, chipped at family life. Flying gave me direction, but it also left wreckage in its wake.

And yet, through the turbulence, came Jax. She was now my constant. With her, the kids were no longer distant passengers in my life. And with Paul, there was Redline, our new operation. Not warfighting this time but protecting skies and lives in a different way.

My flying chapter was over, but it had done its job. It gave me the skills, the grit, and the perspective to start again.

Not in uniform, not in a cockpit, but still with a mission worth everything.

Chapter 123. Good Morning Ghana

On the inaugural manager's course, we had a couple of overseas delegates, a lady from St Maarten in the Caribbean, and the security manager from Ghana Airways. Moses was a model delegate and promised we'd hear from him again.

A month or two later we did indeed hear from Ghana, but it wasn't Moses, it was an acquaintance of his; Pastor Ebenezeer Teteh, a prominent preacher and businessman, well connected across Ghana. Eben wanted us to go to Accra to deliver the same managers course to his security company.

This would be our first overseas trip. We eventually agreed a price and some basic conditions. We were absolutely clear on one point, we weren't getting on an aircraft until we'd been paid.

We met Eben at Gatwick, where he duly handed over an envelope stuffed with cash. I wrote a receipt, just in case.

When we arrived in Accra, we were met at the aircraft steps and whisked through VIP arrivals.

A small convoy sped us through the busy, bustling streets of Accra and took us to our hotel. The first day was to be spent setting up the course, which didn't take long, we changed into shorts and went to the pool for a beer.

Eben arrived and said he wanted to take us to meet some friends. Right now. Still in flip-flops, we boarded the Prado convoy with no idea what was going on. We assumed it meant for a drink, it didn't.

We arrived at the headquarters of the Ghana Broadcasting Corporation and were taken to the religious programming department. Eben had a contact there who was about to go live on air, with us as his special guests.

We had about five minutes with the presenter to work out what in the holy fuck was going on. He asked our names.

Paul jumped straight in. *"I'm Paul, and this is Lord James of Doncaster."*

That was it. Before I could say a word, we were ushered into the "studio". I knew it was a studio because of the microphone and the egg-boxes on the walls.

Ghana's answer to Terry Wogan launched straight into the

introductions.

"A big Accra welcome to the world leaders in airport security, Mr Paul… and hear me now… my main man… Lord Jimmy of Doncaster…"

Neither of us could breathe.

All I could see was Paul's shoulders going up and down as he tried to hold it together.

Which, of course, finished me off.

The next day we started the delivery of the course. I started the introductions. Halfway through the first module, Paul stood up in the questions section and simply said, *"Put your hand up if you can speak English."*

Not a single hand went up. That was that.

We spoke to Eben and only issued '*course attended*' certificates. He was fine with that, the course wasn't the reason we were there, we were a living publicity stunt for Eben.

But it got worse.

The next day, Eben told us we had a TV interview. At least this time we had some warning. Suited and booted, we were taken back to GBC to appear on their version of the breakfast show. We were last on.

After hanging around for a while, we were ushered in for make-up.

We sat side by side in salon chairs. In front of us was a full spread of powders, every shade darker than white. Nothing even close to what we needed.

A runner appeared with a bag of cornflour.

Perfect for gravy. Less so for two pasty Englishmen about to go live on national television.

We were given a light dusting and wheeled into the studio to wait our turn. The set alternated, one segment on air, the next being built in the dark. We sat on our allotted sofa, waiting.

The presenter was Gifty Anti, one of Ghana's most recognisable faces. Smart, direct, and not someone you'd want to get wrong.

She began her link into our segment and walked towards us.

She sat down; the lights came on. I glanced over my shoulder. There was a giant A1 sized poster on an easel right behind me. It was an image of a giant dick, perched on my shoulder.

Ah Bugger!

The previous segment in this part of the studio was about vasectomy.

I nudged Paul and gave him that '*look up*' sign, the same one I used for Ali on that boat in Wroxham. I let him do the talking so that hopefully the camera would stay off me.

We laughed all the way back to Doncaster, or as the good people of Accra would forever know it, the seat of Lord Jimmy's stately home.

115 The GBC Interview with Lord Jimmy

Chapter 124. Gathering Steam

Word was out. The government had placed two MI5 agents into UK civil aviation security in a bid to get the industry where it needed to be.

Or rather, that was the rumour doing the rounds about Redline.

We never confirmed it. We never denied it either. There was no need. The suggestion alone did a lot of heavy lifting for us, opening doors, focusing minds, and encouraging a level of cooperation that might otherwise have taken years.

If the rumour occasionally nudged itself along, that was hearsay.

Training was going gangbusters. We needed to employ trainers to deliver the training that Paul and I were building. As approved providers we could train our own trainers. We brought in ex-military and civilian trainers.

Paul was now full time too. Our original business plan, like any good tactical plan, hadn't survived the first contact; we had to adapt quickly. We were now planning to expand into quality assurance, as well as consultancy. This new model was honed to perfection, each segment supporting the other two.

The first big opportunity came from an unusual source, in-flight catering.

It was Halloween, I was wandering through a '*maize maze*' with the kids when my phone started going mad. It was a security manager from a very high-end, first-class catering company in tears; his entire operation was on the verge of being shut down by the DfT for some serious infringements.

They needed all the help we could provide.

The bat signal in the shape of the Redline logo was glowing in the sky; we were on the way. Paul and I went down to their Heathrow catering facility the next day. We met the management team and realised after a quick look around that they were indeed in deep trouble.

Back at the Premier Inn on Bath Road, our unofficial London office, we went into overdrive. Paul in his room, me in mine, bouncing documents between us all night, drafting a response to DfT that would satisfy them and permit a stay of execution. This method of working became the norm.

We met the managing director the next day and briefed him on our proposed recovery plan. *"Make it happen"* he said as his only response. We did just that. His operation responded to our multiple changes and survived.

We knew this company was not an isolated case. The DfT had recognised that the industry now had access to dedicated support; assistance that could step in when inspections identified corrective actions.

We pre-empted the DfT and hired a very experienced in-flight catering security manager, eased him gently into our way of doing things and we ended up creating a whole package just for these organisations.

Our single office at Directions was no longer big enough for us, we expanded into a further 4 rooms, taking over the whole of the lower floor.

From the administration point of view, we pushed through ISO9001 early. It was painful, but it set us apart. It was also another selling point when we were putting proposals together.

The hours we spent in those offices was unreal. Leaving the offices quite late one balmy summer evening I had a 'training aid' I'd brought back from Germany after being over to see the girls.

It was a tiny, but working pin-fire pistol fitted with a flare.

As we walked out, I took it out of my pocket and said, quite calmly to Paul, "*here watch this*" I loaded a flare, raised it over my head and before the words *"don't"* left Paul's mouth I fired the flare.

An impressively large green fireball left the pistol and very gracefully arced over the nearby airport perimeter fence. It touched down in some long, dry grass.

A thin pillar of smoke started drifting upwards.

Ah Bugger!

"Fuck, fuck, fuck!" I ran back inside the building and grabbed a fire extinguisher. By the time I got back outside, that thin pillar of smoke was now more like a roman candle, growing bigger by the second. I

poked the extinguisher as far through the chain link fence as I could.

It was landing just short of the growing inferno.

After about 3 minutes of trying desperately to put the burning airfield out, from across the other side of the airport we heard the fire engines crashing out.

When Barney McGrew, Cuthbert, Dibble and Grubb arrived, they had the fire out in no time at all. They asked us what had happened.

With a poker straight face I shrugged, *"no idea mate, we just left the office and saw smoke. But I did see a crow flying over with a cigarette butt in its beak"*

"Ah that was probably it", the head fire bloke said *"it happens a lot"*

We quietly sloped away to Paul's car. Got in and watched the fire engine pack up and leave before exploding in fits of laughter.

"A fucking crow smoking a tab!"

Back in the real world, delivery of GSAT was non-stop, whilst it helped get us off the ground, it was becoming a drag. There was a better way. And that was to deliver via the internet as computer-based training. We didn't have the technical capability, but surprisingly, there was a local company based in Grimsby that did, more fish alley than Silicon Valley, but they knew their stuff.

We contacted the company and met up, agreed a partnership and profit share and forged ahead with what would become the first version of the Redline web based GSAT. With fewer overheads, it worked financially. Paul and I could see it very clearly with an entrepreneurial view.

It would become one of many strands to Redline, but, crucially, it was one of only two strands to the tech company; they got greedy.

Inflight supplies was snowballing, from one company in deep trouble, to another; and then to clients that were trying to avoid getting into trouble in the first place. All our trainers rapidly learnt the in-flight supplies syllabus and how security was implemented in a catering facility.

Redline wasn't just growing anymore. It was being pulled forward by the weight of the industry behind it.

Chapter 125. A'hhv Bin ta Leeds

In the spring of 2008, the contract for the provision of airport security at Leeds/Bradford was up for renewal and there was a new player on the scene, circling for the kill. The same security company that had delivered my managers course were the incumbent, holding much of the UK market, but with that dominance came complacency.

A new model was needed.

We were approached by the MD of Omniserv, a company that didn't have any AvSec contracts, but retained a pedigree through their founders. They asked if we could do their training for their staff; our reputation for excellence was growing rapidly.

We could do a lot more than train; we could get into bed with Omniserv and do more than tickle their fancy.

We drafted a proposal that would see Redline assume a sub-contracted role in training, quality assurance and governance. It wasn't the cheapest option presented to the airport.

On the award of contract, the two trainers already at the airport moved across to us and became Redline trainers. We revamped the entire airport security training programme starting with the airport directors and working down to GSAT.

This model wasn't widely copied. We started our covert testing of security, with Leeds becoming a living test bed for how security should work.

Our reputation was everything in those days. Redline was Paul and Jim, our names, our credibility. Every course and every client carried our personal stamp in the early developmental days.

We personally knew every client and forged a partnership with them, not through rhetoric, but through actions.

Every email was responded to immediately; every requested meeting would happen within 24 hours where possible. We spent many nights driving around the country, we were there.

The team we were building was fundamental to our successes; the instructors we brought in were put through their paces with a comprehensive 'Competency to Instruct' package.

This training package was applied regardless of background, experience or qualifications. Although we didn't see it at the time, this upskilling and standardisation package was to drive the standard across the industry for years to come with the regulator eventually making a similar programme a national requirement.

Naturally, not everyone could keep up, or indeed wanted to. Some saw the credibility that was becoming synonymous with being a Redliner, but not the hard work required to be there.

We discussed the team; its composition and how we forged it probably more than anything else. The strategy was easy, it was the people that was the most complex issue. For Paul and me, it came to one thing; we called it '*The Pilot Gene*', that intangible aspect that you can't put your finger on. The drive, motivation and commitment to succeed.

The '*The Pilot Gene*' wasn't restricted to aviators, it was an intrinsic thing that we could see in people. It was an intuition that served us well.

Our obsession with people and standards was noticed. Before long, word reached the DfT, and we were invited to apply for something bigger: the Global Opportunities Fund (GOF).

This programme funded AvSec training in key foreign countries that needed assistance. After the invitation to apply was followed up on, we became part of the programme, sending a male and female instructor to some very far-flung destinations, most of which would have been served by BMED and for the same reasons.

Those GOF trips cracked open the next horizon. Redline wasn't just a UK training provider anymore; we were international operators. Leeds had proved the Redline model, GOF proved the reach.

In Doncaster, the offices at Directions were shrinking around us. The admin team had expanded to cope with both company admin and training. The training team had a full-time manager and a growing pool of trainers. We had the embedded trainer's resident in Leeds.

Things were growing exponentially.

We'd been careful with the money, conservative and retained enough in reserve to survive six months with no income at all. That caution bought us confidence. Then space. Then momentum. We took over the entire upper floor of the office wing and barely noticed the step change until it was already normal.

Redline was no longer just a company name on a door. It had weight.

People had opinions about Redline and us as the leaders. They either wanted in or wanted us gone. The loudest critics were almost always those who'd made a good living doing things the old way, to a standard that survived only because no one had challenged it. Change didn't threaten their values. It threatened their seats.

But outside the classrooms, life was winding up a curveball for me and Jax; one we never saw coming.

Chapter 126. From a Jax to a Queen

Jax was always Jax, it had said so on her BA name badge; that's how I knew. Short for Jacqueline, but it sounded like something a 13-year-old gamer would call themselves. By mutual agreement we had a celebration and a renaming party.

Enter Jaq, a much more elegant and sophisticated handle.

Jaq and I, with only two very modest incomes were struggling to survive. Having done the right thing by the kids, we were back in potato-week territory. Without staff tickets, bringing the kids over was crippling. We couldn't afford to pay attention.

Aware that Jaqs' step-mum had had an inheritance, and given we had a great relationship with her parents, we thought we would broach the subject of a small loan to help us through this period.

Redline was increasingly a safe bet, but not a given by any stretch. Plan 'C' would be to go back to the airlines where I could achieve a descent salary immediately if this failed.

What we got was the third degree, some very intrusive questions about our financial acumen, our priorities and everything in between. Taken aback somewhat, we felt that this was simply a case of tough love; nothing more. We were over a barrel and had nowhere else to turn.

The terms were hammered out, but more on a commercial basis than a family bale out.

A misunderstanding crept in. In essence, Jaq's father thought half of the loan had bought him 10% of Redline, when to us it was only secured against it. The other half came with commercial-bank interest.

We thought we'd agreed terms. He thought he'd bought in. None of us realised the difference would matter.

We thought nothing more of it; we had a wedding to plan.

Having been engaged since Addis, we thought it high time we bloody well went and did it!

Financial pressure had eased off, but we still needed to be very careful. Our first choice was a cheap all-inclusive deal on an island somewhere. It was very cost effective and had a lot of appeal. Raising the idea with the father of the bride revealed that he wouldn't go to our wedding if that's what we chose.

Back to the planning board.

We came up with Sicily for a couple of reasons; it was romantic and affordable. The Sicily plan was met with universal approval; Jaq went into 'military-grade list mode'. I graciously left her to it.

The first week of September 2008 arrived and, courtesy of Ryanair, 21 of us, including the very happy couple landed in Palermo. We put together a small convoy of hire cars to drive to Taormina, our wedding location. Jaq had a found a small resort made up of villas; of the 10 villas, we had 8 of them; it was gorgeous, rustic and totally Sicilian.

The convoy lasted all of thirty seconds before a Squadron of Vespas, with Fiat 500s in mutual support, cut through us like a swarm of angry wasps. Everyone knew the address; every person for themselves.

The last car into the Taormina villas was Jaq's father with the steam from his piss still fogging the car windows, unhappy that we'd left him behind. A few bottles of Birra Messina later and all was sweetness and light again.

The days between arrival and the wedding were glorious; Taormina sits high above the rugged Sicilian coast, a maze of narrow streets and terraces that all seem to point towards Mt Etna. At its centre is the old town hall, the Palazzo dei Giurati, where Jaq and I were to be married. The square outside was the heart of the town, busy but timeless with street musicians and artists.

Below us lay Isola Bella, a narrow strip of sand connecting the mainland to a tiny island. The journey down was part of the attraction. A cable car drifted through the trees before the Mediterranean appeared beneath you in a blaze of blue. The beach wasn't particularly large, but standing in the shallows with the island at your shoulder, it was easy to see why people called it the *'Pearl of the Ionian'*.

The morning of our day arrived. We made our way to the majestic town hall. With Ed, my best man, we walked in to meet the mayor who would be marrying us. She was the doppelganger of Judge Judy and didn't speak a word of English.

I adopted my position at the top of the aisle, excited and nervous knowing that this was the person I would spend the rest of my life with.

When '*Palladio*' by Karl Jenkins started playing, I knew she was here. I didn't look. Ed did and he smiled.

Jaq glided into position next to me. I looked at her and welled up. Jaq welled up. Judge Judy welled up. We held hands, squeezing tightly.

Judge Judy did her thing in Italian; I have no idea what she said but it was beautiful. When it came to vows, I had rehearsed mine with help from the Italian wedding planner, in Italian.

Judge Judy cried so I must have hit the spot. Jaq gave me her vows. I cried, Jaq cried.

At some point, we were declared husband and wife. Grinning like cheshire cats, we walked out of the Town Hall and into the sunlit square. *'Signed, Sealed, Delivered'* playing, and for once, everything felt exactly right.

Jaqs father and step mum beamed with pride, their daughter was a beautiful bride and centre of much warranted attention.

We walked arm in arm through the town to a restaurant overlooking the sea, tourists and locals applauding the flushed bride. A dozen photo stops later, and we managed to grab our first glass of champagne.

The wedding lunch was a full Italian 5 course banquet, and no one there will ever forget just how good it was, everything fresh; cooked like Nonna would have. The wine? Well, of course it was from Gattinara; we had to.

For each person, instead of some crappy old, sugar-coated almonds, we went personal. Before the wedding Jaq and I had gone through the music that had defined our relationship and our journey together, burned the playlist onto CDs, made labels and gift wrapped them.

The playlist started with *'Air Hostess'* by Busted, went through memories of our meeting in Addis, through to those days in Belfast.

Some songs were literal; others evoked our feelings and thoughts about each other.

After lunch, as the day slipped into evening, we strolled together down towards the reception, a champagne bar tucked beneath the castle walls. The sun was setting behind us, painting the town in gold.

Mr and Mrs Termini took to the floor for their first dance to '*Sometimes We Cry*' by Tom Jones and Van Morrison; a song that reflected our quiet recognition that life would bring its share of struggles, but that we would face them side by side, always together.

We'd defied the odds and signalled that we were together, come hell or high water, surrounded by family and friends. It was, for us and our story, the perfect way to sign the deal.

While we were making our life-changing commitments in Sicily, Redline was ticking forward, Paul had been holding the fort in Doncaster; other than some glitches with the website, it was all going well. Now firmly in control of half of Directions office space, we were joined by an international organisation that trained aircraft engineering apprentices. This was a big win for Directions, but we were relegated instantly to second fiddle, which given how much we'd invested personally into the site, was a kick in the slats.

Back in Ranskill another important subject was about to raise its head.

Early on, we'd talked about children. Her boyfriend at the time had two, and whilst they were nice enough kids, she really didn't want any of her own. Which was just as well because just after Amelia was born, I'd had a vasectomy.

Jaq's ticking body clock was playing games on her, and the time came when she confessed, she wanted us to have kids; time and circumstances had shifted her perspective.

It floored me. For years I thought that chapter was closed, but here she was, asking me to reopen it. I agreed; on one condition.

She had smoked those bloody awful, stinky menthol cigarettes since I had met her. I detested them, as well as the fact she smoked. I agreed to put myself through the trauma of a vasectomy reversal if she packed the tabs in.

It was a good plan, but she'd already pre-empted me, knowing that would be my strategy. Miraculously, she somehow managed to stop instantly and permanently. Had I been outplayed?

More than likely.

Shortly, in fact far too shortly, after the reversal operation I was down in Heathrow having discussions with a major in-flight catering company; this was potentially a huge contract win for us.

The surgery had left me with long black lengths of plastic cord hanging out of each reversal wound. They were there so that the stents in my plumbing could be extracted without further surgery; a quick tug, and they'd be out.

These long cords started off neatly coiled and taped to the inside of my thighs. Unfortunately, the tape became unstuck, the cords sprung out and started to catch on the seams of my suit trousers. Given the other end was attached to my *'ball strings'*, every time I moved, they tugged hard.

I had to excuse myself and shuffle to the toilets, walking from the knees down, looking directly ahead and desperately trying not to chunder on the way.

Every step was a sharp flick to the nuts.

I entered one of the traps and gently eased my trousers down, taking great care not to touch the black cords of death.

The only saving grace was that there was enough sticky glue left on the tape, to re-coil them and fix them back in position.

I'm pleased to report my gonads didn't suffer any lasting damage and we landed the contract.

UK, Haxey, North Lincs: Jun 09 – Feb 11

Chapter 127. New Directions

At the end of 2008, Paul and I were invited up to Inverness, the headquarters of the Highlands and Islands Airport (HIAL) group. They were interested in talking about a range of services, from training through covert testing and consultancy.

HIAL was a curious organisation: a Scottish government-owned company tasked with running a scattered network of tiny airfields across northern Scotland and its islands, the largest of which was Inverness. On paper it looked like a collection of loss-making backwaters, but these strips of tarmac, and in Barra's case, a tidal beach, were lifelines.

We developed a plan to provide everything HIAL asked for across their whole network. The geographical spread, from Orkney and Shetland to the Western Isles and Inverness, was a logistical challenge that would have floored most. But we thrived on it; we loved those challenges, and this one was a biggy.

We agreed to embed an experienced security manager within HIAL to help their centralised security team based in Inverness. We had a willing volunteer, a highly regarded manager from another airport.

The covert testing programme started, with all of Redline chipping in. Family and friends joined the growing pool of testers; we needed as many varied faces as we could train and trust. One of these was my mom, who went on to become a very capable covert testing team leader.

The testing for HIAL could mean Tiree one day, a hop over to Campbeltown followed by Stornoway. It didn't pay much, but most testers loved the travel around the Highlands and Islands.

In May 2009, Jaq and I had a weekend away in Andover and Twickenham. We stayed in a hotel on Andover High Street, met up with all our friends early and then got the train into Twicker's for the annual Army v Navy Rugby match. It was always more about reunions than rugby. We'd been a few times and always had an absolute blast.

Early on the Sunday morning as we both nursed sore heads and hangovers that had arrived like a town crier, Jaq took a phone call.

It was the Police, and they were in our house.

We'd been professionally burgled.

When we got home, we found every room had been ransacked, the burglar alarm had been cut off the wall and dumped in a water butt.

All they took was an Xbox and a laptop charger, but bizarrely not the laptop. Another set of disappointed thieves.

We just didn't feel safe in that house anymore.

We let the lease expire and packed up once more, this time heading to Haxey, a quieter village on the far side of Bawtry. Another new chapter.

The house sat directly opposite the local pub, The Loco, which had a Quizzy and a regular cast of familiar faces. It was so close that if nature called mid-pint, it was genuinely quicker, and significantly more pleasant, to nip across the road and use our own loo rather than risk whatever was happening in the pub gents.

The first weekend in Haxey was in June 2009 and was one of the most random couple of days we ever experienced.

Ed was staying with us with his new girlfriend, we were still unpacking boxes, and it was a hot day. I opened a window to see Pete Dougherty from Babyshambles checking in to the pub's accommodation.

As if that wasn't weird enough, standing next to him was Cliff Richard. Things took a turn when we then saw Luciano Pavarotti. *"Ummmmmm, he's dead, isn't he?"*

It was the Haxey Fake Festival. Part of the day was a standard village festival, donkey rides and stalls, stuff like that; but the organisers also had 'look-a-likes' wandering around. It was the funniest thing, sat having a picnic when David Brent or Jim Royle rock up for a quick chat. After the tribute bands; The Fillers were good, better than they had any right to be, we went back to the pub.

I was sitting between Jim Royle, Cliff Richard, Mr Bean (real name Steve, an insurance salesman from Birmingham) and Simon Cowell putting the world to rights.

At work, the relationship with Directions was becoming strained. They'd helped us early on, but things were now becoming difficult ever since the engineering company had arrived.

Parking at the hangar was limited; we had a couple of dedicated spaces close to our entrance. Paul had parked, as usual, in the same space. It was a Monday. During the week he still lived with us. In fact, the house hunt for Haxey had included as a requirement, space for Paul.

The manager of the engineering company had deliberately taken Paul's space.

Words were exchanged.

Loud ones.

Later that morning, Paul was asked to go up to the management suite to see the MD of Directions. He came back with a letter in his hand that he silently passed to me.

Directions evicted us. We had 30 days to relocate a national operation.

We were already a busy company, around 25 full-time staff and close to 100 part-time covert testers spread across the country. Overnight, everything else stopped.

For a CIC meant to nurture small aviation businesses, they chose the multinational and shoved us off the edge.

By dinner, the anger was gone, there was no time for that.

Out came the notebooks.

When RAF Finningley had been purchased, the developers had hoped to build a small business park, and to get that ball rolling, they had speculatively built a two-story office block.

We had driven past that empty building every day for the past 3 years.

Paul called the agents advertising it. The building was owned by a pair of Irish property investors that were very ready to get it leased out, we struck up a deal with them that was beneficial to all parties.

Next up, get our vision for what would become the National Security Training Centre (NSTC) kitted out properly. I contacted key suppliers, one of which was the national sales manager for a major security technology manufacturer. We convened a meeting for later that first week.

Paul was aware that his previous company had two complete classrooms of computerised desks, designed for training Apache pilots, in storage. We were going to need those.

I contacted a manufacturer of security fencing, access control systems and CCTV. We were going to need those too.

On the day of the first site meeting, we gathered downstairs and walked everyone around our concepts and described our '*shopping list*'. That list included 4 x-ray machines, archway metal detectors and an armoury for training aids.

Tracey, from the manufacturer, pulled some strings, made some calls, and came back with *"yep, I think we can help"*. The other reps followed suit. It was an amazing feat from all concerned.

Upstairs would be two fully equipped, modern classrooms, offices, a boardroom and a reception area. There was also a kitchen and dedicated crew room.

For two weeks, Paul and I worked day and night fitting out the site.

Jaq's father chipped in, helping us reinforce the flooring with steel supports for the heavy x-ray machines and building us a swinging road barrier, he was a talented welder and craftsman. The hoarding outside was replaced with a bold NSTC sign and our new secondary branding.

The classrooms were fitted with high-definition projectors and fixed screens.

Less than 30 days after eviction from Directions we moved all our furniture out of their facility and into ours.

The NSTC was born.

Just before the official opening in November by Caroline Flint, the local MP, another challenge hit us, and this one cut far deeper. We had a traitor.

It was the manager we had embedded with HIAL. She had been on leave, so we had sent another manager to provide cover. When he came back, he asked to speak to just Paul and I, alone.

We discovered that she had approached HIAL directly about a

permanent role, along the lines of *"employ me, cancel the Redline contract"*.

In the investigation, we discovered that large amounts of our intellectual property, training courses, testing protocols and operating procedures had been copied without authorisation.

When we raised it with HIAL leadership, they confirmed she would be joining them. For us, the financial loss was painful, but the personal hit was worse. We had brought her in, trusted her, and invested in her development. To see that trust abused was crushing.

For Redline, integrity wasn't negotiable, and whether compromised by clients or team members, it left a mark that stayed with us.

In the meantime, I looked at what files had been stolen; to my horror, she had copied covert testing reports from all our clients. These files were obviously held securely on military grade 'ironkeys', but in her role before going up to Inverness, she had had legitimate access. We were morally bound to report these losses to the police.

The Police deemed it a national security issue and we were bound to a comprehensive non-disclosure agreement.

All I can say is that she went on to work for HIAL.

We didn't.

Chapter 128. Testing Times

Paul and I were spending a lot of time in London with DfT, we'd been invited to join a couple of working groups on various AvSec issues. For Redline, change was good; but we needed to understand it first.

On one of the working groups was a senior representative from our only real competitor. During one meeting, her tone toward us in front of the regulator was pointed and dismissive. It sharpened our focus for the upcoming British Airports Authority (BAA) covert-testing tender, of which her company was the incumbent; and we prepared accordingly.

She'd put herself on the target list.

Paul and I went into overdrive. We didn't outspend them. We out-thought them. We treated the tender the way we'd planned operations for years: understand the ground, work out where people were comfortable, then move where they weren't. Our proposal provided everything that was missing, innovation, technology, drive and commitment.

This wasn't just a contract; it would make Redline unavoidable. Win it, and Redline would stop being a challenger and start being unavoidable. Heathrow, Stansted, Edinburgh; the places that defined UK aviation. But we needed more full-time people, the headcount was growing.

We now needed a full-time office in Heathrow; we found a suitable location and established Redline's southern office. We promoted one of the original ex-military trainers as the LHR contract manager, who resigned almost immediately. It was the clearest feedback we could have asked for.

That's when Paul Cooper, an ex-Coldstreamer that Paul knew well, joined the team as the BAA testing manager, he excelled and took the delivery of covert testing to the next level. He was a machine and simply got on with it.

After HIAL, being able to trust a manager wasn't a luxury. It was oxygen. More importantly, it allowed Paul and I to continue to drive growth at home, and to kick start our international plans properly and independently from GOF.

At home, Jaq's father, having seen the massive growth in Redline firsthand, was continuing down the line that he owned 10% of the company. We maintained it was a secured loan, not equity. In his version, we were thieves.

We knew it was getting bad when Jaq's brother, a person I got on with well, called us. He sounded furious.

Despite one of Jaq's father's loan conditions being that no one else in the family should know about the agreement, he had told Jaq's brother, the person she was closest to, that we were "*ripping him off.*"

We were shocked he knew and hurt that anyone thought us capable of that. By the end of the call, everyone was upset.

A few weeks later I was in Liverpool, discussing a covert testing programme for an airport group that owned three airports, when Paul called me from the office. Jaq's father was there, in a rage and creating a scene. Before I returned, Paul had "*assisted him out of the building*" and he'd gone home.

I called him. In the two hours it had taken him to travel home, Google had apparently made him a barrister, commercial solicitor, and judge rolled into one.

He was taking us to court.

We started receiving letters with various demands that just could not be met, legally or otherwise.

We could fight competitors and regulators; family was different. That was the kind of battle no one really wins. It wasn't about the money; it was about family cohesion and unity.

We won the contract for testing across the group of airports, Liverpool, Teesside and Doncaster. Along with some independent airports, we were now easily the leading provider of covert testing in the UK. Only Manchester Airport Group stood between us and a clean sweep, except Gatwick.

Our rigid application of standards and innovation, combined with trust, was working.

We were reaching saturation in the UK.

Expansion had to be international.

I was in touch with UK Trade and Industry (UKTI) requesting targeted support. We could also receive grant funding to support overseas business development. We selected the US as our first target country.

Research had revealed a private US company that provided aviation security and were also expanding. We had what they needed. I contacted them, they set up meetings with the TSA and airport clients.

Paul and I flew first into Boston. It gave me real delight to take him, my best friend, up to Pittsfield and introduce him to my family there. After a couple of nights, we headed south.

On the way we stopped in Washington to meet the TSA and then detoured to Norfolk, Virginia, to see my dad, who was living just outside Oceana Naval Air Station. He'd moved there from Massachusetts simply to be close to the runway, spending his days on the balcony with a cigarette, watching F-18s thunder in and out.

His apartment was a shocker. The decor was nicotine yellow, ashtrays overflowing with burnt offerings to the gods of tobacco. He looked yellow, thin, and gaunt. Still, I was glad Paul could meet him.

The contrast between our fathers couldn't have been sharper; mine a worn-out costume of a man, Paul's a tall, immaculate, Regimented former Guardsman.

I was always a little in awe of Mason senior. He was a straightforward, no-nonsense bloke who said whatever came into his head; if you didn't like it, that was your problem. The first time Jaq met him we walked into Paul's parents' living room, and it would be fair to say I could have lost a pound or two.

He looked straight at my belly and barked, *"That's disgusting!"* Jaq nearly wet herself laughing, Paul just stared at the floor shaking his head, and I was left speechless.

From Norfolk we flew on to Kissimmee. After pretending we cared about golf, we got down to the real purpose, the proposed partnership. After the meetings, a way forward agreed, we were ready to go home, we'd been away too much.

We were heads down, one foot in front of the other, always looking to the horizon and never pausing to look back at what we'd already achieved together.

The next year ran on two tracks: Redline surged forward while family life fractured. Airports and regulators we could handle; the damage at home followed us everywhere.

Chapter 129. The Birds and the Bea's

2009 quickly became a rollercoaster on two fronts. Paul and I were still flat out, we were usually in the office moulding the company, developing business and dealing with clients.

A normal day started at 07:00. You'd fire up the computers, plough through emails, look up and it was suddenly evening. Do it again tomorrow. Twelve to eighteen-hour days became routine. If we had meetings the next day, we'd then set off to wherever we needed to be.

That working regime clearly takes its toll, but it's the toll on families that goes unnoticed. Days blurred into weeks, the weeks into months and suddenly it was well into 2010. Redline now had a full-time strength of around 30, all of which had families to support and mortgages to pay.

In early March we finally got the news we'd been waiting for. My vasectomy reversal had been successful and Jaq was pregnant. She was overjoyed, happier than I'd ever seen her.

Yet at the same time, I could see the strain the ongoing paternal campaign against us was placing on her. It was the sort of hurt that is difficult to describe and even harder to challenge. We kept the news to ourselves, sharing it only with Jaq's cousin, Meral, and her husband, Mick. They were the only members of her family still in regular contact with us.

Everyone else could find out later, if at all.

Jaq should have been enjoying her first pregnancy, fully supported by her wider family, instead she felt she had to keep it from them.

I was so sad for her.

Most of Jaq's family, now very aware of the breakdown in relations between us and her father, tried to remain neutral as much as possible. We understood it, even though it left us feeling quite isolated.

One quiet exception was her aunt and uncle, her father's sister and brother-in-law, who never took sides but never stepped away either. They stood by us gently and consistently, their love for Jaq never wavering, and that relationship has endured and grown ever since.

In April, I was asked to meet the security team from the Olympic Delivery Authority (ODA) for the London 2012 Olympic Games. They wanted us to train them in covert testing. After a conversation of about an hour, I was asked to give the ODA team the room and wait outside. I could see what was going on.

They needed covert testing, but it was already clear it sat outside their capability and appetite for risk.

I found a quiet stairwell and on a terrible line, called Paul to discuss where I thought this was going, whether we could commit and what the impact on the company would be. This level of testing would include testing the security applied to building materials, working construction sites and in some unusual locations.

Stratford at that time was one of the most heavily policed areas in the country; ISIS was at their peak. Did we really want our people walking around Stratford with simulated threat objects while ISIS dominated the headlines?

I went back in; my hunch was correct. We were asked to provide a detailed proposal to the ODA for commercial assessment. Paul and I fired up the proposal machine, produced a detailed plan, and the ODA accepted it, the plan included testing of all security checkpoints all the way up to and including the games.

We placed an embedded manager with ODA, based in Canary Wharf. His main duty was to coordinate all Redline testing operations with the other agencies involved in covert policing operations; and there were quite a few of them. Having joined us from a covert background, our man on the scene was a perfect fit.

For Redline, this represented a major step forward in diversification, we could now add areas of significant critical national infrastructure in our portfolio.

Even with all this growth, most of our revenue was generated by term limited contracts, what we craved was long term stability for the team. Stability in contracting terms means over performing, not just some of the time, but all the time.

While the ODA project ran smoothly in the background, our focus at home shifted to things far bigger. Life at home would give us moments of celebration too.

In May, Alex and Barry were married at Craythorne Hall in North Yorkshire, it was a beautiful day. We were delighted. Barry was the right man for Alex, we could see reflections of our own relationship in theirs. This wasn't a marriage of convenience just to get allocated an Army house. There was real love, affection and tenderness between them.

From a father's perspective, what more do you want for your daughter other than love, happiness and health.

Alex and Barry had been though more than their fair share of trauma in their relationship, and so to see them closer than ever was wonderful.

I had the bonus of genuinely respecting and liking Barry, I could see he was a hard-working young man with my daughters' best interests at heart. He may have been *"just a fucking squaddie"*, but most importantly he was a good man.

We were approaching B-Day. We'd saved up and went to '*babies R us'* with a long list of baby stuff we would need, and that was just the basics. As we were walking around the shop, I'd find something on the list and put it in the trolley. But then I'd see Jaq putting it back on the shelf!

After an hour, we still had an empty trolley. I simply said to her *"you ARE having a baby"*

Jaq teared up. A heady mix of the nerves all new mums go through and the strain of the ongoing dispute with her father. It broke my heart to see her so upset over the rift. This was supposed to be fun. Although we hadn't told him, we did know that he knew about the pregnancy; and that knowledge made his coldness feel even more deliberate, as if he'd chosen silence over compassion.

I was staggered that any man could treat his own daughter in this way. Our solicitors advised us, and him, that, in their view, his claim was unlikely to succeed in any court.

That should have been the end of it. But he wasn't listening to lawyers, or to reason it seemed.

Armed with misplaced confidence, he sent letters bristling with legal phrases, most of them stitched together without meaning, just hyperbole and personal opinions.

Our own solicitor had the thankless job of not only defending us, but effectively coaching him too, pointing out the basics of law that he was mangling. It was surreal, like playing chess against yourself while the other side insisted they were winning.

We heard that other members of the family were also asking him to drop it. It didn't matter how many voices within his own family urged him to let it go, he dug in deeper each time. What had started as a private misunderstanding became a family spectacle, with people who cared about all of us left bewildered, caught between loyalty and disbelief.

Through it all, Jaq's aunt and uncle remained quietly present, refusing the drama but never withdrawing their support for her, a reminder that love does not always need to be loud to be felt.

His refusal wasn't about the money anymore, if it ever had been; it came across as being about pride, control, and about winning an argument no one wanted to fight.

On September 16th our daughter arrived. I immediately thought of Alex and Jessica, both having September birthdays, but the 16th was also shared with Jessica.

We named our little girl Honey Beatrice Jane; our little Honey Bea. She looked just like her mum, but strangely just like my mom when she was a little girl. I don't know how that happens, but it did.

Honey was born mid-afternoon on a Thursday; by that night we were back in Haxey, watching '*Celebrity Juice*' with our baby goblin asleep between us.

Everyone came to see her. Patrick and Amelia were immediately in charge of the baby. We were determined that the kids would see each other as brothers and sisters first and foremost.

Something had to give with Jaq's father. I couldn't let the constant stress drag on. We extended an olive branch and arranged a neutral venue to meet; we were hoping for reconciliation. I booked a hotel meeting room just off the A1. Jaq and I turned up first, Honey was asleep and bundled up in her car seat, her little face just visible under her blankets.

Jaq's father and step mum arrived and sat down on the opposite side of a large table. They both saw Honey, but neither said anything. Her father's first action was to put his phone on record and put it on the table. There was no recognition that a misunderstanding was tearing our family apart.

There would be no reconciliation, something we had desperately hoped for. He said his piece, I said mine.

He closed that meeting with the last thing he would ever say to his only daughter:

"You've made your bed, now lie in it".

Jaq's step mum stole a glance at Honey, I think her heart was breaking too, and her last words were softer: *"your baby is gorgeous"*.

We left that hotel knowing that we would never speak to either of them again.

Ever.

For a while at the beginning, I'd let myself believe he might be that steady influence in our lives; the father figure I'd missed, the grandfather my children deserved. Watching him now, obsessed and distant, I realised he was never going to be that. The sting wasn't just in his rejection of Jaq; it was in the collapse of that hope.

After about 6 weeks, Paul and I were back in the US working with our new partners. The plan was to extend joint assurance operations across their main hubs, which included San Francisco.

The day before we were to come home, Jaq called.

She was in a terrible state, it was upsetting to hear her like this; normally so calm and steady, she was the one to calm me down. The ongoing situation with her father had snowballed since the hotel encounter. She now had the added stress of a very young baby to look after with no family support; when coupled with me not being there, it was too much.

Paul and I tried to bring forward our return flights, but with only a day to go, it was impossible.

I should have been the person Jaq would have been for me. I tried to be supportive, but I missed how badly she'd been shattered. I told myself tomorrow would be soon enough.

It wasn't.

I shouldn't have left her on her own; I was all consumed with work.

Meral drove down from Teesside immediately to be with Jaq, no questions asked. I was, and always will be, deeply grateful for that. In those first raw days, it felt as though Meral and Mick formed the inner ring of support around us, practical, calm and utterly without judgement.

They never took sides, never offered opinions we hadn't asked for, and never once made Jaq feel she had to explain herself. They simply showed up, stayed steady, and reminded us of what family is supposed to do.

Over the next few days, I drafted a compromise agreement with help from Paul and the solicitors.

The recent large contract wins meant that, financially, Redline was in a safe position to pay a first, significant dividend to Paul and I as sole shareholders.

My share went straight to Jaq's father to stop him. I would have paid anything to stop him sooner.

I had to buy him off to stop him taking us to court.

In my life, there have been three baddies. Thacko was the first; you've just met the second.

The whole episode had one positive impact. It caused Jaq and I to circle the wagons, if anything we were even tighter. But I'll always carry the weight of how money exposed fault lines we didn't know were there.

Only a relationship as strong as ours could have survived that.

As for Jaq's father, where we lost money, he lost his only daughter, a granddaughter, and later a grandson he would have adored.

Chapter 130. Snow Days and Groundhog Days

2011 arrived with a flourish more than a bang. It was a long cold winter and we prepared to propel Redline into New Year. Paul had driven down from Teesside late on the Sunday before we went back to work. Mom and John had stayed with us too, so it was a houseful.

We went to bed that night with darkening, heavy skies.

We woke up to nearly three feet of settled snow. Looking outside, my mother's car was literally buried, the roads impassable.

Paul and I set off, the only idiots on the road to get to the office. I hadn't seen so much snow since Finland. By pure luck we arrived in an undamaged car and in one piece. The snow was so thick, the gates wouldn't open, we had to dig them out. Then, one by one, the team started to arrive; we'd fully expected everyone to call in as a snow day.

Soon enough though, there were enough Redliner's to clear much of the car park and the paths. We took everyone to the Ramada over the road and ordered two dozen bacon sandwiches and then knocked everyone off until the main roads were at least cleared.

That morning was real teamwork, people digging out for each other; not because they had to, but because they wanted to.

That was Redline.

The NSTC had also evolved with the company. We now had cargo screening x-ray machines, a purpose-built hold baggage screening system and a hugely expanded range of firearms, training aids and hundreds of covert testing objects. The training hall was now so realistic, we had TV producers using it as a substitute for a real airport.

The relationship with the Grimsby based tech company had run its course, we had ideas for growth in the CBT and threat image recognition areas, but we couldn't be limited by narrow minded purists. We wanted to expand CBT into inflight supplies and other neglected risk areas. We needed our own tech team, but for now, we'd make do with subbing some aspects out.

We also knew that our final major covert testing opportunity was coming up; Manchester Airport Group (MAG). Winning this would seal our grasp on covert testing in the UK with roughly 85% market share.

Did we win it? Of course we did. Attention to detail, a comprehensive proposal, and a hard-earned reputation won it. The only losses were HIAL and, painfully, one of our first contracts: London City. These losses served to nip any complacency in the bud with the larger clients.

As with flying incidents, we went straight into self-examination to ensure we would never lose a contract through circumstances we controlled again.

We look back now at 2010 and see it as the calm before the storm, but at home, finally everything was easing off. Paul and I could pay ourselves correctly. Debts were slowly paid off. We were still having Patrick and Amelia for every school holiday, and we were becoming a closer family.

Jaq and I investigated the possibility of buying the house we rented, but the landlord wasn't keen, giving us a stupid price. So that was that.

Hoping the company would enter a period of consolidation, we went into the second half of the year unsure of what was next and aware that we only had about 18 months left on the Olympic contract. What might replace a contract of that magnitude?

In October 2010, the US express package companies UPS and FEDEX, had flown multiple Improvised Explosive Devices (IEDs) concealed within printer cartridges. The devices had originated in Yemen, passed through Dubai and were making their way slowly but surely towards Chicago and other destinations in the US. One of the devices was eventually intercepted in the UPS facility in East Midlands Airport; the others in Dubai, due to Saudi intelligence.

We knew this would become a game changer for all cargo related aspects of AvSec.

As a result of the printer cartridge devices, I was talking to the security manager of UPS, a friendly Canadian guy by the name of Mike Priest. I was trying to convince him to start a covert testing programme across their UK network. It seemed like the obvious thing that Regulated Cargo Agent who had just flown a live bomb, would consider.

Out of politeness, he invited Paul and I to his office at East Midlands Airport for a coffee and a chat.

He offered to read a proposal. We sent him one. He wasn't very interested, with his feedback being *"too expensive"*.

As the year ended and we rolled into 2011, I kept in touch with Mike and his team, but with not much coming back our way we thought it was dead in the water.

What happened next not even Nostradamus could have predicted. Our world was about to be turned upside down. Everything we thought we knew would be put to the test.

UK, Hatfield Woodhouse, North Lincs: Feb 11 – Feb 12

Chapter 131. Wedding Cakes and Doughnuts

The rental lease on the house in Haxey was up, the landlord wanted to sell; probably because we'd planted that seed in his mind, but we were in no position to buy at the price he was asking. As much as we loved Haxey, its history with the ancient game of the 'Haxey Hood' and the friends we'd made in The Loco Pub, we had to look further afield.

Jaq found a lovely, converted barn in Hatfield Woodhouse. On the downside, it had no garden to speak of and was right on a main road. Yet again, we packed up everything we had and moved house; this was our seventh house move.

The barn had a large attic space above its' garage; I built us a bar in it, and we made our own little pub. It even had the required quizzy, a dart board and plenty of other entertainment. I was surprised how much I enjoyed living there.

In March 2011, a hoax device disguised as a wedding cake was shipped from a UPS office in Camden, North London, bound for Istanbul. Inside the ornate box was a fake bomb; complete with timer, wires, and a detonator. It made its way onto a UPS cargo flight and only came to light after landing in Turkey.

The police later confirmed it was a hoax, not terrorism, but the headlines didn't care.

For UPS, and for air cargo security more broadly, the damage was done. It showed in the starkest way how porous the system remained.

The DfT declared it was taking the incident *"very seriously"* and had already begun an investigation deploying one of the world's toughest cargo-screening regimes. MPs, including the chair of the Home Affairs Committee, called for urgently tightening security rules on cargo and private flights, warning that similar failures could have devastating consequences for communities below flight paths.

We knew 'shit' was about to get real.

A few days later, on 21st March, our family lost my grandmother. She passed peacefully in Southampton General Hospital at the age of 92. She had lived through the war not just as a witness, but in uniform, serving in the ATS. She'd been strafed by Messerschmitt's and lived with the constant reality that each day could be her last one.

She carried on anyway, standing her post, doing her duty, never flinching. That steel never left her.

When my grandfather's illness struck, she carried that burden too. She raised her family with the same quiet strength that had got her through the war. No drama, no complaint, just a soldier's resolve in civilian life.

Years before, when Redline was nothing more than an idea scribbled on a 'post it' note, she had been the one to give me the spark to get it going. Not a fortune, just enough to light the fire.

She never asked questions about business models or contracts; she simply believed in me. That faith grew into something she lived long enough to see succeed. For us, it was never about the money, her belief was worth more.

At the beginning of the year, Ed had joined Redline, initially as Training Director. Given he had recently left the Army as Captain and Officer Commanding the Army Education Centre in Cyprus, it was a shoe-in. We just had to get him up to speed in aviation security.

Then, on Thursday 28th April it came. As we were getting ready to go home, the phone rang.

It was Mike Priest, and not the calm, pleasant Mike we knew; this was Mike Priest, the security manager under significant pressure. *"We need to speak to you guys urgently"* he said, his voice an octave higher than usual. *"OK mate, we'll be there first thing on Monday"* the next day was a one-off bank holiday for a Royal Wedding.

"That's not gonna work, 09:00 tomorrow morning"

"OK Mike, we'll be there, see you in the morning" I answered slightly excited that UPS had eventually responded, but nervous of the urgency.

The next morning Paul, Ed and I drove down to the UPS Tamworth facility to meet Mike and his team. We arrived and were met in the car park by Mike personally. As he signed us in to the site, he was briefing us; the UPS UK district senior management team were waiting for us in the boardroom. This was starting to feel a little too 'ambushy'.

Our instincts were rarely wrong. Waiting for us was Tony Carpachio, an Italian New Yorker and the UPS operations director, Harry Stern, the district security manager and the district transport director whose name I forget.

Tony was the senior person in the room; he reclined back on his giant American leather boardroom chair; hands laced behind his head. *"So, what is it you think you can do for UPS?"*

"No idea" Paul replied, *"you wanted to see us"*

Harry silently passed Paul a letter to read. After a few moments, Paul passed it to me. It was from DfT. This letter effectively informed UPS that unless they responded to the DfT's concerns by a certain time, they would have their Regulated Agent status withdrawn.

"What do you think of that?" Tony asked. *"Have you responded?"* Paul asked back.

Another piece of paper was passed to us. It was a copy of an email with a response letter to DfT, distinctly inadequate in our experience.

"OK, you're fucked. You're about to be shut down at all 34 regulated sites"

The look on all their faces told us everything we needed to know. Tony sat bolt upright. *"we'll just pay the fine"* he said.

"Tony, you don't understand mate; there is no fine to pay, you're being removed from the list of Regulated Agents, IF we have time to draft comprehensive response, we might buy some time, but that's the best outcome here"

"Do you want us to help?"

Without Regulated Agent status a company cannot screen or apply any form of mandated security to cargo that is being flown. The hundreds of thousands of packages handled by UPS would have to be tendered, at significant cost, to another cargo agent that was regulated for screening.

On that bank holiday everything changed. While the rest of the country toasted Will and Kate, Paul and I were booked into a nearby hotel, we rolled up our sleeves and immediately started the most comprehensive rectification plan that the UK, possibly the world had ever seen.

Paul and I went into standard mode, drafting a high-level rectification programme. Ed headed back to Doncaster to keep Redline running. Paul and I knew we wouldn't be back for a while.

The UPS management pledged every support to us, and they meant it. Anything we asked for, they provided immediately.

The next day we met Carly Mueller, a tall American lady and the UK and Ireland District Manager. We sat in her office with her senior team. We briefed her on the rectification plan and the response documents to DfT.

What occurred next was nothing short of corporate meltdown. She literally screamed at her staff, then turned to us: *"and you"*, she paused, *"you'd just better make sure this works"*.

And then just as quick as she went stratospheric, she calmed down, composed herself and turned into one of the best leaders I have ever worked with.

Over the following days and weeks, I would have followed her into war.

I took on redrafting the entire UPS security programme. I adopted an unusual format, but it worked. Instead of just scribing paragraphs, I noted the legal requirements in one font, and then the UPS response to that requirement underneath, in a different font. Whilst unconventional, we had to do something that graphically demonstrated compliance and a reformed approach to compliance.

Paul took on more of the training and education aspects; we had to evidence that hundreds of individuals had the correct records.

We need a much larger team.

Carly drafted in dozens of managers and supervisors from around the company.

People arrived from all over Europe to assist us. We had them at our disposal within 24 hours, for as long as they were needed. The manager of the Stansted Gateway hub was an Italian; Giovanni promptly had his mother send us cases of the most delicious homemade limoncello, and ironically, homemade cakes.

After 72 hours, we'd barely left the building in Tamworth. We were living on doughnuts and pizzas. But then the big guns started to arrive from UPS corporate HQ in Atlanta, Georgia.

We were introduced to Wendy Hayler, a Yorkshire lass from Dewsbury but now the Vice President of corporate security. Joining her was Marty O'Connell, another New Yorker, but with an ego a horse could ride, sounding like Anthony Scaramucci.

Wendy was superb from the first hello, very supportive and understanding. Marty on the other hand was disruptive, trying to make a name for himself. He was arguing about everything we were trying to achieve, he just wasn't seeing the big picture.

As I was drafting the security programme, he was adamant that we didn't need one, *"UPS has a Master Operating Plan (MOP) for that"*.

"Marty," I said, *"DfT don't give a crap about your MOP, they need a Cargo Security Programme". "Send them the MOP"* he replied.

"*OK, but you're going to sign for this Security Programme and an email telling me to stop"* I responded. "*Sounds like CYA* (cover your ass) *to me."* He shot back.

"*Correct, because it is"*. I snapped.

Wendy stepped in, took him to one side, and he backed off.

Then he started interfering with the triage system Paul had set up to review records. Again, he was arguing about things he had no clue about. He stood in front of a flip chart, pens in one hand, knees slightly bent and scratching his chin.

From nowhere he suddenly bellowed that he needed some *"peace and goddamn quiet"* so he could think properly.

I signalled Paul over for a chat. We went outside, leaving Marty to his ways. *"Mate, they're going down, we're not going down with them"* Paul said. I agreed, we could still be paid for what we'd done; they clearly didn't recognise the shit they were in, and this clown was just making everything worse.

Mike Priest had seen Paul and I leave the management suite and came down to the car park to find us.

"Mike, sorry mate, but we're done, we're leaving. We can't and won't work like this, it was all going so well" Paul told him.

"Please don't leave, I'm pleading with you, please don't go"

Wendy joined us.

Mike explained to her that we were leaving. With a *"please don't leave, I'll handle it"* she went back upstairs to the bustling rooms.

"Give us a minute please Mike".

Paul and I took 5 minutes to discuss the situation alone.

We decided that Wendy was trustworthy, we'd give her a chance to deal with the New York motormouth.

She was faithful to her word.

Marty O'Farrell went back to Atlanta very soon afterwards.

He never returned to the project.

We completed the document set, revamped every record and drafted a bulleted ministers brief; fully cross referenced and board ready. Our hopes were pinned on saving the most critical sites.

Of the 34 sites, we shut two immediately and presented a plan to ratchet back to 17, including the two 'gateway' sites of East Midlands and Stansted Airports.

We'd done everything possible, and a few things that weren't. All we could do now was await a formal response from DfT.

We had a 7 day wait. In that intervening period, the UPS Chief Operating Officer (COO) flew over from Atlanta, at Wendy's request, to meet us. We still had to formalise the commercial arrangement and the framework for us moving forward.

We had discussed beforehand that we wouldn't take advantage of any company in difficulty, including a company of UPS' size and stature. Trust was still the way forward. We had a daily rate in mind as a starting point for negotiation.

When we started talking numbers with the COO, he simply said *"nope, that's not going to work for UPS"*. We looked at each other, confused and unsure of how to progress, thinking we'd been very reasonable.

The COO then gave us a figure that was double what we had in mind!

Wendy grinned behind him.

He asked us if there was anything else he or UPS could do. Half-jokingly I replied, *"well you could send my wife some flowers, we have a baby, and she's barely seen me"*

The next day both Jaq and Catherine received a huge bouquet of flowers and a substantial cheque; made out to them personally in recognition of their support to me and Paul. It was a beautiful touch and very UPS.

The biggest challenge facing Redline was then finding and recruiting the nearly 30 members of staff we would need to service just the UPS contract. Phone calls were made and feelers put out. With UPS financing the posts, we could attract some real talent.

One of the senior positions we had to fill, and critical to success, would be the UPS contract manager. Paul called on an old friend of ours, Ed Macy, who had recently published his own book and was a decorated AAC pilot.

We weren't just patching cracks; we were bringing in serious, proven leadership.

For Redline, this was more than a contract. If we could help pull UPS back from the brink in the UK, it would prove that our little company in Doncaster could take on the biggest problems in global aviation security; and win.

Chapter 132. The Three Musketeers

The DfT responded. They accepted our rectification programme, but with significant caveats. We requested 17 sites; they gave us 10 and denied both airport gateways. The 7 remaining sites would have to be re-listed as Regulated Agents sites, and the original 10 would be subject to audit.

Those caveats lit the fuse on the busiest, most relentless six months of our lives. Relentless in the way Bosnia and South Armagh had been: minus the death and destruction, just the chaos, long hours, and the creeping sense of impending doom.

From that point on, Paul, Wendy, and I became the core of the UPS rectification effort: a three-person engine room fighting a two-front war: stabilising ten live sites while clawing back seven more, including the two gateways.

Days blurred together into an endless loop of conference calls, site visits, spreadsheets, checklists, and midnight rewrites.

Every UPS screener would be re-trained in Doncaster, and every site reviewed, audited, and supported by a Redline Facility Security Coordinator (FSC). These embedded supervisors were critical in providing impartial security management to UPS. Each site would undergo a mock audit as part of the preparation for the re-listing.

The mock audits we ran were designed to be every bit as brutal as the DfT; in truth, often worse. The DfT audits were already legendary: hours of relentless questioning, paper trails picked apart line by line, no let-up from start to finish. Every procedure, every requirement, every process had to be comprehensively evidenced.

Tippex wasn't a correction; it was fraud. A missing signature? Practically a capital offence.

At this stage of the plan, Sleep was optional. Mistakes weren't.

The first site up was Tamworth, one of UPS's biggest and most important facilities. We got off to a poor start before we'd even opened a file; the DfT inspectors arrived to find no parking spaces reserved for them and had to walk some distance in.

From that point, the mood soured and never really recovered. The audit became death by a thousand cuts; the site failed on minor infractions that, in isolation, seemed trivial but collectively painted the wrong picture.

It was a hard knock, but we learnt fast. The second site sailed through, setting a higher bar for everything that followed.

The three of us formed a formidable team, fully backed and supported by Carly. We eventually got to the point where we were ready to present the first gateway for listing; it was fundamental to the UPS operation that we recovered these sites.

Stansted was first up.

When the 'three musketeers' arrived, we were met with a surprising attitude from the site manager. Wendy was furious with this guy, took him outside and gave him the hair dryer treatment. I don't recall him coming back in!

Stansted was successful; we moved over to East Midlands and again put the site through their paces, rehearsing every person in a barrage of questions. We weren't training them to pass a listing inspection; we were bringing them up to speed so that they had the confidence to hold their own during the actual re-listing, as we were forbidden to answer questions directly.

We no longer felt like consultants from Doncaster; in the eyes of UPS and the DfT, we had become the team holding a global giant together.

It took everything we had in those 6 months to help UPS recover from a period of significant regulatory challenge in the UK and move to 17 highly efficient Regulated Agent sites.

The role that Carly and her UPS team played was extraordinary. Her drive and leadership were exemplary, but even Carly's superhuman effort paled beside Wendy's.

Wendy didn't know us from Adam when she arrived in Tamworth; her instinctive trust in us, and ours in her, created a team that I was proud to be a part of. We worked twelve-hour days without pause. Wendy never complained and was as vital to success as anyone in Redline.

The team back in Doncaster had kept the home fires burning, it enabled Paul and I to get on with our role in supporting UPS. The recruitment for UPS had gone better than we had expected; as well as bringing Macy on board to manage the whole contract, we had some exceptional managers join us.

We thought we'd earned a breather.

UPS had other ideas.

UPS corporate was suitably spooked by events in the UK and now asked if we could review their international operations to safeguard against similar regulatory action. Paul and I were invited to Atlanta to meet the US-based team.

As our aircraft touched down at Hartsfield–Jackson, our phones lit up with messages from home. The same night as our flight, the UK Airport Operators Association held its annual dinner and awards in London.

A group of Redline managers were there representing the company, enjoying a well-deserved and rare evening off. To our surprise, Redline had been voted '*Best Small Company*' by our client base. A humbling award, and the timing could not have been better; it felt as though the wind was firmly at our backs.

We spent a few days meeting Wendy's security team, along with the management of various supporting functions. They then flew us down to Louisville, the home of UPS Worldport.

This place was unbelievable. Home to the UPS airline, it sat near the top of the global pile: number two worldwide by cargo moved, and one of the largest airlines by fleet. In the UK we were used to seeing packages being screened at a rate of about 10-15,000 per hour in the largest sites. Here in Louisville, they were processing hundreds of thousands of packages an hour; the facility was gargantuan.

Louisville showed us the scale of the game we were now in. Redline wasn't just recovering the UK operation anymore; we were being pulled into the very heart of one of the world's largest logistics empires.

We were to become close friends with Wendy and her family, as well as her deputy, Matt James and his family, forming quite deep bonds forged through friendship, trust and respect.

The international plan saw at least one representative from Redline accompanying UPS Internal Audit (IA) on their highly developed quality programme. Matt, (taking Wendy's place), Paul and I visited their most crucial sites around the Far East. We did a bit of a tour, starting in Incheon, Korea, through to Hong Kong without IA in tow to establish what our routine would look like.

We realised that a single Redline representative was sufficient to work with IA. I was allocated Cologne in Germany, followed by Singapore. Both sites were critical; Cologne was the main European sorting hub for UPS and second only to Worldport. Singapore was the gateway to all of Asia Pacific.

Chapter 133. Sing Sing

UPS only has one standard: the highest international standard. Sorting facilities are spotless, administration faultless, management meticulous. After the challenges in the UK, there had been corporate soul searching: how had IA missed the near collapse of a high-priority district? UPS IA overhauled their entire global audit schedule to ensure this never happened again.

Wendy and I went Cologne to provide IA with technical AvSec expertise.

Given the UK experience, the rewritten IA audit schedules assumed that a screening facility regulated by any EU member state would be regulated to the same standard as the UK; they weren't. This assumption would prove to be the flaw in the plan.

We started the audit. Each evening, we gathered with coffee and doughnuts to review the day's findings. The internal system categorised issues by levels of severity, and by the end of the audit there were several findings rated among the most serious. Some of these were reconsidered after discussion, but ultimately a few significant issues were recorded.

It caused chaos; a senior manager was removed from post on the spot. Apparently, the UPS 'C' suite was 'unhappy', as IA's findings painted their main European hub as being in disarray. While there were areas needing attention, as there always are, the situation was not as bleak as the audit suggested.

Back in Atlanta, the audit was assessed and a few of the calculations used to determine severity were adjusted. We had gone from one extreme to another. The next shot had to hit the target in terms of accuracy and objectivity.

A couple of weeks later the same team assembled in Singapore. The regulatory environment there was very different again. This was multilayered and boiled down to government-controlled screening; UPS had little control over their own standards.

Reading through the national regulation that applied to foreign air carriers and regulated agents, it specified that in the quality assurance programme, UPS must conduct covert tests on the screening. OK, I thought, we can run a covert test as part of the audit. I went and bought some everyday components to construct a viable test object.

UPS sub-contracted its cargo shed operations, who in turn sub-contracted the screening to a government approved and owned entity. We asked for a meeting, and I described what I wanted to do. I presented the homemade test object, and they agreed a screener should detect it.

What occurred next surprised me to say the least. For a variety of reasons, the test object wasn't detected. After a debrief with the screening team, I went back to the audit which was on its last day; we were flying home the next day.

That evening, I left the hotel and went shopping to buy some presents for Jaq and the baby. A police car pulled up next to me; the cop asked if I was James Termini. *"Yes, what's up?"* I replied casually.

I was immediately detained and taken to the headquarters of the Airport Police Division (APD) and placed in a room by myself; but crucially I still had my phone. I called Wendy and told her what had happened. I then called Paul back in Doncaster to get UPS legal cued up ready to bail me out. I still had no idea what it was I'd done.

Eventually a police inspector came to talk to me; pleasant enough, the 'good cop'. They alleged I'd carried out an unauthorised covert test, but I knew I was working within their regulations, I hadn't done anything of the kind. Next, *'bad cop'*, whom I now know as 'Dave' arrived to give me a grilling.

The routine went on for about an hour. During that hour, my phone was continually buzzing away on the table next to me. I looked down at it. There was a picture of my head superimposed on the poster of *'The Usual Suspects'*, complete with me holding up a prisoner number. The first thought on Mason's mind was to send me photoshopped pictures to take the piss!

I ended up quoting their own regulations to them to defend myself. 'Dave' eventually found a piece of UN guidance stating that simulated threats required explicit clearance, and took me back to the hotel, but retained my passport.

Ah Bugger!

The next day the police arrived to escort me to the airport. They gave me my passport back and subjected me to every kind of security they had available, as if to say, *"that'll teach him"*.

In a twist of fate, years later I would meet '*Dave*', who turned out to be a great guy. We have a laugh about him doing me over and the story gets a little more embellished every time it's told.

We're good friends now.

When I arrived back in Doncaster, I'd missed Honey's first birthday, exactly what I didn't want.

Before I'd gone to Cologne, I'd written to the headquarters of the International Civil Aviation Organisation (ICAO) requested their endorsement of Redline as an Aviation Security Training Centre (ASTC).

On my return I had their answer. *"We regret to inform you..."* The basis of the refusal was that ICAO only endorsed state operated training centres, and any request for endorsement would have to come from HM Government, and DfT in particular.

They said no. The trigger word. I wrote to the appropriate manager in DfT to request they submit the endorsement application on our behalf. After 2 weeks we got their reply. *"DfT doesn't want an ASTC in the UK, so we regret to inform you..."*.

I wrote back to this manager and explained very clearly the benefits of endorsement to all concerned, not least the DfT itself. There followed a backwards and forwards exchange of thoughts that ended with our application for endorsement as an ASTC finally being accepted by ICAO. We would be subjected to an assessment from the appropriate section from the ICAO secretariat in Montreal.

A lovely Polish lady arrived to assess us. We'd prepared the death out of the NSTC, everything was pristine, every training aid laid out, every aspect we would be evaluated against rehearsed. Just before Christmas we were officially endorsed as an ICAO ASTC, with a formal 'flag ceremony' scheduled for May.

Redline was the first private company in the world to achieve endorsement from ICAO.

For a company that had started above a hangar in Doncaster, it was validation we never imagined asking for.

We were the only centre in the UK permitted to fly the ICAO flag.

I had the honour of being the first Director of ASTC Doncaster.

The UPS contract was settling down, the covert testing programme across all sites was underway. Macy was spending a fair amount of time in Tamworth building relationships with his counterparts in UPS.

Airport testing operations were going well; evolving all the time with a digitised reporting system. Courses were all full.

That Christmas of 2011 was special. UPS had paid a substantial invoice, and although we'd expected it, Paul and I quietly took our first large dividend without mentioning it to Cath or Jaq. They found out in the ladies at the Redline Christmas party; judging by the shrieks of laughter, they weren't upset.

2012 promised even more: the Queen's Jubilee, the London Olympics, and finally, some stability. Debt was behind us, we were house hunting locally, and to top it off; Jaq was pregnant.

It was hard not to reflect on the irony. If her father had trusted us, stood by us through the hard times, and shown even a little compassion, he would not only have grown his financial worth substantially, but also kept his daughter, granddaughter, and soon another grandchild due the following August.

UK, Epworth, North Lincs: Feb 12 – May 18

Chapter 134. Home Sweet Home

In February 2012, after four years of marriage and eight together, Jaq and I stepped into our first home with Honey. We'd found a rundown farmhouse on the outskirts of the village of Epworth. Bridge House was set in 4 acres of woodland and had a small paddock.

There were outbuildings and garages, everything I'd always wanted. On the downside, the kitchen was barely usable, the plastic conservatory an eyesore, and the whole place needed serious work. But it was ours.

The winter wasn't as bad as it had been in 2010, but there was still a lot of snow; it was bloody cold. The day after we moved in, the heating oil ran out. Not great with a baby. It was hardly the stately country retreat I'd imagined, more like camping indoors with central heating as a rumour. I found a random bloke who sold me a couple of cans of oil. That set the scene.

Jaq and I had great plans for Bridge House; this was our forever home. We wanted it to be the place that all our family called home.

The Olympics' covert testing contract was ramping up as we approached the start of the Games in July. The athlete's village was now coming online along with the main London venues, but another piece of critical national infrastructure was about to request a proposal.

I was invited by the security manager for the Palace of Westminster to present some options for covert testing of both the House of Lords, the House of Commons and Portcullis House, the large office block opposite Parliament.

After the allowed 60 minutes the only question was *"when can you start?"* As we prepared to mobilise the contract, we were provided a private tour of the estate.

We were to provide two sets of reports using our digitised system; one set to the Serjeant at Arms for the Commons, one to Black Rod for the Lords.

With such a heavily armed police presence, we had to have rigid protocols that eliminated the possibility of one of our testers being arrested, or worse, injured.

We put it in Paul Cooper's hands. He took what we'd built and made it better. Macy was still tied up with UPS, but we were already planning the next layer of leadership.

Early in the year, DfT had an unusual overseas task for us. GOF had transformed into a new programme; now called Overseas Aviation Security Training, or OAST. OAST required us to place competitive bids for each proposed delivery. Each bid had to include one male and one female trainer; this bid was for a single trainer experienced in delivery of manager training.

The task was to deliver an abbreviated manager's course in Kuala Lumpur to the South East Asia Regional Counter Terrorist Team (SEARCCT) in Kuala Lumpur.

SEARCCT, on this occasion, was being supported by New Zealand and the UK. I was representing the UK's contribution; representing NZ was Murray "Hugo' Porter.

Over a beer one evening towards the end of the course, Hugo and I were talking. He was the Director of security at the CAANZ but held additional roles directly with ICAO. He was the Chairman of the ICAO Working Group on Guidance Material (WGGM), and in the aftermath of the printer cartridges bombs the group was overhauling the recommended policies and procedures contained in its Aviation Security Manual.

He explained that the group was attended primarily by representatives of government authorities with some representation from international organisations such as IATA and ACI, but that operational expertise was lacking. He invited me to attend the next WGGM taking place in Zermatt, Switzerland in a few weeks' time in the capacity of 'observer'.

I arrived in Zermatt with no idea what to expect. There I met Stefan Wenger, deputy chairman of the working group, and Alec Doazan from the ICAO secretariat, a Frenchman and the document owner.

Even in the relatively relaxed setting of this working group, I quickly learned a lot about international diplomacy and thoroughly enjoyed the experience. I would stay with this group for the next thirteen years, contributing operational expertise. Stefan, Alec and Hugo remain some of my closest friends to this day.

Throughout my life I had kept in loose contact with my father. I'd been to the States a few times to see him, in fact Jaq and I had been over to Pittsfield and stayed with my cousin Amy and her husband Lee, so Jaq had met him.

At the start of April 2012, my Aunt Fran called me to say that I needed to be over there; my father was in a hospice back in Pittsfield with lung cancer and short on time. I spoke to Paul and told him we were on our way to the US. Ed didn't want to go for his own very valid reasons.

Jaq and I arrived to a sombre mood and went up to the hospice straight away. He lay frail on the bed, barely conscious, with only the dignity of a sheet to cover him. I sat with him that night.

After a few hours' sleep at my cousins Amy and Lee's house, I returned the next morning. Then the next. And the next.

On the fourth day I asked the nurses if they could help wash and dress him. To my surprise he was lucid and alert. For the first time since we'd arrived, he looked like Dad again.

The three of us talked about Jaqs pregnancy and that if this was a boy, we would call him Samuel; after him. He managed a smile and asked a few questions about the baby.

That was the last conversation we ever had.

A couple of nights later, the lung cancer finally claimed what was left of him. His last hours were absurd; lying there, miming a cigarette between his fingers, the habit was burned so deep into his soul, he couldn't let it go, even as he died.

I sat beside him, listening to his breathing waiver and stagger. I held his hand tightly, felt the faintest squeeze back, and then, nothing.

After a few minutes, I lay his hand down and went to the nurse's station. *"I think my dad's just died"* I whispered, a lump in my throat. The nurse came back to his room with me and listened to his chest. *"I'm so sorry, he's passed"* she said, *"you can stay with him if you like"*. I sat down, picked up his hand and held it, kissed him on the forehead and said goodbye.

The undertaker arrived within 45 minutes. I helped him put my dad in the body bag and zipped it up. I didn't cry, I don't know why, but I didn't when I should have.

Our relationship had always been thin, more absence than presence. Sitting with him at the end, I felt the weight of what I'd missed, not just a father, but a guide. It made me more determined to be present for my own children.

The next few days in Massachusetts were spent making his funeral arrangements and putting his few affairs straight. He had nothing to leave anyone; I kept his wallet; an empty wallet apart from an expired credit card, a metaphor that summed up his life neatly.

His sisters started bickering about everything. Jaq and I made the decision to leave them to it; If they wanted to turn his funeral into a circus I couldn't stop them; but we wouldn't be there to witness it.

He had a funeral with military honours. Some of his cremains, his cornered stars and stripes and the cartridges from the rifle salute were dispatched to me back in the UK.

We had some of his ashes built into rockets by a specialist company. Our family gathered one night later in the year to say goodbye, Ed and I lit the blue touch paper together, and the old man was launched into the night sky, exploding in a huge shower of different shades of blue, before fading into the darkness. It was the most fitting tribute imaginable. I could picture him watching, lip curled in that trademark half-smile.

He was a good man in ways that mattered, but he never worked out how to be a father. Maybe I inherited that flaw. I hope my sons do better, I know my daughters have. But he was kind, generous, and in his own broken way I think he cared. I tried to make him proud.

I'll never know if I succeeded, but I like to think, in that flash of blue light, he knew.

While I'd been away in the US, the Redline team had been preparing the NSTC for the ICAO Flag raising ceremony. Normally, the Secretary General of ICAO, at the time Raymond Benjamin, would formally present the flag to much fanfare.

Maybe it was because we were a private company, but we were informed just before the ceremony he would be replaced by the Europe/North Atlantic Regional Officer for Aviation Security and

Facilitation, Connie Ludorf.

Like most ICAO staff, she started off quite sceptical about having a private company in the network. The overall perception was that we would commercialise ICAO and drive standards down. Nothing could be further from the truth.

In May, Paul and I headed off to Dubai to attend the mandatory annual ASTC Directors conference. This week was to be a turning point in the history of Redline.

We were met before the meeting even began by the Chief of Implementation at ICAO's AVSEC Branch. He was clearly not pleased to see us and gave us a warning about what was expected of us and what the sanctions were should we displease ICAO. I felt he was itching to get us out of the network.

There were others in the room that felt the same way, but crucially, there were some very supportive directors too, who saw the benefits we would bring.

We would walk miles around the Mall of the Emirates on an evening just talking about where we were in the evolution of Redline; we realised we had a choice.

Option 1: strip profits and ride Redline until natural death, or;

Option 2: reinvest the profits, constitute a board of directors and build the company for a sale to a larger corporate at some point in the future.

We went for option 2.

We had a notion for who we wanted to be Chairman. It had to be someone we both respected, trusted and liked. Someone that knew us, who we were as people and how we operated. There was only one choice, and we knew he was about to retire from the Army as 'Lieutenant General Sir Gary Coward', our old CO from 1 Regt AAC and Bosnia. For us, it felt like closing a circle.

Additionally, we would bring in a Training Director and move Ed to Managing Director, given the job he'd done when UPS kicked off. We would recruit a Technical Director and build a full tech department for software and e-learning development. Macy would become Operations Director, with his position backfilled through a promotion. From the start, Paul had assumed a MD role, he would move up to CEO and I would be COO, which shortly changed to Commercial Director.

Paul had ideas on suitable people, he'd met individuals he thought would fit in from his days on the Apache programme. I trusted his judgment.

The key was finding a way to pay the right people a salary that would attract talent; and retain it. The answer to that would come from our accountants who had been with us since day 1.

During the week we got quite close to the ASTC management teams from Dubai and Kunming in China. Both sets of unlikely directors got on well, and we went home with plans to forge closer partnerships in training and best practice.

In the summer work started on Bridge House in earnest. The existing kitchen and plastic 'conservatory' were being demolished and a tailor-made, 2 storey oak framed extension was being put up. The entire downstairs would be a huge new kitchen with a dining area and open plan lounge. Upstairs was a master suite, a massive bedroom with a walk-in wardrobe and ensuite. It featured floor to ceiling windows, a cathedral ceiling and a cast iron slipper bath. It would be stunning.

The structure was manufactured by David Salisbury, one of the country's leading oak framed building providers.

Unfortunately for us, the builders they used were cowboys, utter charlatans. I called the designers up from Somerset to inspect the shambles. Steel work held together by a single bolt, levelled with broken tiles and the remains of their lunch in the cement, not concrete, foundations.

All the foundational work had to be taken down and rebuilt by competent local builders. But to be fair, David Salisbury threw in a load of extra oak work to compensate us.

The kitchen was expertly designed and fitted by Mick, Jaq's cousin. The finished extension was truly a thing of beauty; we were so happy.

But as we went into July, Jaq was having some real problems with the pregnancy. She was admitted to Bassetlaw hospital for observations; and then couldn't leave. I had Honey to look after, and work wasn't going away. After a week or so, my mom came up to help look after Honey, but Bassetlaw was a good hour's drive away and not getting any closer.

Now confined to bed and forbidden to even stand up, she was going stir crazy, I needed to be with her as much as possible. We thought a move to Scunthorpe hospital might help; it was much closer to home, and I could be there longer. Because of her precarious position, a midwife had to accompany her in the ambulance from Worksop up to Scunthorpe – just in case.

Scunthorpe didn't have the same good reputation as Bassetlaw, and I felt awful for suggesting the move in the first place. We'd made a mistake. We then requested a move back after a nurse tried to give her the wrong medication.

We were too late.

I got the call in the very early hours of the morning: I needed to be at the hospital.

Right now.

Jaq was being prepared for an emergency caesarean section. Still tugging on a T-shirt, I strapped into my rocket car; a 5-litre supercharged V8 Jaguar XKR.

It was 02:30, the roads were empty, and the M180 had never looked so inviting.

The Jag howled like Concorde on take-off as I tried to balance two competing urges: to arrive before the surgeon made the first cut, and not to arrive via the Accident & Emergency entrance myself.

I broke the land speed record in the first quarter mile drag. After 5 miles the supercharger was screaming, it had had enough, I kept pushing it.

Approaching the motorway exit at exactly 70 mph (tops), I slammed the brakes on, they started smoking; I could smell it. I arrived at the hospital like the pizza delivery kid in Home Alone, hit a sign in the car park and abandoned it.

I raced into the ward to see Jaq being wheeled away; the nurse instantly knew who I was, even though we'd never met; the 1950s racing driver look on my face, and the smell of burning brake pads must have given me away. She took me to a changing room, and I put some of those green scrubs on.

In the operating theatre I stood by Jaq's head, stroking her hair and straining to see over the small curtain where all the work was going on. I was barked at and concentrated on talking to Jaq.

What they pulled out first didn't look much like a baby; just a greasy, bloody bundle, but it was Sam. Then the bundle was whisked away, I followed.

Samuel Rocco Termini was here. 3 weeks early.

After he'd been cleaned up, his tiny face was screwed up like he was in pain. His lips and fingers turning blue. He was taken straight to the special care baby unit and handed to a junior doctor who was incapable of getting a canula in his arm. Getting frustrated, I went and found a nurse, who came and took over the job.

Once I knew Sam was OK, I went and found Jaq.

She looked at me and tried to smile. She was on a different planet, still drugged up the ying yang. She slept for hours.

I headed straight back and checked in on Sam first, his fingers were a weird shade of purple. the cannula had cut off blood to his fingers, and no one had noticed, they did when I got hold of a nurse!

His hand was sorted out, and I took him to meet his mum for the first time. Poor Jaq could barely register that I was there. Sitting in a wheelchair she held him gently, I don't think she knew he was there at first. Slowly a dawning recognition spread over her face, and she smiled very softly at him.

I put Sam back in his NHS wheely cot and Jaq went back to sleep.

Looking back, that first year at Bridge House was a blur of extremes. We built a home, but just as we were laying down roots, we flew across the Atlantic to watch my father fade away.

One moment I was holding his hand as the last breath left him; the next, I was racing through the night to hold Jaq's hand as our son fought his way into the world.

In between, Redline was thriving, winning contracts others could only dream of. In the space of a few short months, I said goodbye to one generation and welcomed the next, learning that everything; love, loss, ambition and fear all belong to the same story.

Chapter 135. Boom and Bust

2013 arrived. Over a couple of weekends Mick and I built a proper pub in the back garden. It started as a three-bay garage and ended up as a fully functioning boozer with a fireplace, quizzy, pool table and a small lounge area.

The walls were covered in mementos of our travels and my time on Lynx. The ceiling fan was 4 Lynx tail rotor blades I had bought from eBay. If Carlsberg built pubs in your garden, this was it.

Honey had a pony, Sam had mud pools. Patrick was helping me with cutting and splitting logs and Amelia was becoming a talented little artist. Jaq had lots of friends from the mums at Hill House School. We had a beautiful family home and were so very happy.

Fran had decided that she wanted a break from Germany. After some conversations I borrowed a neighbour's clapped-out Bedford horsebox and coaxed it all the way to Gutersloh.

After collecting Fran and her horse, we made the 30mph drive back to Calais for the ferry. I was fully armed with a paintball gun just in case we picked up some unwanted passengers at Calais!

Bridge House became a much livelier place with Fran; the kids really loved her.

The downside was that I was away an awful lot of the time; sometime in the UK, most of the time in other countries building the company.

The travel was exciting, but even then, it was starting to become routine. Paul and I were taking part in some high-level conversations about the regulatory framework for aviation security.

We had started the reshaping of Redline to better reflect where we wanted to go. Sir Gary joined as our chairman. We reformed the way in which we paid ourselves as company directors.

On the advice of our accountants, we adopted a more tax-efficient remuneration structure. It was a common approach for small businesses at the time and allowed us to reinvest more money into the company while helping us attract and retain the right people as Redline continued to grow.

At the time it seemed entirely unremarkable. The arrangements were professionally advised, widely used and fully disclosed. None of us imagined they would return years later to cast such a long shadow over our lives.

We had recently won most covert testing contracts we had bid for, so there weren't that many left to find. The Royal Collections Trust (RCT), the body that manages the Royal Palaces requested a proposal for the testing of Buckingham Palace and Windsor Castle.

When we were awarded the contract, we had a private tour of the palace by the security manager.

In this period of change, Paul and I were keen to build on the relationships with the Chinese and Emirati ICAO ASTCs. I arranged a trip out to China via Dubai and Hong Kong. In Dubai we met Maj. Marwan Singel and Capt. Jamal Ali Hassan, the centre directors. We hit it off instantly.

Straight away there was an instant rapport; the kind that only comes when rank, nationality and language get thrown out the window and you're just four blokes laughing at the same daft jokes. That friendship would become the foundation of a long-lasting exchange programme. We remain in contact as friends to this day.

From there we headed to Hong Kong, where we linked up with Bill Yeung, a former LA cop we'd met at an earlier ICAO directors' meeting. Bill, being of Chinese heritage, was our unofficial fixer and cultural coach. He talked us through what to expect in Beijing; the customs, the style of meetings, and most importantly, the drinking, *"They'll try and drink you under the table,"* he said.

"Whatever you do, don't let them."

We strode into Beijing full of misplaced confidence.

At the airport, still looking for the taxi signs, we were approached by a tall Western man in a trench coat. *"Are you James Termini and Paul Mason?"* Instinctively we replied, *"Who's asking?"* He turned out to be the Post Security Officer for the British Embassy. Let's call him Nigel. That wasn't his name, but it fits him.

Nigel slid into the taxi with us and, over a green tea at the hotel, came clean. He knew we were there to meet the Deputy Director General of the Chinese CAA; a man the Embassy had been denied access to for three years.

He wanted to know who the hell we really were. Two nuggets from the Northeast being hosted at a level beyond his reach?

He was suspicious. And he wasn't joking when he reminded us that bribery was an offence. We explained it was nothing shady, just ICAO ASTC business, and invited him along to prove it. To our surprise, he accepted.

The meetings went smoothly, until dinner. Just as Bill had warned us, out came the Baijiu, China's own liquid wrecking ball. Potent, fiery, and accompanied by the cry of *"Ganbei!"*. Dry the glass. At first, we held our own. The suits were sharp, the ties Regimental, and we toasted with the confidence of men in control.

Then came the Wolf of Wall Street moment. One minute we were dining like diplomats, the next we were out in Beijing at 3am, in jeans and shirts, with Paul stealing cigarettes from a Chinese policeman and riding bronze rickshaw statues. The next morning my phone was full of incriminating evidence.

I rang Paul, head pounding:

"What do you remember from last night?"

"Sweet fuck all. What happened?"

We compared photos over a coffee, both of us terrified about what we might have done. Doing what good Sergeants do when the situation is utterly unsalvageable, we packed up, changed our flights, and legged it back to the airport.

As for Nigel, he rang me later. What we didn't know was that he wasn't exactly built for nights on the Baijiu. We hadn't just nudged him off his stride; we'd yanked the eject handle and fired him straight into the mess below.

His wife gave him both barrels when he got home. Poor bloke copped the hangover and the domestic fallout.

We only had the hangover.

Chapter 136. A Tale of Two Cities

Just before Christmas 2013 I was in the Premier Inn at Westminster Bridge for two nights, in London for meetings that could have been emails. I was lying in bed late on the first night, grinding through the day's emails, disillusioned with how much time I was burning away from home. I ended up doom-scrolling Facebook to bore myself to sleep.

I saw a thread about a family that lived aboard a boat, they just travelled the waterways of Europe on this floating council house.

On the second night I looked at boats for sale. I messaged Jaq. *"How do you feel about taking 3 months off work and travelling Europe on a boat?"* her reply intimated that she thought I had lost my mind.

I wasn't joking. With her usual *"F'kit"* attitude, Jaq agreed.

After Christmas I discussed the plan with Paul and Gary. They accepted it would work. And so, Jaq and I went about planning the expedition of our lifetimes; along with a 4- and 2-year-old.

I found a suitable boat in Swansea marina that was for sale. Kimya was a 40ft Cranchi Atlantique motorboat with 2 cabins and enough space for the 4 of us. She'd been a gin palace for some time, so the engines had hardly any hours on them. We agreed a price.

Having NEVER been on a motorboat before, this was a bold move. We needed training and licensing before we could do anything. I found an ex-soldier that was now an instructor and arranged for a full course using Kimya to qualify Jaq and I as 'Day Skippers'.

Then a bombshell landed.

Swansea marina was about to undergo extensive work on its lock doors, meaning the marina would be unavailable for 6 – 8 weeks. I had 4 days to get her out.

Michelle, the Redline admin managers husband Mark was an avid sailor, but sail boats; not engine powered ones. I called him, a busy man himself: *"mate, we're driving down to Swansea tomorrow morning and sailing a boat from there to Cardiff – happy?"* Being upfront worked because he agreed – thankfully.

We powered up in Swansea and headed for the lock. I took the wheel with the confidence of a man who'd never been on a motorboat in his life. We entered the lock like drunks in a shopping trolley, ricocheting off both walls and kissing the gates at each end. Mark stood there with his head in his hands, speechless.

It was a miracle that we got out of Swansea, let alone into Cardiff alive and with a boat. Michelle picked us up and we headed back up to Doncaster crying with laughter.

Things were simple after that. Take Kimya 'around the corner' from Wales to Southampton, have her lifted out of the water and do some deep servicing.

Jaq and I got a train to Cardiff and met with Cliff, the instructor. Day 1 we motored to Padstow, then day 2 around Lands' End to Falmouth, day 3 to Exmouth and finally on day 4 to the marina up the Hamble River.

We arrived in Eastland's boatyard as newly qualified day skippers, complete with an international licence, the ICC. Kimya was lifted out, and the clock started. I serviced both diesel engines, changing all the filters and consumables.

Who knew a Chieftain tank driver's course would be useful for boating?

Patrick and Amelia came over and helped me with the cleaning and odd jobs.

I was stressed out, they were bickering, and I'd been barking at them both about standards of cleaning. They both insisted they'd done a good job on the glass. I spun round to storm out of the boat and promptly nearly knocked myself unconscious on the very door I'd claimed wasn't clean. That made their day, nothing like watching me nearly break my nose again to prove their point.

With the engines being water cooled, I couldn't fire them up to run any leak checks until she was back in the water, so I was nervous when it came to starting them up as we'd only have one chance to get it right.

Jaq drove down to Southampton with the little kids and dropped them at my mother's house in Hythe. She was then dropped at the boatyard. The giant boat lift gently lowered Kimya into the murky freshwater of the Hamble river and I turned the ignition switches.

I'd left the engine bay open so I could check for leaks and/or smoke!

Both Volvo Pentas' ran as sweet as a nut, no leaks.

We would now move her over to the small marina near my mother's house at Hythe, have a couple of days with them before setting off.

As we powered gently down the Hamble, the depth sounder started showing shallower water. It couldn't be right; we were in the middle of the marked channel.

Then we hit the sandbar that had been shouting at me for 5 minutes.

The right engine stalled.

Ah bugger!

Jaq freaked out, "*FUCKING FUCK A DUCK*" she shouted. Our momentum carried us over the shifting sands and the engine restarted first try. 30 mins later we moored in Hythe, still shaking like shitting dogs and nervously laughing.

How would we manage for 3 months in international waters if we couldn't go 5 miles in England without crashing.

My mother still had Nan's ashes. She'd grown up in Portsmouth and used to swim out towards the Isle of Wight as a girl. We took Kimya out, found the Mary Rose buoy, laid daffodils on the water, and let her go in quiet contemplation of her extraordinary life.

At the beginning of April, we set off with intent. Our routing over the next day or two was to Eastbourne then a direct channel crossing to Boulogne-Sur-Mer, where Kimya would be again lifted out of the water and then trucked to St Jean-de-Losne on the Saone River near Dijon.

That would take 4 days, so we went by train to Paris, had 2 days in Disneyland Paris before meeting the boat in her new marina.

We stocked up on wine and food and departed with no real plan other than to meander down the river Saone at 5 knots, stopping when we needed to and enjoying the amazing French countryside.

It's not like we could get lost!

After a week or so we arrived in Lyon, our first big city. Kimya was a delight, not a thing went wrong with her. The daily routine we'd developed saw a breakfast of fresh bread, cheese and cereal.

All of us would then go up to the flybridge, the kids fastened on with sailing harnesses and lifejackets.

Meander slowly down the river.

Sam soon christened crows as '*baby dragons*'. If we went past somewhere that looked interesting, we'd stop. We navigated some huge river locks, concrete cathedrals up to 20m deep in some cases.

We stayed in the centre of Lyon for a couple of days, did some clothes shopping and enjoyed the city. Heading south through the confluence of the Saone and the Rhone, we decided to leave the giant Rhone and instead detour through the Camargue on the Petit-Rhone. We saw the fabled white horses and stopped in Avignon.

We needed a non-boat shower, and a porcelain sit down by then, so we got a hotel for a couple of nights. It was quite nostalgic being so close to Orange, I briefly thought about paying the ATC tower a visit.... I'm sure they would remember me!

Another stunning overnight stop was Aigues-Mortes. It was like something lifted straight from a medieval film set; walls, gates and towers still standing as if the next crusade might march through at any moment. Inside the grid of streets were cafés spilling out onto the cobbles, the smell of bread and strong coffee cutting through the heat.

Beyond the gates, the pink shimmer of the salt flats stretched into the distance, surreal against the blue sky. It was the kind of place that made you stop, look around, and think not much has changed here in hundreds of years, except the tourists now wear Ray-Bans instead of armour.

We had planned a week in Port Camargue; I wanted to do some servicing on Kimya and get her ready for the transition from fresh to salt water. Jaq had contacted Ciara who now lived in France with her French husband. We had a fantastic catch up with her and celebrated my 47th birthday in style.

We knew our destination would be Denia in the south of Spain, Mark and Michelle had an apartment there and highly recommended the town. It had a couple of marinas to choose from.

But how to get there?

Option 1: turn left and go down the French coast, down the Italian coast and cross the Mediterranean, island hopping across the

Balearics.

Option 2: turn right and take the Spanish coast all the way down to Denia.

Option 2 was the least risky; less open water crossing and so that's what we did; turned right and handrailed the coast all the way to Denia. The plan was basically the same though; stop when we wanted, stay where we liked.

The next couple of weeks saw us slowly bimble down the coast with extended stays in Barcelona and Valencia before arriving in Denia. Mark and Michelle were right; Denia was perfect, sitting between the mountains and the sea, a town that felt both ancient and alive.

The castle watched over the marina from the hill, its weathered stones catching the last light of the day, while below, the narrow streets twisted with tapas bars, bakeries, and the sound of Spanish voices spilling into the evening air.

No 'brits on the piss' was a big attraction.

We settled into the marina and signed up with a young Anglo/German couple, Sam and Anna Spencer-White, to look after Kimya for us once we left. To complete the adventure, we had to do the Balearics.

We set off, discovered Ibiza and stayed for a night or two in the old city. The next day we headed off east again and went over to Mallorca where we met up with Fran and her boyfriend from Bawtry, who were in a villa there with his family.

We spent 2 nights in Mallorca, the second was to recover from the first night. It's fair to say that Fran's boyfriends' family were expert drinkers that made us look like rank amateurs.

We arrived back in Denia having had many close encounters of the dolphin kind. They would arrive in a pod, play with us for a short time in Kimya's wake, and then wave a cheerful goodbye.

It reminded me what life could be without the constant hum of Redline in the background. Sadly, Redline wasn't about to give me a break.

With great sadness we packed our cases and headed back to Doncaster after our 3 months was up. This spur of the moment, wild and stupid idea was also the best idea we ever had.

For 3 full months it was just us, our family together with hardly any email distractions, phone calls from crying security managers or discussions over the definition of a word.

Throughout the trip Jaq had adopted the role of diarist/blog creator. She issued daily Facebook updates and our progress and then, when we returned, had a specialist format her posts into a single book. That book is one of our most cherished keepsakes.

It was nice to be back in Bridge House, but it took a little while to get back into the swing of work.

Before we'd set off on Kimya, I had worked with the Airport Operators Association, of which we were platinum sponsors, and the DfT to form a non-partisan UK Covert Testing Working Group. Sir Gary was the inaugural chairman. The idea was to create an industry standard for the delivery of covert testing.

I had also been talking to an ex-pat Irishman who worked for Aeroport's de Paris (ADP) who were closely following our work at Heathrow. I invited him to attend the CTWG.

Whilst I had been away, our proposal for the delivery of testing across Paris Charles de Gaulle, Paris Orly and Le Bourget had been submitted, we were now awaiting a result.

The government's Finance Act 2014 landed like a depth charge.

We knew the dividend-replacement rules were going to change. What we didn't expect was backdating.

That was the killer.

Worse still, we had only 90 days to cough up whatever figure the taxman plucked from the air.

These weren't polite "gentle reminders" either; they were Accelerated Payment Notices.

No debate, no delay, just *"pay now, argue later"*.

Arguing meant taking the government to court with money we no longer had. It felt like getting a parking ticket from God: absolute, unchallengeable, and ruinously expensive.

As with the Directions eviction notice, we had a grossly inadequate period to get a working contingency in place. We would have no choice but to sell a significant portion of the company and seek a financial injection just to pay the taxman.

When the APNs arrived, we were on the hook for millions.

Paul and I spent hours on spreadsheets, working numbers and arriving at a solution.

An intermediary matched us with a handful of potential investors, and we started doing the rounds meeting these people and answering all their questions.

At a meeting with one option, Janus Asset Partners, my phone buzzed. I glanced at the text.

"Sorry", I said, looking back at the assembled money men, *"we've been awarded a 5-year contract with Aeroport's de Paris".*

An English company had just been awarded a prime security contract with the airports of the French capital.

I let that fact sink in.

The terms of the investment were agreed, due diligence conducted, and signatures exchanged. Paul and I were no longer masters of our destiny, but we had high hopes. We liked the investor assigned to Redline; we trusted him when he said that they were there to help us grow. Robert Bowles was very personable, likable, a bit of a geek, but likable, nonetheless.

Payday arrived and our salaries were wrong. Not a bit wrong. Wrong enough to make my stomach drop. I rang Robert.

Then the mask slipped.

Robert performed a volte face and insisted that although our salary would be corrected, the fact we had raised it had gone down very badly in London.

Robert was moved to another investment after about 6 months. We didn't see the warning flags; we were too busy preparing the company for what would be a very different 2015. New structure with investors now on board and the evolving threat environment in global aviation security. Our magic 8 ball was focussed on work.

In October 2014, ICAO held its first symposium on Innovation in

Montreal. Organised by Stefan; Paul and I were due to take part as speakers; Sir Gary was asked to be a panel moderator.

Paul dropped out; partly due to the cost, partly because of the looming circus with JAP, so it was Gary and I who made the trip.

The symposium itself was the usual alphabet soup of acronyms, PowerPoint slides that could have doubled as chloroform, and a lot of people congratulating themselves on being innovative, while sticking rigidly to the echo chamber of nonsense. The real value was in the evenings.

One night Gary suggested we convene in his apartment with a decent bottle of wine, he chose, naturally. He knew his wines and chose well.

There we were: a Knighted General and a former Staff Sergeant, sat in an apartment in Montréal putting the world to rights.

We talked about Bosnia; his lived experiences as CO, mine on the line. Tail rotors, fences and decisions that could have gone differently; nothing was off limits.

It wasn't therapy, but it was close. Gary will never know how much that conversation meant to me. For years I'd carried those memories from Krupa like a stowaway, quiet most of the time, but impossible to forget.

I slew a demon that evening over a bottle of wine I couldn't even name.

Chapter 137. Performance under Fire

JAP replaced Robert with a new fellow, a young South African called Roland Du Preez. Roland was under the supervision of an older, but no less arrogant handler called Elliott Wedge.

Elliott reminded me of Michael Douglas in 'Wall Street'. He'd sit back, hands steepled, radiating that Gekko certainty: *I know best, you just don't realise it yet.*

The problem wasn't arrogance; unlike Gekko, Elliott wasn't playing the markets. He was fiddling with aviation security; a game he didn't even know the rules of.

The divide and conquer tactic was obvious. Private conversations. Different messages to different people. Pressure applied just out of sight, just plausible enough to deny.

In March 2015, the world of aviation was rocked by the Germanwings disaster. An Airbus A320 left Barcelona for Düsseldorf and never made it. The first officer, Andreas Lubitz, waited until the captain stepped out, then locked the door and set the aircraft on a steady descent into the French Alps.

One hundred and fifty people were killed in minutes.

The shock wasn't just the crash, but the fact it was deliberate.

Airlines scrambled to introduce "*two in the cockpit*" rules, and suddenly the talk wasn't just about terrorism or bombs in cargo holds, but about the fragility of the people flying the aircraft.

For us, this wasn't abstract. It was exactly the kind of risk that doesn't show up in a spreadsheet until it's already killed people.

Following the MH370 incident, I had become the face of Redline. I was interviewed by multiple global news agencies. Every time I appeared as an expert witness, the tool in beers owed just grew!

Gary had given me a crash course in media handling based on his experiences as UN spokesman. Following the tragedy of Germanwings 9525, I was approached by Channel 5 to take part in a documentary.

The setting for my piece was inside an A320 simulator. I had Patrick with me for some reason, I don't recall why. The interviewer, Matt Frei was easy to get along with and asked some sensible questions about the state of mental health support for commercial pilots.

I was honest and open about my own experiences with BMI.

The simulator instructor had set the visuals up to be on the same route as the actual aircraft. That was fine, but I was then asked to demonstrate the actions probably taken by the FO that crashed the aircraft. I refused. It wasn't about sensitivity. It was about ethics.

You don't rehearse someone else's suicide for televisual entertainment.

Away from the cameras, we had the French contract to mobilise. The legal and corporate aspects were exceptionally complex; I knew France was a socialist state, but we were dealing with bureaucracy of a scale I hadn't thought possible outside a scene from Orwell's '1984'.

Next level in terms of what we could, and more importantly, could not do. We hit gold with the manager we employed.

Frank Charbonneau was a former French secret service spook and Gazelle pilot before that. He was very pro Anglo and knew the French systems well enough to always find an answer. Macy had another superb manager reporting to him. In no time he had collated a significant pool of testers, agreed testing protocols with ADP and was up and running.

Paul and I had been invited to the French Embassy in London to attend a wide-ranging discussion with much larger UK companies wanting to expand into France. The meeting was chaired by the French Ambassador, but attended by their Minister of the Economy, Industry, and Digital Affairs, a certain Monsieur Emmanuel Macron.

Being the minnows in the room, we sat back and watched Macron take a full ear-bashing from the bosses of several UK giants. They ranted about French employment laws making it impossible to serve customers properly; complaints we'd soon understand for ourselves.

Listening to them complain, I realised the way to survive in France was not by fighting the system, but by learning how to move inside it.

Redline now firmly occupied the position of the leading provider of aviation security support services in Europe. Our French team would rapidly expand to include another 10 major airports in France.

JAP seemed never to be satisfied, continually arguing that nothing mattered more than the numbers in the bottom right corner of their spreadsheets. The initial spiel from Robert about them investing in us as people unravelled quickly, but we had no choice other than to bend over, time and again.

They didn't understand Paul and me as people, or as partners. They had no sense of the bond between us, or of the things that had brought us together and forged that trust. With the constant turnover of JAP representatives, no one ever stayed long enough to even try.

The company was building steadily, but our model relied on bulk, especially in covert testing. To make the numbers work, we had to hold a large share of the market. That was never going to be a straight line of sales targets; most of our growth came from word of mouth and a reputation for doing the job properly.

JAP never understood Redline because they didn't understand risk. We weren't selling polystyrene packaging or logistics, like their other investments. We were selling trust. Our product only proved its value when nothing happened.

Paul and I had understood that from the days of the 'post-it' note wall.

They never understood that our partnership wasn't contractual. It was much deeper.

In my opinion, they lacked "the pilot gene" as well as any spark of entrepreneurship, the very things they claimed to be investing in.

Paul described it perfectly as them having *"a long-handled screwdriver"* that would appear from time to time to tweak things that didn't need tweaking.

Chapter 138. Lights, Camera, Action

March brought a brief reprieve from JAP's meddling. A letter appeared on the doormat at Bridge House; it was from the TV production company, Channel 4 Creative who produce the daytime TV show 'Couples Come Dine with Me'. Apparently, someone had driven past the house and thought, Perfect, this'll look great on telly.

To be honest, I couldn't be bothered with it. Jaq was giddy and thought it would be a lot of fun.

Before I could say *"meh"* she was on the phone to the researchers. We went through a couple of phone interviews and then someone visited the house; as soon as they saw inside it was a done deal.

Couples Come Dine with Me follows three couples who take turns hosting dinner parties, secretly scoring each other on food, atmosphere, and hosting.

At the end of the week, the scores are tallied, and the couple with the highest total wins a cash prize. The format mixes cooking with '*unscripted*' humour, personality clashes, and the kind of awkward dinner-table moments that allegedly make for compulsive viewing, so we knew what to expect; or so we thought.

We were scheduled as hosts for the second night. At around mid-morning on the first day a taxi picked us up and took us to our hosts who lived a good distance away in Lincoln.

We genuinely had no idea about them at all. I knocked on the front door after hearing a booming laugh from inside.

The door opened and we were met by Mel, a huge South African guy and owner of the earthquake inducing laughter. His partner was Bianca, a nurse with Italian heritage; between them they were as mixed as we were.

Initial impressions of Mel & Bianca were very positive, they were both very likeable, interesting and engaging. It wasn't until we saw the show for the first time that their impression of us was less unsure.

Then the third couple arrived, Chris and Liz from South Yorkshire. Chris was of average height, loud with a shaved head and sharply dressed; Liz was much quieter and reserved, letting Chris do most of the talking; a serious amount of talking.

Chris thought he was some kind of comedian. Just not the funny sort.

I knew he was going to be a TV producers dream and a nightmare for everyone else at the same time. He would make good TV, and that's what mattered.

The *'unscripted'* part means there's no script, but you do exactly what the director tells you. The first night it all became apparent; someone would say something at the table, and the director would jump in with *"that's great, but save it for later"*, or *"try saying what you just said, but like this..."* it was all very frustrating.

It was obvious a confrontation was being engineered between me and Chris. There were two obvious angles, one was that they wanted a duel for *'funniest man on Channel 4'*; and secondly that Jaq and I were the rich, posh couple from Harry Enfield's sketch show.

We eventually got home after the first night at about 2am having been driven around in a taxi to give our scores on Mel & Bianca. We thought it had gone well and scored them highly.

The next morning the crews arrived at Bridge House at about 08:00 to start the 'behind the scenes' filming. Our menu included potted pheasant as the starter, with the hook that I had caught the pheasants myself with my Harris Hawk 'Gypo', named after the famous Red Arrows flying display manoeuvre.

We spent all morning filming me with 'Gypo' flying her around the local area.

We then got into the cooking.

When the director arrived, I took him to one side and explained, in no uncertain terms, that if the contrived effort to make me and Jaq look like some kind of Lord of the Manor meets Hyacinth Bucket combo continued, they would be Gypo's next target. I had a professional reputation that didn't need any damage.

He backed off.

Our dinner party guests arrived and were met with a glass of bubbly. We deployed our strategic reserve early.

Fran.

After some banter in the living room, we had to get on with the cooking, so we tasked Fran to 'entertain' our guests in the pub, off camera. Alcohol is pretty much off limits for a daytime TV show, except for a glass of wine with the meal.

Fran didn't get that memo and morphed into a girly Tom Cruise from 'Cocktail' and had them half-cut before the starter.

We kicked off with potted pheasant, which, after some raised eyebrows, went down surprisingly well. Then back to the pub it was for our guests, while we tackled the main event, confit of duck with the best organic vegetables in the area.

We had purple carrots, parsnips, and potatoes that were in 6 inches of mud that morning. On the plate, all that freshness magically transformed into fifty shades of brown. Tasty, yes, but not exactly photogenic.

Fran wheeled in our guests who must have been so pissed they didn't notice, because it was a culinary delight.

Back to the pub again while we started the dessert course. This was, according to Jaq, 'Decadent chocolate Torte'. It was an upside-down affair, so the choreography from the TV crew was to have a great reveal, 'Bake off' style, as the torte would be placed delicately on an ornate glass serving dish and presented with a grand flourish.

It was going so well.

Jaq carefully placed the serving disk over the torte, turned it upside down and gently tapped it to release the chocolatier's masterpiece.

The glass dish instantly exploding, sending shards everywhere.

A stunned silence and then, looking straight into the camera, she whispered "*ohhh shit, that wasn't supposed to happen*".

The torte had glass embedded across about three-quarters of its surface, so we went about cutting away the good part and ensuring there were no glass landmines in it.

There was just enough left for a sliver of torte each.

The guests resumed their seats, Fran now guiding them into their chairs. Jaq went full on-air hostess mode and made out like nothing had happened. She slid a plate with a wafer-thin slice of torte, with the immortal excuse of *"It's very rich. Too much would ruin the experience."* Translation: this is all you're getting.

Another 2am finish and another 08:00 start; not what we signed up for at all. At least we didn't have hangovers, the others must have.

The third and last night was at Chris and Liz' house in Sheffield. We all knew each other better by now and there was some genuine conversation, but Chris was still a pain in the arse, albeit more relaxed and less hyper than the previous evenings. He dialled it down from "stand-up audition" to "slightly overexcited dinner guest."

Chris was probably a good person at heart, and we all saw a different side to him.

The meal was a take on some Yorkshire classics, and was good, but the main event was Mel asking Bianca if she would marry him. We had seen it coming, but it was still lovely to be a part of their moment.

After some producer-invented bollocks about comedy socks, the final voting occurred, and we herded into the living room to hear the results.

But before I announce the winners, Chris and Liz deployed the same tactics we had, they fed us wine and beer all evening when we weren't on camera.

By the time the scores were in, we were all nicely lubricated. For a moment, it looked like their liquid bribery might just swing it.

The prize arrived on a silver tray: a wad of £50 notes and a bottle of Sainsbury's Cava; daytime TV at its finest.

By a single point the winners were… Jaq and Jim.

I was truly amazed having served three courses of brown, albeit tasty brown. I leaned back in drunken incredulity on the sofa and clapped my hands over my head in celebration.

I missed my own hands and flailed about like one of those inflatable car-lot mascots, all arms and no coordination.

When the show aired, none of the Hawk flying footage was shown, apparently it was unethical to serve pheasant that had been caught in that way.

There was still an attempt to showcase me and Jaq as pretentious, but nowhere near as bad as it could have been.

A decade later and every time the episode gets repeated, we still get ribbed by friends, family, and random strangers. And fair enough; it was ridiculous, but Jaq was right: it was fun, exhausting, and the perfect reminder of what 'reality TV' is all about, more TV than real life.

After a week of forced smiles, exploding desserts, and "reality" that was anything but, it was a relief to get back to our own brand of madness. No producers, no retakes; just family, friends, and a Fourth of July party with no producers, no retakes, and no one telling us what to say.

Chapter 139. Independence Counts

In July, I took some leave for a very specific reason; we'd invited Amy, Lee and their kids to spend the Fourth of July with us at Bridge House. Lee didn't share the rest of our enthusiasm for flying, so just getting him on a plane across the Atlantic felt like a minor diplomatic victory. I think he'd heard about some of the parties Bridge House had thrown and couldn't resist.

The damn Yankees arrived at Heathrow T5 where I picked them up in a hired minibus that would become the party bus for the next 10 days. First on the itinerary was obviously the Independence Day party; all the family was there plus about 20 of our closest friends.

Bridge House became the White House for the day; star-spangled banners everywhere, red, white and blue on anything that would hold still, and a marquee in the garden. During the day we'd hired the full bucking bronco set-up.

Remarkably there were no broken bones, though several egos took a battering.

The highlight of the culinary day was the full hog roast we'd prepared, but to be honest the gallons of beer probably came a close second.

As it got dark, the deck was converted into a stage, and the hired singer took over the entertainment; she was brilliant and had the crowd up and dancing. Just before she delivered the finale, I snuck off with Ed to get the ridiculously huge stockpile of fireworks prepped. Thanks to a mate at a local firework factory, we had enough fireworks to concern a small Balkan state.

As the singer bowed out with a flourish, Ed lit the first of the 1000 shot fireworks and I lit a rocket; not one of those tiny Finland fireworks, but more of a rocket you could launch a kid into space on.

Unfortunately, this is where it went sideways, literally; the 1,000-shot *"Four Horsemen of the Apocalypse"* failed catastrophically, firing all one thousand shots at once.

The sky lit up like the opening credits of The Simpsons. Stray missiles rained down on the crowd. My mother was last seen sprinting into the barn, pursued by a screaming banshee of fire, before it detonated beside her. Not ideal in a hay barn, though she did manage to avoid immolation.

The mayhem continued outside, with the dogs chasing spinning wheels of sparks everywhere. When it eventually came to an end, a haze of smoke drifted across the garden, broken only by thirty people laughing, coughing, and checking their hearing.

After a period of recovery involving a lot of coffee, we boarded the minibus and headed off to Swinton Park in North Yorkshire.

Swinton is a stately home now operating as a high-end hotel; Jaqs self-described 'happy place'.

On the way, I took the opportunity to drive through Ilkley to show Amy and Lee the culture shock I'd been through as a kid moving from the US.

The difference between a wooden duplex and a Yorkshire stone terrace, between American sprawl and English compression.

Swinton Park looked like a period drama that hadn't realised filming had finished. Inside, chandeliers swung gently, staircases creaked, and the whole place smelled faintly of history.

For Americans, used to hotels built from glass, steel and plasterboard, it was a cultural jolt. Dinner didn't help their confusion: foie gras on brioche, venison served rare, and desserts that looked like they'd been plated with tweezers.

They loved it, but you could see them silently recalibrating everything they thought they knew about food and buildings.

Amy and Lee were given the Turret Room, a suite that spirals through three floors of the castle's tower, a circular sitting room below, a curved bedroom above, and right at the top a bathroom with a freestanding bath and windows that make you feel you're bathing in the treetops.

We strolled the grounds, went to the local pubs in Masham, had a few pints of Old Peculiar and some top-notch food. It was easy to show off the Dales; it's just so beautiful.

All aboard the bus again, this time to change to a train and head off to the smoke. We did all the tourist things, Madame Tussauds, the London Eye etc; but then we had something special lined up.

I had arranged with an old friend to visit the Warders Bar in the Tower of London and join the private tour that takes in the ancient Ceremony of the Keys.

The Warders' Bar, known simply as 'The Keys,' is one of the most exclusive pubs in Britain, hidden inside the Tower of London and closed to the public. Behind the fortress walls, it feels part officers' mess, part private club with red leather benches, polished wood, and walls crammed with Tower memorabilia: plaques, portraits, even the odd weapon.

After the gates are locked and the tourists gone, it becomes their sanctuary, a place where the uniforms come off, stories flow, and history feels less like a museum and more like a living room with very good beer.

Dressed accordingly, our guests were treated with tremendous hospitality and kindness. The session after 'the keys' in the bar was topped off with a raffle of unique items from the Yeoman Warders private collection.

We bought enough tickets so that we were guaranteed a prize; how could we not?

It was a truly special week for all of us; one that we still talk about. And just as the smoke cleared, it was back to aviation security; fewer fireworks, but no less risk of things exploding.

In the wake of Metrojet Flight 9268's devastating bombing over Sinai in October, British security consultant Control Risks Group (CRG) was swiftly enlisted by the Egyptian government in December 2015 to audit security at three crucial airports: Sharm el-Sheikh, Marsa Alam, and Cairo.

We were surprised, to put it mildly. CRG were a big name in security consultancy; oil fields, kidnap insurance, that sort of thing, but aviation security wasn't their patch. I was straight on the phone to find our part in this.

Control Risks were tasked with the diplomatic and strategic side of the response. Redline was brought in for what happens on the ground. We deployed two of our most experienced and capable managers, Dave Openshaw and Neil Coull, to support operations at Sharm el-Sheikh, Marsa Alam, and Cairo.

Redline supported the sensitive task of risk mitigation at airport security layers in Egypt that had, until then, remained porous due to systemic issues.

The collaboration with CRG worked well, and it wouldn't be the last time our paths crossed.

At Bridge House, Jaq and I still had a happy and idyllic setting, albeit the travel wasn't decreasing. Fran was loving being with us, she was very popular in Bawtry with a lot of good friends. Her and Jaq were close.

Fran was now working directly for Redline in the covert testing department as a coordinator; and she was very good at it. In the offices Fran was the life and soul. Laughter drifting down the corridor usually meant she was setting up another practical joke.

And then there was Mick Harper; my old mate from Berlin and the 15/19H who joined us as a tester when the London contracts expanded to include the Palaces.

Having Mick back in my life was a gift I hadn't realised I'd missed. With him around, there was always room for one more story, one more laugh.

Chapter 140. It Ain't Half Hot Mum!

The rebrand to Redline Assured Security came in early 2016. On paper it looked like marketing polish. In reality, it was how we escaped the "aviation-only" pigeonhole. Aviation security had been our bread and butter, but we were already dipping our toes into critical national infrastructure, and the new name gave us the credibility to do so without everyone assuming we were just the *"airport people."*

It made investors happy, and it impressed regulators enough to get us a seat at new tables. Of course, changing the name wasn't just about strategy; it was also about forking out a small fortune to redo logos on everything from business cards to coffee mugs.

2016 was when JAP finally got to see how Redline actually operated. A regulator in the Civil Aviation Authority of Bangladesh (CAAB) had been in touch with us asking some vague questions about our Threat Image Recognition Training (TIRT) system.

I replied promptly, expecting a modest £10k-a-year support contract.

They wanted to see a full demonstration; I arranged a remote session but was met with a barrage of near abuse about us not taking Bangladesh seriously. They wanted an in-person demonstration the next week.

I called the Bangladeshi manager and detected that familiar trace of panic in his voice. Even Paul was sceptical about this one at first, but true to our credo, I flew out to Dhaka. Within a couple of hours, I got to the bottom of what the real issue was.

They were in deep shit with the UK DfT and multiple other international partners.

For a country that relies heavily on air cargo, having your only system shut down hurts.

Australia, Canada and the EU had banned the introduction of cargo from Bangladesh, the direct Heathrow flight was under significant threat too.

I called Paul, briefed him and got him out on the next flight to Dhaka, along with a trainer who could commence some immediate interim cargo screener training.

We fired up the Paul and Jim machine rapidly.

Together we established the magnitude of the problem, arrived at a solution and talked it through with CAA of Bangladesh, the UK DfT and other interested parties.

A single trainer wasn't even close to fixing this epic mess.

Confident we would win this work, we headed back to Manchester on a flight via Dubai. As we taxied into the terminal at Dubai; a message from CAAB. This had to be awarded through competitive tender, and they needed a formal presentation.

The day after tomorrow.

Paul called Michelle in the office and had us booked on the next flight back to Dhaka; the same aircraft that took us to Manchester, also took us back.

In Dhaka, we met the team from another UK company. We gave our presentation, they gave theirs. And that was that. No feedback, just a *"we'll let you know"*.

Back on another flight home.

Confident in our proposal, we started the mobilisation process from FL380. From the transit lounge in Dubai, we had both Eds, and the rest of the Redline team on a conference call.

We asked JAP if they wanted to be on the call – they declined, which was disappointing, I really wanted them to see us 'in character'. The call ended with a *"get to it lads, see you in a couple of hours"*

Our mission had two high-level objectives.

Secure the Dhaka–Heathrow passenger flight by rapidly improving aviation security to UK-required levels.

Lift the international air cargo ban by overhauling procedures, facilities, and regulatory compliance.

This meant a complete re-drafting of national aviation security regulations, creation of an aviation security master plan and transforming Dhaka's cargo facilities into fully EU compliant sites.

Every screener in Dhaka had to be re-trained to EU standards.

To do that we sub-contracted a UK general security company to provide us with people; people we could train to EU standards in Doncaster and then deploy them to Dhaka for up to 3 months, so that the Bangladeshi screeners could be released for retraining.

On award of contract, we were able to hit the ground running. Paul and I deployed back immediately to get the systems, processes and procedures in place. This included everything from accommodation, travel, team security, administration and even the procurement of office and training room equipment. We bought computer systems for TIRT and enough desks chairs and furniture to get a full training centre up and running.

We needed covert test teams, supervisors, administrators, and a management structure that could survive international scrutiny. We spoke to CRG and went into an Egypt style arrangement, bringing in 2 of their senior specialists.

We flew in over 50 individuals to help Bangladesh rebuild its security framework.

All of this was done under intense international scrutiny. Paul and I were attending meetings where we briefed Bangladeshi ministers, High Commissioners and EU representatives. The US TSA was heavily involved, as was the Japan International Cooperation Agency (JICA), who were about to build a new terminal in Dhaka.

At every meeting, all eyes were on us.

Representing the UK was the same inspector, now in a different role, that had nearly shut down UPS. He personally knew what our capability was and expected nothing less than a similar transformation.

Management of expectations was an important feature, culturally this was a different ball game and at first, we struggled to understand the cultural mindset of the Bangladeshis; being colonial Brits, everything was still our fault.

It was a tough line to walk, and at least once I strayed well past diplomacy into full "*Who the fuck do you think you're talking to?*" mode with senior management, delivered with all the restraint of Sgt Maj Williams barking *"Shut up!"*.

Thankfully, Paul kept me balanced, and between us, we got the job moving.

Paul and I met the first flights of teams coming in and welcomed every person into country. A couple were clearly shell shocked on arrival, despite the Emirates Business Class treatment.

We sent one individual home; he was missing his dog too much…

Mick arrived.

Mick was one of the team members that would be staying full time, for the whole 2 years. He was an instant hit with the locals and the wider Redline team, as I knew he would be.

After a few weeks, Paul and I began to hear whispers of Mick's antics; we'd clocked him avoiding us in the operation. He wasn't looking great. Mick had strived to blend with the locals and had decided to go for a train journey 'Bangla' style; on the roof.

Some basic research, like timetables, would have stopped him from climbing onto the roof of a train with hundreds of locals if he'd known that the next stop was hundreds of miles away in Chittagong.

Left with no option and no emergency pull cord, he jumped off, severely hurting his back in the process. He went to a local pharmacy, not well known for their high standards of pharmaceutical excellence and bought some strong painkillers.

It must have been close to morphine for the effect it had on him.

He was useless for days.

The practical joker in him was well and truly out; I was happy to stand by and watch from a distance as he tortured the younger former soldiers in the team, blokes that should have known better. Mick would write songs and sing about other team members; a couple of them were even repeatable.

Over the next 2 years Redline also saw the other side of Mick, the side I knew. The kind, generous and thoughtful side. He would buy cricket sets for the local kids and arrange matches in the grounds of the local mosque. He would organise boxes of clothes and other essentials to be sent out from the UK.

Mick was a lynchpin in keeping morale high in people far from home and in a strange land.

Eventually, once the screeners had been retrained, we were able to move out of hotels and into the best apartments available in Dhaka.

Each apartment came with a housekeeper who would do all the cooking, cleaning and washing for the occupants.

The food was magnificent; fiery, fragrant curries that put the British takeaway version to shame. The traffic, however, was less fun. Sitting in a minibus in Dhaka rush hour was like being trapped inside a live-action game of Tetris, with rickshaws and buses and even elephants slotting into spaces you were sure didn't exist a second ago.

I lost track of the number of near-misses, but every driver seemed serenely convinced it was all under control.

Paul and I handed over a robust operation to Emma Mason, the Redline Director in Bangladesh; she had full authority for all matters Redline in country. We weren't handing the keys to a stranger; Emma had been the contract manager for our Heathrow testing operation, so she already knew how Paul and I worked and, more importantly, we knew her eye for detail.

Nothing slipped past her. She had the kind of quiet authority that achieved results without ever needing to raise her voice. In a place like Dhaka, where small errors could snowball into international headlines and where women weren't always taken seriously, Emma was exactly the steady pair of hands we needed.

In Dhaka, competence mattered more than volume, and Emma had it in spades.

Paul and I had other fish to fry.

After the report into Egypt was submitted (and largely ignored) we were approached by an Egyptian company to put together a Dhaka style proposal to the government to cover Cairo and the other affected airports.

It was highly speculative and so didn't really get off the ground. Disappointing, but that could have been a project too far at that time.

With the work done and the pressure briefly off, we did what everyone does in Cairo.

Before our late flight home, we hired a local guide, pretty much because he wouldn't go away until we did; but he guided us around the Egyptian museum and did seem quite knowledgeable.

Omar, might not be his real name, then took us out to see the great pyramids at Giza for a camel-back tour. A simple routine, we would mount the camels and go to a stop, be shown something, buy some tat and then mount up again.

Paul's camel turned out to be the grumpiest bastard since Moses was presented with the Ten Commandments. It would scream and shout at everyone within a 10m exclusion zone.

After we'd been inside one of the pyramids, where we marvelled at the graffiti by Shane from Luton, who had allegedly shagged Leanna down there, and taken in the unique bouquet of stale piss in 40 degrees heat, we went to get back on the humpty backed beasts from hell.

Mine was OK and stood up with me on its back. Mr Angry however objected to Paul's very existence, let alone him climbing on his back. In the most heinous act of defiance since Bobby Sands redecorated his cell, it pissed all over its own tail and accurately flicked about a gallon of piss all over Paul.

"ah, the bastards pissed aaal ower us!" he yelled whilst wiping the most foul-smelling piss from his face, and to top it off, I caught the whole thing on my camera. Paul getting a good camel piss soaking was manna to many eyes.

The worst part was sitting next to Paul smelling of eau de piss for the flight home; no amount of airport toilet scrubbing was shifting that without significant amounts of bleach.

Back in Doncaster, just as the team in Dhaka were getting into their stride and things were settling down, Kuwait Airways called us in for what sounded like a routine audit.

Their deputy CEO asked if we'd take a holistic look at aviation security at Kuwait International.

Deja vu.

I headed out to Kuwait on a staff ticket the airline provided and realised very quickly that there was more than met the eye. This wasn't about a polite health check; it was about saving their flagship New York route from being chopped off at the knees.

The Americans had dropped the hammer this time. The TSA had issued an Emergency Amendment (EA) that gave Kuwait ninety days to clean up its act, or the JFK flights were finished. By the time Redline got the knock, there weren't ninety days left, there were barely twenty.

It was less "consultancy project" and more "defuse the bomb with whatever's in your pocket."

I asked for another ticket to get Paul out; it was game on. Again.

Just to keep things interesting, it was Ramadan. That meant our hosts spent the day running on empty stomachs, short tempers and gallons of mint tea at sunset.

Meetings were at 03:00 every night/morning when the locals were awake, and we should have been sleeping. We ran on empty, writing up the documents we would rely on by day, and attending meetings all night.

The CEO put us in touch with all the right people. We were there to prevent international humiliation. It was hot (it hit 54 degrees one day), tense, and frustrating in ways only the Middle East can manage, but it was also exactly the kind of firestorm Redline was built for.

After 10 days of meetings and drafting extensive, costed proposals we left to go home.

At the airport we knew it was between us and a major player in aviation security who were well connected within Kuwait. Still, we knew we had submitted a comprehensive programme that would satisfy the TSA and was cost effective.

Same drill as Dhaka, we held a conference call from the airport, so that the ops team, led by Macy, were ready to spool up as soon as we got the green light.

Arriving back in Manchester we did indeed get the light; only this time it was a red one.

Our competitor had been awarded the contract. Stunned, I called to get the low down. In short, I was told we'd been done over. Even with the full support of Kuwait Airways behind our bid, we'd lost out.

Allegedly, someone had provided our proposal to the competitor, who simply knocked some numbers off and used our operational plan.

For us, this was highly disappointing; JAP seized on this as proof that Redline was faltering after Egypt. This was just another example of playing in a dirty, political game. We weren't failing, we were learning. We were fighting fires most of our critics wouldn't dare get close to.

Roland appeared to think that the best course of action was to increase the pressure on us.

At Bridge House, the best possible Christmas and New Year happened. Apart from Jessica, everyone was there; the house was full of siblings, aunts, uncles and cousins. This was the very reason Jaq and I poured our hearts and souls into Bridge House.

The kitchen was decorated in our traditional woodland style, strings of fresh Ivy, holly and sprigs of pine from our woods. The warmth from the Aga keeping a strong earthy fragrance that will always be Christmas for us.

In the living room, the massive log burner was well stocked and permanently roaring. Toys everywhere and kids running around screaming and laughing was just perfect.

That's how memories are made.

I knew 2017 would be a difficult one for Redline. The situation with JAP had to come to a head, I'd tolerated plenty in my time, but I could feel the red mist rising with JAP; and once that switch flipped, there'd be no walking it back. I also knew Paul felt the same, but he was managing a more balanced thought train with them.

Frictions were creeping in across the board. JAP had brought their preferred finance director to the board; at first there was no issue and although he was a nice guy, he too just didn't get the way that Redline had been run prior to 2014, he was focusing only on numbers while ignoring operational realities.

Macy was increasingly doing his own thing, he was going out to Bangladesh and making changes we didn't know about. On the face of it, not a problem with the levels of mutual trust we enjoyed, but when it came to higher level meetings, issues were raised by CAAB that we didn't know about.

Chapter 141. Bless the Rains Down in Africa

If Dhaka had been a baptism of international fire, Nigeria was confirmation that Redline was now trusted in the most challenging environments. In January 2017 we were contracted by the United Nations Counter-Terrorism Centre to support the Nigerian CAA.

Abuja and Lagos were the focus, but the issues were national. Cargo, screening, regulation; the whole system needed assistance.

We deployed a team of eight of our instructors, splitting them between the two cities. They trained thirty-five Nigerian Master Trainers in everything from passenger screening to covert testing. This wasn't just classroom based; there was operational mentoring, exercises, and the introduction of Threat Image Recognition Training.

On top of that we re-drafted their regulatory frameworks, the kind of dry paperwork that kept ICAO and the EU happy, but which meant the difference between banned flights and open skies.

The heat was one thing, the traffic another, but the real challenge was the sheer scale of what Nigeria needed in such a short window. We had six months to complete the programme.

Paul and I never set foot in-country, the team there did the heavy lifting, but we made sure team support and oversight were in place. By July the new framework was in place, the Nigerians had their cadre of trainers, and the UN could chalk it up as another success.

We continued our highly successful partnership with ASTC Dubai, at least once a year they would send delegates to us for training and we'd go out to Dubai for business development meetings arranged by Marwan and Jamal.

Jamal took us down to Abu Dhabi and introduced us to the General Authority of Ports, Borders & Free Zone Security (MANAFTH). He knew they were responsible for covert testing at all Emirati airports. They weren't looking for any training or actual testing but were interested in a technical solution.

The project was called Al Hosn 2, and on paper it was a complex Security Management System (SeMS) with all the trimmings: digitised QA and covert testing, integrated CCTV, GIS mapping, and enough acronyms to make your eyes water. All in Arabic.

The first invitations had come in 2016, but in 2017 things got serious and the scope was expanded. MANAFTH now wanted us to train their inspectorate at every level; inspectors, team leaders, covert testers, trainers, analysts, the lot.

This wasn't to be just classroom-based either, but included a lot of fieldwork with live testing, operational planning, and systems that would stand up under ICAO's and international scrutiny. That was the carrot. The stick was that Abu Dhabi was deadly serious about this. Their reputation, and by extension the UAE's, is tied to aviation and border security.

It was ambitious and, truth be told, bloody exciting. Compared to haggling with European airports over covert test dates, Abu Dhabi felt like the big leagues. It was also highly lucrative.

A win here would have vaporised the pressure from JAP overnight.

The problem was that the Arabic way of working is very different to ours. Arabs do business through getting to know people, in the same way that we had gotten to know Marwan and Jamal; it was personal.

Emails meant nothing; coffee, eye contact, and a two-hour chat about your family did.

Arabs traditionally did business like Bedouins, ships passing in the desert night. You do business with people you trust, knowing you may not see them again. The phone arranges meetings; trust is built face to face.

Chapter 142. Sandyman to Sandyland

Ever since leaving Kimya in Denia with Sam and Anna, we went back at every opportunity. The Marina de Denia became our escape, even if only for a weekend.

Kimya was more than a toy, she was a reward and reason for all the hard work and separation.

The only issue was that she had just two cabins: one master and one twin. The long and short of it was that we wanted to spend time there, but with our large family we were strictly limited on space. The only obvious answer was that, to borrow Chief Brody's famous linc from Jaws, *"You're gonna need a bigger boat."*

I started looking for boats that had more cabin space but retained the things we loved about Kimya: a flybridge, reliable engines and safety. Eventually I found a Fairline Squadron 52 that fitted the specs.

The Squadron 52 designation felt like a good omen. It was close enough to 652 Squadron to matter.

Sandyman was also berthed in Spain, up the coast near Barcelona in Sant Carles de la Ràpita. She had 3 full size cabins and a crew cabin that could accommodate an extra 2 people. Perfect on paper.

Fairline marked their boats with nautical flag emblems for the make and model. Sandyman carried the flags for 5 and 2 on her mast. It didn't take long to find the matching flag for 6. With that, she was rebadged as 652, marking her firmly as ours. She probably still carries it today.

I arranged to have Sandyman lifted out of the water for a marine survey and made sure that I would be there. Not being an expert on these things, I asked Sam if he would come with me and help with the survey with his opinions, which I respected.

To make it appealing, I also mentioned that we could do the Barcelona F1 weekend taking place at the same time.

Sam's only job was to book a hotel; not hard for someone fluent in Spanish.

We met up in Sant Carles and conducted the survey along with an engineer. Sam and I then took her for a 'sea trial' in other words, a blast around the coastline. The survey uncovered a small issue, but no showstoppers and she handled like a jet ski in open water.

Sam and I headed up to Barcelona in our respective cars, with me following him despite his attempts to beat Lewis Hamilton to the start line.

As we entered Barcelona, Sam navigated to the hotel and parked up, I had to drive around the block twice to find a spot but then met him in the lobby. We queued at reception to check in. Whilst in the queue I noticed some key combat indicators of what kind of hotel Sam had booked.

There were no women in the queue, just pairs of men. Some holding hands. Then I noticed the artwork. Multi-coloured abstract images of very hairy men wearing leather shorts and caps. I nudged Sam. *"don't look around, but I think you've booked us into a gay hotel"* I whispered. *"don't be daft"*, he laughed it off.

We checked in, the receptionist giving the wink that 'wasn't I the lucky man to have such an attractive young partner'.

We went to the room; the twin room Sam had booked.

The twin was, in fact, a double room with one bed and a large, shelved headboard. Each shelf thoughtfully stocked with a range of lubricants, double thick condoms and antiseptic wipes.

We made some very masculine comments about the situation, compared notes on the F1 and headed downtown for a shot of manly entertainment; we made do with a tapas bar overlooking the old marina where Jaq and I had moored Kimya when we'd passed through on or way south.

The F1 was a great race, we both had a blast and Lewis won the race. We headed back to our hotel as late as possible, just to avoid the prospect of sleeping in the same bed. The hotel had a rooftop bar, so we headed up there.

We ordered some Estrella Damm and another selection of tapas; Patatas Bravas, Pan con Tomate, Croquetas and Jamón Ibérico. All very civilised until Sam spotted some head/groin bobbing action going on in a dark corner of the rooftop.

Liam from my Maersk days had taught me a lot in Stuttgart; exposed me to many things, but not that much.

We went to our room with lots of jealous men eying me up and down, wondering what I had to offer such an attractive boyfriend!

We put an offer on Sandyman, which was accepted, and arranged a handover in Sant Carles. The plan was that Jaq and I would pick her up, and then take her down to Denia, via Valencia, and leave her with Sam who would use her for charters.

After a night on board, we prepared for the trip down to Denia. Sandyman had a complex set of engine controls that could be alternated from the flybridge to the main helm. This was controlled by a switch that had been exposed to sunlight and had faded.

I didn't know that part.

I started the engines from the flybridge, Jaq went down to untie us and detach the ship/shore electrical power cable. Just as she untied us, the engines burst into life, the switch assumed I was at the main helm and those engine levers must have been forward, demanding power. We lurched forward off the jetty, Jaq taking a face full of sea water.

The electrical shore cable jolted forward, tearing the control box clean off the jetty and into the salt water. I slammed the engines backwards; they immediately went into reverse, and we smashed into the jetty.

This was worse than the lock in Swansea by a considerable margin.

Luckily there was a safety cutout in the control box that cut power the instant it detached from the jetty.

The fenders absorbed the impact of Sandyman against the jetty, the only real damage was my pride and the electrical control box. The harbour master was on the scene in about 15 seconds. Mortified and shaking, I asked how much the repairs would cost.

He just laughed, shrugged his shoulders and said it was OK, it happens.

After composing ourselves, when my hands stopped shaking, we set off without any further damage limitation required. We had a stunning evening in Valencia, the sundowners on our huge new flybridge were made for memories.

In Denia, we moored next to Kimya. I felt like a total traitor; like I was having an affair. Cross loading all our stuff from Kimya to Sandyman made us realise just how much larger she was. Kimya's new owners came and collected her. She was going to be trucked back to the UK.

Kimya gave us the best 3 months of our lives, she genuinely meant more to us than fibre glass and engines.

Later that evening we were driving into the town when, at a roundabout, we saw Kimya loaded on the truck on her way back to England.

Queue wobbly bottom lips all round.

Sandyman gave us space, but she also came with problems. Hydraulics failed. Toilets leaked. At one point we dragged mattresses onto the flybridge just to sleep when the AC failed.

She was a money pit, but she gave us what we wanted: time together.

We'd take Sandyman over to Ibiza or Formentera every time we were in Spain. We would always have 10 or 15 minutes with the dolphins; I'd put the boat into a hard turn to generate a huge, circular wake in which flipper and his mates would give us an aquatic display. They never failed to impress.

They say there are two moments in a boat owners' life that are the most satisfying; when you buy it and then when you sell it. That was kind of true for us, selling Kimya was very bittersweet. Sandyman did provide us with what we wanted, quality time with family and friends. She was a money pit, but we loved being on her even when things weren't working as they should.

I loved the simplicity of being in the engine room, tinkering and keeping those huge engines sweet; it took me back to keeping Daisy on the road.

Most of all, our time on the boats was a relief. They were our sanctuary from the real world, a place where the stress of work couldn't reach us, and where memories were made that will last far longer than the boats themselves.

That contrast was sharpest when I turned fifty. I spent my birthday in meetings in Abu Dhabi and Dubai; Jaq had come out with me, leaving the kids with my mother. We made a point of being together, refusing to let another milestone slip past in an airport lounge.

While we were both there, we met up with Mireya from our BMED days. She was now married to Dale, a former RAF aircraft engineer working for Emirates. The four of us picked up like nothing had ever been different. Still the fiery Spaniard we knew, still capable of detonating over a crooked fork at dinner, but still our Mireya.

The call of MANAFTH was powerful, every call from them was to arrange another meeting in Abu Dhabi, we always seemed to be one meeting away from a signature. We did try the tactic of bringing them to London.

Gary arranged for us to use The Rag; the Army and Navy Club on Pall Mall. The Rag gave the meetings weight. Heritage, privacy, and ritual mattered to them.

Collectively we did everything we could think of to develop this critical business relationship.

I was spending so much time in the UAE I was starting to wear a kandura and flip flops to coffee meetings that would last all night and tossing "*inshallah*" into the conversation like a native.

For a few months Paul had been talking to Keith from our 652 Bosnia days. They had caught up at Paul's villa in Cyprus and started kicking around the idea of him joining Redline. When Paul finally ran it past me, I was genuinely excited; I couldn't wait to see Keith again, let alone work with him.

He joined Redline in the second quarter of 2017 as business development manager, which meant we'd be working side by side. It felt like a reunion and a fresh start all at once. From muddy airfields in Bosnia to boardrooms and contracts.

Chapter 143. Sandyland to the Land Down Under

My work with ICAO was now ongoing. The sixth meeting was in Brisbane. Paul should have been there with me, but JAP pressure kept him back again. Instead, Keith came with me. It would be a baptism of AvSec fire for him, and a clear signal that Redline's reach was not rhetoric.

That meeting, from a work perspective went very well, but Brisbane was to leave a lasting impression on me. Keith and I tracked down Dan Perren, our Australian flight commander from 651 days, and finally made him buy us the beer he probably owed us.

Whilst things were starting to show signs of fracture at work, at home there was a development. Jaq had received a text message from Jessica in Germany. She was in grave danger from her partner and with a young baby; it was serious.

Jaq showed me the text asking for help; her initial thought was to leave her to it; not unsurprising after the way the relationship with Jess had gone sour and this was the first real contact in a very long time.

My reaction was to message her back; ask for her address and tell her I was on the way. I looked at Jaq and said, *"I wouldn't be the man you think I am if I didn't rescue my daughter"*

I packed a bag and booked the next Hull–Rotterdam ferry. Jessica's partner was a dangerous criminal, in and out of prison, and he wouldn't simply let her leave. I was given a time when he wouldn't be there.

I arrived early and waited a couple of miles away so that my car, on English plates, wouldn't be spotted. Once he was out of the way, I drove to Jessicas house; she was ready and waiting with bags packed and the baby wrapped up.

We had about 15 minutes to load up and get the hell out of Dodge.

I headed for the autobahn.

The next issue was that the baby didn't have a passport. We went to the British Consulate in Dusseldorf for help with an emergency travel document. Recognising the urgency, they pulled every stop out. Twenty-four hours later, we had the documents.

Jessica and the baby moved in with us at Bridge House for a few months as she went about rebuilding her life; This time, we were part of it. That made me very happy.

The personal highlight was being able to take a 16-year-old Patrick to New Zealand for the British & Irish Lions Tour. Given the amount of flying I'd done, I was loaded up with 'skywards points', enough to get me and my boy to Wellington and back in business class.

We stayed with Hugo for the rugby test matches but made a special trip up to see my cousin Geoff from Eastern Hill farm days. Geoff had emigrated to New Zealand and now had his own dairy farm with a herd of Jerseys.

His farmland was extraordinarily beautiful; we hadn't seen each other since his father's funeral a couple of years prior; it was great to catch up.

Watching the haka with Patrick, surrounded by red shirts roaring as Sonny Bill Williams was sent off, I felt something rare. I was just a dad. For once, my air miles bought something priceless: time with my son

That trip remains one of the best things I've ever done. For me, that's what 2017 was about.

117 Keith, me and Dan Perren

Chapter 144. Administering the Last Rites

We pushed through the madness of 2017 and into a pivotal year, 2018 marked a significant change in our company structure and the defined relationship between Paul and I; a relationship that was forged in Germany, strengthened in Bosnia and then Northern Ireland and gave birth to Redline.

This relationship had endured being shot at, changes in domestic status and way too much Baijiu. There was nothing we wouldn't do for each other.

At a board meeting just after the start of the year Roland had taken me to one side and asked for a private meeting, we were joined by his mentor, Elliott. They started asking questions about Paul; his style of leadership and management and what I thought about my position compared to his.

I replied that they didn't know him, and never would, unless they started to get a measure of the man, they would never succeed.

JAP unilaterally decided that Paul needed "help" as CEO and put it to the board. They gave us no choice but to accept some bloke as a mentor to Paul. We also had to pay him, as they wouldn't be.

Paul quite rightly took this personally, as would anyone.

JAP had pored over every decision since Kuwait, seemingly desperate to pin the loss on us, rather than accept the reality of how the game was played. They needed a number, something more tangible than *"we just lost out through politics and brown envelopes"*.

It was never said, but I knew they were trying to divide and conquer, to get rid of one of us. They didn't want the cost of two founder directors, and they were after my position.

At every board meeting they would chip away at us, setting one director against another. In my opinion, they behaved like parasites, feeding off a business they didn't understand and couldn't build themselves.

Gary offered counsel on several occasions to just accept them for what they were. Paul and I discussed getting rid of them, buying them out, doing whatever it would take. We just didn't have the legal ability or the finances. All we could do was keep up the hard work and prove them wrong.

The Bangladesh contract was due for completion in a couple of months. We'd beaten our own deadlines in restoring international cargo and securing the DAC – LHR route.

There was little hope of a contract extension, although we tried.

The UPS contract was under constant pressure as senior positions rotated and the corporate memory of what the UK went through was waning, they were looking at ways of cost cutting.

The pressure, perceived or otherwise, was ramping up like some kind of evil ratchet.

There were a couple of wins, we won a tender to build a security management system for the 2018 Commonwealth Games in The Gold Coast. ISS was an Australian security provider that wanted to look at TIRT; the pipeline from Australia was building quite nicely.

I had also been trying to build a portfolio for Redline in South Africa with their ICAO training centre and their CAA.

But MANAFTH was still the goal; it answered every question. The issue was that chasing it was costing us a fortune in flights and hotels.

In March, me, Paul, Cath and Jaq went for a quick holiday to Zermatt. We met up with Stefan and his family for a couple of days skiing and peace and relaxation. That's what the 3 of *them* got; I got a badly damaged knee and had to be rescued off the mountain.

At least I got to experience first-hand the *thrill* of being taken through airport security in a wheelchair.

It was at the April board meeting when Roland said something, I don't recall what he said, it might have been innocuous, but tensions were running high, but whatever he said, that was it for me. I knew then I had to get away from JAP, and especially from Roland with his brand of slick condescension.

I put a plan to Paul and Gary that would see me relocating to Dubai to establish Redline as an Emirati company, apparently the final legal hurdle to being awarded the MANAFTH contract.

I would also expand into business development for Middle East and North Africa; with a keen eye on Asia Pacific. We went through the numbers, and it made sense, but of course JAP eyes had to be cast over the plan and approve it.

They set some conditions on contract awards, and I went home to Jaq to drop the bombshell. We were moving to Dubai for 12 months minimum, probably 2 years. She took it in her normal style of '*F'kit*', let's do it.

I had surgery on my twisted knee on my birthday before flying out to Dubai the next day; Jaq and the kids would follow as soon as I had a house and domestic arrangements sorted.

Until then, I bummed a room with Dom Giles, now a senior captain in the Dubai Royal Flight; along with many other former AAC pilots. Dan Brodie was there, Baulfers and many others. It was like being back in the Aldergrove mess, only a lot more expensive and a bit less 80's style décor.

Being with these boys meant the world to me.

I found a house in a Dubai housing estate known as 'The Sustainable City', this place may have had solar panels and golf buggies to tool around in but was far from financially sustainable. Jaq and the kids arrived soon after.

Whilst Jaq and the kids settled into new schools and the social life that goes with living in Dubai, I set about jumping through the vast quantities of red tape required to set the company up as a legal entity that could do business with a government body.

Mireya and Dale were there helping with settling us into the expat life, Jaq was loving it. She had loads of friends from BMED days and was making new ones daily.

There were accreditations required from multiple government bodies, each with its own maze of prerequisites. Every office was in a different part of town.

We even needed a licence to buy a beer.

I was rarely at home, and if I was, it was in the maid's room now being used as an office. I had project plans all over the walls, just to help me keep track of where we were and where we were going. I had business development streams growing in Oman, Kuwait and Saudi Arabia as well as South Africa and Australia.

We didn't stay in The Sustainable City long; it was way too expensive, so we found a nice maisonette apartment in a much cheaper, but still nice area of Dubai.

As each lead in the pipeline developed, I would then have to write a comprehensive requirement back to Macy and the Ops team; but things were getting lost in translation from what a potential Arabic client wanted, and what Doncaster would *interpret* as the requirement, they were continually putting a European spin on everything.

It became monotonous: describe the requirement, correct the interpretation, start again.

One example was a straightforward provision of TIRT, well straightforward in the context of working with Arabs, as always, it dragged on until eventually Roland even floated the idea that I'd invented a contract to justify staying in Dubai. To me, that said more about his mindset than mine.

I took it personally and was furious.

In haste, I blasted off an email with the confirmation letter attached from the client. But the letter was in Arabic, so I bluntly suggested he would have to get it translated if he didn't believe me. I could have been more diplomatic, but even with the distance between us, he was still throwing shit my way, and I'd had enough.

I took up golf; Dale talked me into it. We would go to one of the many golf courses in Dubai, more for a *'walk and a talk'* than twatting golf balls. I needed to unwind and de-stress; this was a start, but sadly not enough.

One glimmer of hope was from a French guy, a friend of Alec from ICAO, now working in Washington DC for a major US defence contractor. They had been awarded a contract by the US State department for aviation security capacity development in 3 Libyan airports, but were struggling as, ironically, the State department had also banned US contractors from setting foot in Libya.

This potential contract smashed the financial target I had been set by JAP and the board, so I went all in for it. Before anyone could say no, I booked a flight from Dubai to DC and met with the leadership team from the defence contractor.

I got along with their leadership straightaway, and over a couple of days hammered out a much wider ranging partnership to deliver this key project with them.

As was the standard protocol for us, I called Paul and had him fly out to DC. The pair of us further expanding the scope and worked out a pricing strategy that worked for everyone. We then brought out some training course developers and Macy to take over the project.

Ed arrived and entered the room like a bull in a china shop thinking he was well received by the Americans; he wasn't, and some firefighting was required.

Signing this contract was a relief, it would put JAP back in their box and stop them from distrusting me and Paul. For me personally, I saw light at the end of the city boy tunnel, they might just let us get on with our jobs and be supportive.

Back in Dubai, I could see signs of extreme stress manifesting in me by the way I was responding to standard comms from Doncaster, I was starting to see everything as a direct dig, even when it wasn't. JAP were firmly under my skin, and I saw challenges to my work everywhere I looked.

On board meeting calls I was abrupt and on the verge of being rude, even to Gary and Paul. On one hand I knew I was acting poorly on some occasions, on the other hand, Gary and/or Paul would grip me: wouldn't they?

Just before I was due to fly back to Doncaster for a quarterly board meeting, I had bid for and won a landside security assessment at a major international airport. They wanted an independent review of landside security arrangements so that, if our report was well received, it would lead to a lot more work. I had single handedly won their trust.

It was critical that we put our best foot forward and only use the best team available to us for this audit. Those auditors and their CVs had been part of the winning bid, in the face of some major competition.

We were sitting around the board table; I was opposite Roland when he said something to me that caused my peripheral vison to start closing in. He then said something else to Paul that questioned a decision he had made; nothing to do with me, but that was enough.

The jab at Paul was the trigger. Not just for what he'd said, but for the years that had led to this point, the long-handled screwdriver treatment, the constant undermining, the subtle little digs dressed up as questions.

It had been a sustained campaign.

I'm not a violent man; I never have been. But in that moment, in that room, I was going to cause him damage. Something inside me snapped, not slowly, not subtly. Like a circuit breaker reacting to an overload, just click, and everything went quiet.

I wasn't thinking about professionalism or optics or outcomes. I was thinking about grabbing his throat and applying the Vulcan death grip with no thought for consequences.

I leapt up, my chair skidding behind me, and went for Roland, stopping just before I made impact. I looked at Gary who looked back, not in horror or anger, but with disappointment. I muttered an inadequate apology to Gary and walked out of the board room, going over to the Ramada to breathe and have a coffee.

Macy followed me over after 5 minutes or so and told me I needed to go back.

I went back with a clearer head. The Ramada's finest instant coffee and five minutes of deep breathing had taken the edge off, just enough to stop me imagining Roland heading back to London in an ambulance, complete with green tubes and a C collar.

I apologised to Roland and Elliott. No excuses. I acknowledged the moment for what it was. To their credit, they apologised as well. Roland even tried to clarify what he'd said, but no one was really interested in a playback. It wasn't about revisiting it; everyone just wanted the temperature to drop.

The meeting concluded and the JAPs left, I thought the worst of the day was over.

It wasn't.

I then needed to talk to Ed and Paul about the landside assessment. The sticking point was that we had named experienced international inspectors that would conduct the review in the tender. Ed was determined that he knew better and wanted to send a pair of capable, but young and inexperienced trainers to do the job.

It wasn't about capability; it was about the optics of sending junior people when we'd promised senior.

We went around in circles.

"Ed, fuck off will you, we need to send these guys," I said, exasperated.

He went off like a grenade. *"That's it, I can't and won't work with you anymore,"* he shouted, storming for the door.

I tried to pull him back the way only old mates could. *"Ed, for fuck's sake, I've seen your cock! Talk to me!"*

Normally that would've cracked him; we'd laughed about worse in the hot tub; nothing had ever been off-limits. But this time he didn't even blink. He kept walking. Out of the room, out of the NSTC, and out of my life.

I followed him all the way to his house, trying to talk to him friend to friend.

He wouldn't even open his front door to me.

Back in Dubai, Jaq and I sat on the balcony with a glass of wine, the city humming around us. I told her how I felt. Every betrayal, every door that had quietly shut behind me. She didn't interrupt, just sat with her hand on my knee, steady as a rock.

Then '*Sit Down*' by James came on the Sonos; a song for the bruised and bewildered, for people trying to hold themselves together. It felt like it had been written for that moment.

The only person sitting next to me feeling ridiculous, supporting me, was Jaq.

And that's when it all spilled out. I'd smashed through my own limits without even realising it, I was in freefall. Feeling isolated, ostracised and eliminated from my own company.

I broke down right there on the balcony.

Without Jaq talking me down, I dread to think where my mind might have taken me.

The fallout from that boardroom blow-up hit me harder than I expected. My relationship with Macy was dead in the water. I felt distant from Paul and Gary, even they seemed unsure what to say to me. The years of relentless graft, the late nights, the pressure to deliver, it had all backed me into a corner.

And in that corner, a voice was waiting for me. Not shouting, not even stern. Just calm, cruel, and matter-of-fact: *"You're not up to this. You've blagged your way this far. But now they've seen it. You're a fraud and you're not needed"*.

Later, I found out later that Macy's first call, his first move, had been to brief JAP. Not to check in with me, not to de-escalate, just… use it. No context, no caution.

He went straight to JAP HQ behind my back with the dirt.

It felt like a calculated betrayal.

I hadn't laid a hand on Roland, but it didn't matter. Macy had thrown the switch and lit the whole thing up. I felt like I'd lost someone I thought was in my corner.

There was no coming back from that.

Within days, Paul and Gary were summoned to London to explain themselves, not to Elliott, but to their boss. The message was clear: this was no longer an internal wobble; this was now a matter of record.

I don't know exactly what was said, but it was made clear I was to be got rid of. I did know that Paul and Gary would be on my side; and I knew that Paul was under the same stresses as me, but for different reasons.

What were Macy's motives? I'll never know or understand.

I agreed with Gary that I would seek some professional help. I thought I knew what my issues were; I just didn't have a coping mechanism. This made leaving the Army seem like a party.

I arranged to see a doctor in Dubai, but had said to Jaq before we stepped into his office three things;

1. If there was a certificate on his wall, purportedly from '*The University of Islamabad*', I was out.

2. If he said I had PTSD, I was out, and;

3. If he tried to medicate me, I was out.

I walked in. Strike one: the certificate. Strike two: PTSD. Strike three:

sleeping pills.

Life had to go on; the business had to go on.

One foot in front of the other. Keep going. Push it out.

Chapter 145. Walk Towards the Light

Christmas 2018 contrasted with the way the year had panned out. We spent it with Dale & Mireya and her Spanish only speaking parents. Between Jaq and Dale, the most delicious Christmas dinner ever was served, party games played and lots of singing and laughter.

It was a great way to draw a line under the year, and by the end of it we were fluent in Spanish; Mireya's parents were fluent in bollocks.

After the debacle of seeking professional help, I pondered for a while about how to deal with stress, from any quarter.

I knew I was still stressed to the hilt. Heading north on the E300 motorway, you drive under what feels like half the UAE national grid. I glanced up and, just like that, Bosnia came crashing back.

The pylons and wires.

My chest tightened before I could blink. Sweating is normal in Dubai, but not in an air-conditioned Land Cruiser.

I was still doing the rounds of Dubai's golf courses with Dale, who, like me, needed to decompress and recalibrate. Around the same time, I struck up a surprising friendship with an Austrian guy whom I met in The Sustainable City. Together, we'd load up our 4x4s and head out into the desert, just to light a fire, share a few drinks, and talk nonsense under the stars in a hammock.

We talked about everything and nothing; work, life, women, the Austrian Alps and dead cows, solving the world's problems one cold beer at a time. It was gloriously aimless, and for a while, exactly what I needed.

As 2019 came in, I was feeling more optimistic for the future. The leads for Australia were converting. the initial plan to service ISS with TIRT had grown legs and they were now more interested in a full blown SeMS system. They'd asked me to go down under and do a tour of their clients, showing them their new direction and innovation they wanted to bring.

The Arabic clients were coming in, the fruits of being constantly available and at their beck and call for the previous 14 months.

The work in South Africa was taking shape; albeit slowly. To operate in South Africa, we had to comply with the 'Broad-Based Black

Economic Empowerment' regulations, and that meant having a black majority shareholder. I had located a suitable person, the chairman of the black business council and very keen to work with us.

I felt there was a lot of suspicion and lethargy coming out of Doncaster, I needed the board on side for this to succeed. I arranged for this guy to fly up from Jo' Burg to talk to the board in person. The JAPs were rude to him, flat out. Roland seemed to think a South African accent gave him an advantage. Shame it didn't come with manners.

Despite smashing the targets set by the board, by a significant margin, it was clear that JAP had disregarded the defence contract in Libya due to it being a US based company.

There was no winning.

There was no room for me in Redline, all I could do was the best I could. And that was no longer for the board, it was for the Redline team; it was for Paul, Michelle, Dave, Mick and the rest of the team.

Around the middle of the year Paul told me he'd been approached by a UK aviation company with a view to an acquisition. They say that timing is always everything, and in this case, they weren't wrong.

My reaction was *"Just sell the thing,"* like it was a dodgy second-hand car.

When he raised the subject with JAP management, they were of the attitude that if anyone was stupid enough to buy Redline, then they could have it. It was JAPs professional opinion that we were a delinquent company and were a liability.

In September, Gary and I attended the Emirates Group Security AVSEC Global symposium and awards. Gary presented a keynote address and Redline were awarded Best Training Organisation of the Year. Despite the judging panel being international and included the Edith Cowan University, there was little pride or joy in the award. There was a perception that I had somehow contrived the award.

Final proof that I literally could not win.

Paul managed all the negotiations with the buyer, with only minor input from me. We went into due diligence knowing that we weren't the company or people JAP thought we were. Knowing I wouldn't be staying with the company after acquisition meant my salary was suddenly clear profit that added significantly to the sale price.

One of my last trips from Dubai was to Japan with Dale to the 2019 Rugby World Cup. Whilst I was there, the phone rang, I was still pissed having watched England thrash the Wallabies. It was the EGM of ISS in Australia; one of their sites was in serious trouble with the regulator and they needed immediate support.

I arranged for an experienced and versatile trainer to meet me in Brisbane where we would then travel onwards to the site in trouble. We were there in 72 hours.

We got stuck straight into working out what was going on and how to fix it. I discovered that the national training manager for ISS was leaving and asked the EGM privately if they might consider me for the role once I left Redline.

He declined, but only because it was too junior a position. I was then interviewed to be the national QA manager for ISS, covering all their 9 sites in Australia.

I spoke to Paul and arranged for the lawyers to draft a clause in the sale agreement that allowed me to work for a Redline client.

In Dubai, the conversation with Jaq was easy. We both knew I wasn't continuing with Redline, so I simply said, *"how do you feel about Australia next?"* a big ask seeing as she'd never been.

We agreed that we would go over on a recce, I had the choice of anywhere in Australia that ISS operated, a fantastic position to be in. We narrowed it down to either the Gold Coast or Brisbane.

We maxed out the credit cards and in November 2019 we had 10 days scoping possible locations and homes. We started on the Goldie and its surrounding areas then moved up to Brisvegas. We'd spent hours on house purchase websites and had pre-arranged some viewings, mostly clustered around the village of Samford, to the northwest of the city.

Jaq wasn't keen initially on Samford, believing it was going to be like some *"little house on the prairie affair"*. The first house we walked into had an immediate and emotional impact on me. I don't know why, but I had an overwhelming feeling of belonging and peace.

I welled up, not unnoticed by the real estate agent and Jaq. It may have been the similarity of the countryside to New England, or the wild Wallabies grazing in the backyard; I don't know, but it felt like exhaling after years of holding my breath.

The next house we went to was the one I'd liked from Google. We all loved it and went for a family vote on which house we should buy if we moved over. Sam chose the first one but was outvoted; we put a conditional offer in on our new home.

ISS were handling the work visa for us, so we went back to Epworth for Christmas, letting the family know what our plans were.

The sale of Redline completed just a few days before Christmas 2019, marked by muted celebrations that felt more like a wake than a wedding. And then, like an unexpected punch, the world changed. COVID hit.

The deal went through on the last viable day before the aviation industry was flattened by the pandemic. A single day's delay, just one more boardroom postponement or a missed email, and it would have been the holidays, then January. And by then, the deal would have died.

Macy turned up late, for the formal signing, cloaked in some Equaliser–style Barbour trench coat, looking like a man auditioning for a part he didn't deserve.

We signed the paperwork with few words.

Before leaving he offered me his hand, I refused. Not out of spite, just a quiet refusal to fake closure where there was none. That was the only honest ending.

I said my goodbyes to the other board members in the lawyer's office, some well-meant, others obligatory, and then, like Elvis, I slipped the building. No fanfare. Just a hollow ache that settled in deeper than I expected.

Earlier I said there were three baddies. You've now met them all. Each left a different scar. The last one just took longer to show.

After I left, I started hearing other people's views of Macy. Overbearing. Sidelining. Things I hadn't seen, or hadn't wanted to see, while Paul and I were looking forward instead of behind us.

We'd trusted him implicitly. Learning that others felt pushed aside left a bitter aftertaste. It wasn't the culture we thought we were building.

Not long after, Macy left Redline too. "*Mutually agreed terms,*" they said. Maybe. All I knew was the silence that followed must have been something.

Chapter 146. Far, Far Away

On 10th January 2020, we left Manchester with suitcases big enough to smuggle livestock and expectations to match. Redline was in the rear-view mirror; sold, boxed, and shipped off to someone else's inbox.

I exhaled properly for the first time in months.

Maybe years.

On the flight, I was scrolling through news apps, just passing time. A few headlines jumped out: some bloke had been bitten six times by a brown snake and died; bushfires were ripping through parts of Queensland; floods hammering the rest; and to really complete the picture, someone had photographed sharks swimming through a children's play park.

I glanced at Jaq. She was reading the same stories. We looked at each other, wide-eyed, and just burst out laughing.

"What the actual fuck have we done?"

We arrived thinking that we'd be stepping over snakes and other nasties; that never happened although it didn't take Jaq long to say *"oh, it's only a carpet python, nothing to worry about"* like that's a perfectly normal sentence in everyday life.

In the weeks that followed, I started to find my rhythm. Slowly at first, like trying to dance in bare feet, but it came. Then, the best thing ever happened: I was stood down by ISS because of COVID-19.

At first, I thought I'd go mad with boredom. Instead, I found the unexpected joy of doing absolutely bugger all. Days spent in shorts with questionable elastic, burning snags on the barbie and calling it part of the Australian Combat Readiness Check.

Stress is a sneaky bastard, like Japanese knotweed. It grows quietly in the cracks, throws down roots where you least expect, and before you know it, it's wrapped tightly around your lungs and tapping on the inside of your skull.

Getting rid of it isn't easy. But somewhere between the early morning swims and the hum of the ride-on mower (with beer holder), I started to feel human again. Not all at once, but enough to know I was on the way back.

Queensland takes some getting used to. The humid air is thick with eucalyptus and wild garlic from the firefly-lit creek just beyond the trees. The coffee is dangerously good, and the local reptiles look like they've wandered straight out of an Attenborough documentary and decided to squat on the back porch indefinitely.

There's so much more to where we live, the national parks are sprawling, wild, impossibly ancient and open to use. I can drive for hours along sandy beaches with the ocean crashing on one side and bushland whispering on the other.

Stradbroke Island slows the world down; the headlands are perfect for whale watching. The sand squeaks, the tide sets the pace, and I can almost feel the stress evaporating. Southbank in the city becomes our lazy-Sunday walk, markets, buskers and some amazing cafes.

The Brisbane River, affectionately called the 'Brown Snake', allegedly teeming with bull sharks, winds its muddy way through the city like it's up to something.

Queensland isn't just a place. It's a lifestyle choice, with better seafood, louder birds, and zero tolerance for whingers.

The light at the end of this tunnel wasn't a train after all, it was the fresh air and countryside of the Sunshine State. In the three parts of my life, Redline was by far the most difficult, but it bore the highest rewards financially.

The emotional toll is harder to calculate, but the view from where I'm writing this makes it all worth it.

After everything, I don't feel like 'Unlucky Jim' anymore. Sure, things went sideways often, near misses, full-blown disasters, and enough bureaucratic bollocks to fill a skip, but looking back, I see more than just the chaos. I see the grit, the laughs, the people who mattered, and the moments that somehow made it all worth it.

I've faced my baddies. Different uniforms, different eras, same result. Each one tried to knock the compass out of my hand. For a while, they succeeded.

What they left behind were scars. I've decided to keep them. Scars tell you where you've been, and they remind you what matters when it counts.

Each one taught me a little more about trust, loyalty, and the importance of calling bullshit when you see it, even if it costs you. Especially then.

Paul and I built something really bloody special in Redline. When it soured for me, it wasn't because the dream was flawed. It was because trust broke. That happens more often than people admit. I'd like to say I saw it coming. I didn't. I don't regret it, though.

Regret is a luxury for people who never took risks.

That chapter ended in a sterile lawyer's office with a handshake I refused and a silence that said more than words. I walked out feeling like the last song had been sung. But looking back now, it wasn't an ending. It was a release.

Those baddies? They didn't break me or Jaq. They made us.

Paul and I? We're still standing too. Roland didn't divide us, though God knows he made the mistake of trying. Many better men have tried and failed too. We may have wandered different paths since then, but the bond never frayed.

Writing this has stirred memories and truths for both of us, and in doing so, it's been quietly healing. We don't speak every day, but when we do, it's still effortless. Spiritually, we're still close, fire-welded by moments that shaped us, like the day in Sipovo when everything turned in a heartbeat, or the South Armagh cross border shooting incident, when calm heads and trust made the difference between going home and not.

It's that kind of history you don't walk away from. You carry it with you, like scars or medals.

Roll of Honour

Family

Seaman James M. Termini, USN, killed in Vietnam, 1966.

Army Air Corps

Sgt. Les Berrisford, AAC, XZ650 crash, Germany, 1994.

Cpl. Andy Beck, AAC, XZ650 crash, Germany, 1994.

Sgt. Martin "Ozzy" Osbourne, 14/20H, XZ676 crash, Croatia, 1995.

Cpl. Iain MacDonald, AAC, XZ676 crash, Croatia, 1995.

Air Trooper Roger Willingdale, AAC, XZ676 crash, Croatia, 1995.

Air Trooper Graham Witherstone, AAC, XZ676 crash, Croatia, 1995.

SQMS Mick Gwinnel, AAC, motorcycle accident, Germany, 1995.

WO2 Andreij Prenczek, AAC, XZ199 crash, Tilton-on-the-Hill, 1999.

SSgt Stu Donnan, AAC, XZ199 crash, Tilton-on-the-Hill, 1999.

SSgt Pete Clyne, REME, XZ199 crash, Tilton-on-the-Hill, 1999.

The Light Dragoons

Lt. Richard Madden, LD, Bosnia, mine strike, 1996

Tpr. John Kelly, LD, Bosnia, mine strike, 1996

Tpr. Andrew Ovington, LD, Bosnia, mine strike, 1996

Northern Ireland

Cpl. Gary Fenton, RGBW killed in action, 1998.

L/Bdr. Stephen Restorick, RA killed in action, 1997.

L/Cpl. Simon Ware, 2 Coldstream Guards killed in action, 1991.

In Remembrance

This roll of honour names those who are part of my story.

But this reading stands for all Army Air Corps helicopter crews lost in training and on operations. Their courage, professionalism and sacrifice will never be forgotten.

Every crew who strapped in, launched into danger and never returned is remembered here, alongside those whose names are known to us all.

Low Flight *(Inspired by the spirit of "High Flight")*

Oh! I have cut the stubborn skin of land
And skimmed the dust on rotor-beaten air,
Through wire and tree and ridge that test both eye and hand,
Where maps grow vague and judgement rules there;

Nap of the earth, where shadows race the ground,
I've bent the wind to follow where I lead,
Through valleys tight where only fools are found,
And those with just enough skill or need;

I've felt the thump of blades bite heavy sky,
The beat of a machine that wants to live,
Held fast in hands that will not let it die,
And asked of it far more than it should give;

Skids off, time on target set, by day and night with eyes half-blind,
I've hovered downwind, judging drift and line,
Set down in places no one sane designed,

Then lifted out, leaving it all behind;

And under green-lit glass I've worked the night,
Where depth is guessed and distance tells you lies,
Each flicker turns a shadow into shape,
And wires appear too late for mortal eyes;

I've chased dim ghosts across a starless land,
With engines screaming, silence in my head,
Trusted the glow to guide a steady hand,
Where one small lapse would write the pilot dead;

So, when I land, and silence finds the ground,
And blades wind down from fury into rest,
I know it's not the heavens I have found,
But brought a few poor souls back from the test.

James Termini, 2026

Made in United States
Orlando, FL
04 November 2026

90203572R10312